BASIC TECHNICAL DRAWING

HENRY CECIL SPENCER

A.B., B.S. in Arch., M.S.
Formerly Director, Department of Engineering Graphics,
Illinois Institute of Technology

JOHN THOMAS DYGDON

B.S., M.B.A.
Associate Professor, Department of Engineering Graphics,
Illinois Institute of Technology

BASIC TECHNICAL DRAWING

Second Revised Edition

The Macmillan Company, New York

Collier-Macmillan Limited, London

About the Authors

HENRY CECIL SPENCER was Professor and Director of the Department of Engineering Graphics, Illinois Institute of Technology, from 1941 to 1962. Previously, he taught engineering drawing at Texas A&M University and mechanical drawing at Ballinger High School, Ballinger, Texas. He is the author or coauthor of ten books in the field of technical drawing. In 1958 he received the Distinguished Service Award of the Division of Engineering Graphics of the American Society for Engineering Education, and in 1948–49 he was national chairman of the Division.

JOHN THOMAS DYGDON has been a member of the Department of Engineering Graphics, Illinois Institute of Technology, since 1952. Previously, he held positions in industry as an engineer and designer, and he continues as an engineering and management consultant. He has coauthored two books of engineering drawing problems and is a member of many engineering and management societies, including the Division of Engineering Graphics of the American Society for Engineering Education.

The Macmillan Company, New York
Collier-Macmillan Canada, Ltd., Toronto, Ontario

Printed in the United States of America

PREFACE

Basic Technical Drawing is intended to be used as a classroom textbook or reference book for the beginning student of technical drawing. Its main purpose is to provide essential information as clearly and completely as possible in straightforward language and with ample illustrations.

Anyone can use this book. No prior training or skill in technical drawing is assumed. *Basic Technical Drawing* is primarily designed for the beginner—be he in high school, technical institute, or college—who is interested in technical drawing because he wishes to be a draftsman, an engineer, an industrial designer, an architect, or a member of any one of the numerous other professions which demand a knowledge of technical drawing. However, even the student in fine arts will find much information about freehand sketching, perspective, and mechanical drafting technique which will be helpful to him. For everyone this book gives some insight into modern industry and technology. It is the authors' conviction that technical drawing is so interesting in itself, and its value in a technical civilization so great, that side excursions, extra frills, or extraneous motivating devices would detract from rather than further the students' interest. Consequently, *Basic Technical Drawing* is truly devoted to teaching the fundamentals of its subject matter, so that the student can achieve real skill, solid accomplishment, and sound understanding.

This book is as nearly self-teaching as possible. Thus, the teacher is relieved of many tedious explanations of small details, and may devote himself to teaching the more important principles in an effective manner.

The organization of *Basic Technical Drawing* was dictated by the subject matter. While each chapter deals with a different aspect of tech-

nical drawing, the chapters progress from the simple and elementary to the more difficult and advanced. The order follows the topic sequence of most courses. Each teacher, however, can easily adapt this book to his own preferences in organization. For example, if dimensioning is not taken up in class until the chapter on Dimensioning (Chapter 9) is reached, no drawings will be dimensioned up to that time. But the instructor may wish to teach dimensioning gradually and have students apply dimensions to problems from the beginning, with occasional references to Chapter 9, as necessary. Complete explanations and frequent cross references in every chapter allow the teacher to use this book effectively, no matter what the special emphasis of his course, the order of topics, or the individual requirements of his students.

Important terms are defined, explained, and illustrated in detail throughout the book. The technical terms include not only those concerned with drafting, but those associated with the related shop processes. A special feature of the book is a comprehensive dictionary of technical terms in the Appendix (pages 445–450). Thus, *Basic Technical Drawing* should be a valuable reference guide for the trained craftsman as well as for the professional draftsman or engineer.

Every point is amply illustrated. There are approximately seven hundred and fifty illustrations in this book. These drawings and photographs show methods and processes step by step, indicate what to do and what not to do, and provide examples for the student to follow. Every illustration was drawn by the methods and with the tools explained in this book.

Many practical problems are contained in this textbook. No drafting text is complete without an ample supply of good practical problems. Such problems, carefully arranged from the easy to the more difficult, are given after each major chapter. Every problem is an actual industrial problem which has been thoroughly tested through extensive class use in all parts of the country. Early problems in each group are given in sheet-layout form, so that the beginner can get started without difficulty. Most of the problems are designed to fit the standard 8½″ x 11″ sheet. The use of large sheets, partitioned into parts for individual problems, is avoided because such large drawings are awkward for the beginner to handle, become soiled before they can be completed, and tend to discourage the student.

The text material has been substantially revised, and all illustrations and problems have been brought completely up to date in accordance with the latest standards of the United States of America Standards Institute, particularly the drafting standards. The old American Standards Association (ASA) has been superseded by the United States of America Standards Institute (USASI). Standards formerly approved as American Standards are now designated USA Standards.

New chapters have been added on Descriptive Geometry, Electrical Drafting, and Architectural Drafting, to make this text even more comprehensive than heretofore.

The authors appreciate the cooperation of many industrial firms, schools, and individuals who have contributed material for this book. These firms are listed in the Acknowledgments and also, in many cases, in footnote references in the body of the text.

The authors wish to express thanks to Mr. John Kolpak, who prepared the new chapter on Electrical Drafting; to Professor Frank M. Hrachovsky, who prepared the new chapter on Architectural Drafting; to Professor Robert O. Loving, who prepared the new chapter on Descriptive Geometry; and to Professor I. E. Wilks, who prepared the revised chapter on Structural Drawings. Appreciation is also due to Professor Peter Andris and Mr. Charles C. Sands, who reviewed the new chapter on Electrical Drafting; to Mr. Warren A. Koerner, who reviewed the new chapter on Architectural Drafting; and to Dr. Shizuo Hori for preparing the new material on Automatically Programmed Tools. The authors also wish to express thanks to Professor I. L. Hill for his helpful suggestions and cooperation.

The authors extend appreciation to Mr. W. T. Jaycox for his excellent airbrush work, and especially to Mr. Eugene Mysiak, who assisted the authors in preparing most of the drawings in the book. In addition, the authors are indebted to Mr. Philip Burness for valuable ideas in the chapter on Developments and Intersections.

The authors hope that both teachers and students will continue to feel free to write in their suggestions and criticisms.

Henry Cecil Spencer
Waco, Texas

John T. Dygdon
Chicago, Illinois

ACKNOWLEDGMENTS

Ajax Flexible Coupling Co., Inc.
Allen Bradley Company
American Concrete Institute
American Council on Education
American Institute of Steel Construction, Inc.
American Welding Society
Bethlehem Steel Corporation
The Bassick Co.
The Boeing Company
Boston Gear Works
Brown & Sharpe Mfg. Co.
Charles Bruning Company
Buffalo Forge Co.
Caterpillar Tractor Co.
Chicago Public Schools
Cincinnati Milling Machine Co.
Cincinnati Shaper Co.
Eugene Dietzgen Co.
Dodge Mfg. Corp.
Eagle Signal, Div. E. W. Bliss Co.
Eastman Kodak Co.
Federal Housing Administration
General Aniline & Film Corp., Ozalid Division
General Motors Corporation
Ross F. George
Gramercy Guild Group, Inc.
Hardware Products Co.
R. G. Haskins Co.
IIT Research Institute
Illinois Tool Works
Ingersoll Milling Machine Co.

Interstate Drop Forge Co.
Joint Industrial Council
Keuffel & Esser Co.
Kinney Mfg. Co.
Koh-I-Noor, Inc.
Lockheed Aircraft Corp.
Lufkin Rule Co.
Machine Design—Penton Publication
Minnesota Mining and Manufacturing Co.
Morse Twist Drill & Machine Co.
National Bureau of Standards
National Screw & Mfg. Co.
National Twist Drill & Tool Co.
Pennsylvania Railroad
Power Fan Manufacturers' Assn.
RapiDesign, Inc.
Robert Reiner, Inc.
E. Sorensen Tool Co.
South Bend Lathe Works
Sundstrand Machine Tool Division
Templeton, Kenly & Co., Ltd.
Texas Highway Dept.
Toledo Pipe Threading Machine Co.
Tru Point Products Inc.
Twentieth Century Fund
United States of America Standards Institute
U.S. Gypsum Company
Walker-Turner Co., Inc.
The Warner & Swasey Co.
Wettlaufer Mfg. Co.
Wood-Regan Instrument Co.
Wyatt Metal & Boiler Works

USA Standard tables in the Appendix are extracted from the indicated publications with the permission of the publisher, The American Society of Mechanical Engineers, United Engineering Center, 345 East 47th Street, New York, N.Y. 10017.

Draftsman at Work with Modern Equipment.

CONTENTS

BASIC TECHNICAL DRAWING

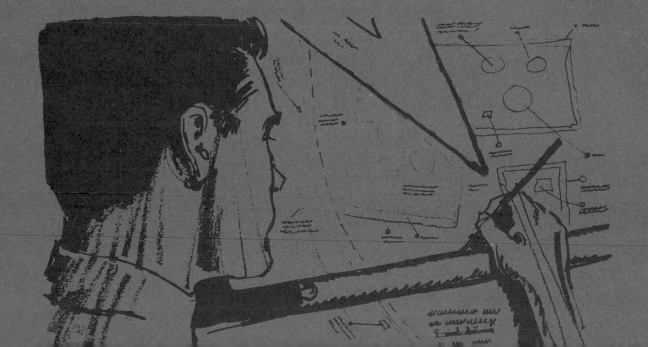

CHAPTER 1

THE GRAPHIC
LANGUAGE

1.1 Your Life's Work. Have you decided yet what your life's work will be? If not, you should start thinking very seriously now so that you can intelligently prepare for it.

The first rule in selecting a vocation is to find the line of activity in which you are deeply *interested*. Some like commercial work, some like building construction, some like scientific work, some like professional work, and so on.

The most successful people are not those who set out "to make money," but those who enter a type of work which they like better than anything else and then perform so well that people are willing to pay well for it. Do a good job, and the money will take care of itself!

Don't make the mistake of assuming that you will necessarily be successful in a given field just because your father, your uncle, or your cousin is successful. The line of work for you is what *you* like to do!

The second rule is to select a type of work that you can *do well*. Everyone has some special abilities. There are some things that you can do better than the next fellow. Your problem is to learn as much as possible about a wide variety of jobs or professions and to select the one that you like best and can do best. But how? Your library has some good books on vocational guidance. Read some of these. See your school counselor; he will give you expert advice. In the larger cities there are vocational testing agencies where you can take scientific tests to find out what your abilities and interests are.

Seize every opportunity to learn by first-hand observation any time you are around a business house, a construction job, or a manufacturing plant. Talk to people. Ask questions. Learn all you can about jobs that interest you. Continue this until you find a line of work which you like and which you can do well. Avoid types of work which clash with your known weaknesses. For example, if you are consistently poor in mathematics, do not plan to be an engineer.

Don't forget for a moment that, regardless of your interests and abilities, your success will depend very largely upon your *personality and character*. You must have the knack of getting along well with other people. You must be absolutely honest. If you are a "square shooter," everyone will know it, and if you are not, they will know that too. You must be dependable at all times. No one will ever have confidence in you or give you any important responsibility if you cannot be depended on to do what you are supposed to do. You must be a hard worker. Remember, Thomas A. Edison said, "Success is one-tenth inspiration and nine-tenths perspiration." *Think* on the job! Do things that ought to be done *before* the boss reminds you. Be cooperative. Be a team worker. Industry has no place for people who cannot work with others.

In this wonderful age of science and invention, opportunities are greater than ever before. Each new discovery, or invention, or improvement opens up new horizons for scientists, engineers, designers, draftsmen, mechanics, technicians, salesmen, business men and women, teachers, lawyers, manufacturers, and many others to make still further advancements. There is a job waiting for every person, regardless of his particular abilities, if he will only find it.

1.2 The Graphic Language. In a technical civilization such as ours, there are many fine opportunities in technical or mechanical fields or in allied fields. If you are interested in planning or building things, you may find your life's work somewhere in this area. To help you "find yourself," many schools offer a wide variety of shop or technical courses, such as woodwork, machine shop, electrical work, drafting, and so on. Drafting is considered basic to all of these. It is the principal means of expression of ideas in a technical world—a *graphic language* which has its own alphabet, grammar, and penmanship. It has been truthfully said that the industrial history of the United States has been written in terms of the graphic language. If you do not understand this language, you will be in a real sense *illiterate*. Even if you find yourself in a field only indirectly associated with industry, a knowledge of the graphic language is essential in order for you to read blueprints. Regardless of your future vocation, you will have many uses for your knowledge of drafting.

Drawing is the oldest type of written expression and is understood the world over. A word is an abstract symbol representing a thing or an idea, but a picture represents an object as it *is*. "One picture is worth a thousand words," said Confucius. To understand the truth of this, try to tell in words how to build a footstool, a house, or a machine. Imagine telling someone how to build a six-jet bomber! You will find that you cannot accurately and completely describe in words how to make even a simple screw or gear. But no object is so complicated that it cannot be drawn. In fact, if it cannot be drawn first, it cannot be built!

Drawing, entirely apart from its uses in industry, has great value to nontechnical people simply as a means of expressing ideas effectively. Usually this is done by means of a freehand sketch on the back of an envelope or on a piece of scratch paper. How many times have you heard someone say, when words have failed, "Oh, I guess we'll have to draw him a picture"? If a person has good ideas, but no effective way of expressing them to other people, the ideas are apt to get nowhere.

1.3 Industrial Drafting. Aside from general idea sketches, there are two main classes of drawings: (1) artistic and (2) technical. The artist expresses philosophic or aesthetic ideas or emotions. When he draws *things*, he draws them as they *appear* to him emotionally, and every artist sees things in his own peculiar way. The technical man is concerned with actual objects, and his drawings show not only how they appear but how they *are*. Technical drawing is an exact means of expression, and accuracy is the main objective.

Every new invention or development starts with an idea in the mind of the originator. If he is an engineer, an inventor, or a designer, he will make the drawing himself because he is the only person who can express exactly what he has in mind. Such a person is usually well trained in drafting anyway, and finds it easy to set an idea down in this way. The difficulty is getting the new idea! Usually such an idea develops through several stages, starting probably with a *freehand sketch*. This is followed by one or more *mechanical drawings,* or *layouts,* drawn accurately with instruments.

Whenever the designer is satisfied with the general scheme, the *working drawings* are made to be used in the shop. The designer may make these himself or he may turn the work over to a *draftsman*. In the course of making these working drawings, many details of construction may be worked out, and the draftsman may be required to be somewhat of a designer himself. Salaries of draftsmen depend pretty largely upon how much "headwork" or designing they are able to do. The young draftsman is in a very favorable position to learn all about the products of a company and to advance rapidly to key positions if his ability warrants it.

Let's do one thing at a time. Your job now is to learn how to make drawings skillfully, correctly, and rapidly. Later on, if you have any ideas to express, you can express them. These ideas will come as a result of further knowledge and actual experience in industry.

1.4 Working Drawings. A *working drawing* is a complete drawing or set of drawings such that the object represented can be built from it alone without additional information. Such a drawing is a *description* of the object and is composed of two parts, the *views* and the *dimensions*.

Just as in written language, the graphic language has its grammar and its symbols. The *alphabet of lines* and the rules of presentation are examples of this.

By means of *views*, the mechanic can visualize the object to be built; if the drawing is right, he will visualize exactly what the designer has in mind. This ability to visualize or "think in three dimensions" is essential to the designer, the draftsman, the mechanic, and all other people in technical work. It is one of the principal values to be derived from the study of industrial drafting. It is also one of the best means of developing the "constructive imagination" so essential in all original designing.

By means of *dimensions,* the mechanic can tell exactly *what size* each part is to be. This applies not only to large dimensions or rough measurements, but in many cases to extremely accurate measurements often down to a tenthousandth of an inch or less.

Thus, the working drawing gives a complete story, and *industrial prints* made from it can be mailed to factories in any part of the country or of the world. Also, all the separate parts can be made exactly as planned and then shipped to a central assembly plant where they will fit into the assembly as intended.

1.5 Technical Drawing and Specialized Branches. The term *mechanical drawing* is widely used, especially in the public schools, to describe industrial drawing. However, this term does not cover *freehand sketching,* which is an important part of the subject. *Technical drawing* is rapidly becoming the accepted term, because it more accurately suggests the broad scope of drawing for industry.

Technical drawing is composed of many specialized types of drawing applied to the various fields. *Architectural drawing* is used in the building industry; *machine drawing* in the machine industries; *structural drawing* in the construction industries where structural steel is used, as in large buildings and bridges; *sheet-metal drawing* in the heating, ventilating, and air-conditioning industries; *electrical drawing* in the electrical industries; *aeronautical drafting* in aircraft manufacturing;

marine drawing in ship construction, and so on. In all these the term *drafting* is often used instead of *drawing*.

The terms *engineering graphics*, *graphic science*, and *graphics* are generally used to describe course work in drawing at the college level. These terms characterize accurately the scope of work in drawing at this level, which includes graphic solutions to engineering and mathematical problems as well as the description of objects for manufacture.

1.6 Aims in Technical Drawing. *First*, a detail drawing must be *accurate*, Sec. 3.23. The requirements of industry are exacting, and the draftsman must have or acquire habits of accuracy in everything he does. A drawing which is not accurate may be completely worthless or may lead to costly errors by those who are depending upon it.

Second, a detail drawing must be executed with the proper *technique*, or good workmanship, Sec. 7.2, which means that the lines must have "sparkle" or "snap" and exhibit good contrasts of lines. A "sloppy" drawing does not have good technique and is likely to be incorrect or unclear.

Third, a detail drawing should be *neat*. Neatness is a habit which can be acquired. It is promoted by observing orderly arrangement and handling of the equipment and by

taking positive measures to keep the drawing clean, Sec. 3.18.

Fourth, a detail drawing must be made with *speed*, for "time is money," and the slow draftsman will soon find himself looking for another job. Speed in drafting comes from mental and physical alertness; it comes naturally as a result of concentration on the job and of intelligent planning, and not from haphazard hurrying. Slowness is the inevitable product of a dull or disinterested mind. Often a slow draftsman would be a fast draftsman if he worked continuously and did not waste valuable time talking, daydreaming, or trying to work solutions out by trial and error without first learning the fundamental principles.

The following quotation from the *Chevrolet Draftsman's Handbook* illustrates the attitude of industry toward the draftsman's work:

"Our drawings are considered by the management, the buyers, the outside sources, the pattern and die makers, the inspectors, and the shop foreman as the last and only word in specifications. Our drawings must stand alone in conveying the ideas of the Chief Engineer to the thousands of people who use them. They must tell all that needs to be known about the parts they represent. They must be so clear and complete that every one of the thousands of users arrives at exactly the same interpretation."

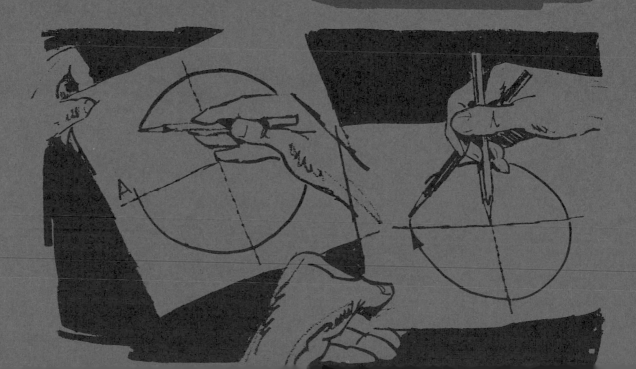

CHAPTER 2

FREEHAND SKETCHING

2.1 Why Learn to Sketch? The importance of freehand sketching to the draftsman, the engineer, or the average person in technical or nontechnical work cannot be overemphasized. It is a valuable means of expression to anyone—an effective way to get an idea across when words fail. In this way, the graphic language becomes a valuable aid to the verbal language.

Most original mechanical ideas or inventions are recorded for the first time in the form of a sketch, Fig. 2-1. The sketch helps the de-

Courtesy The Boeing Company

Fig. 2-1. New Ideas Are Developed Through Sketches.

7

signer to clear up his ideas and to recall from day to day what was figured out before. The sketch also is used to show others what the designer has in mind. In case of a lawsuit over a patent, the original sketch may connect the idea to the inventor.

Sketches are often used instead of complete mechanical drawings where changes of design must be made in a hurry, or where time is not available for a finished mechanical drawing to be made. The greatest use of sketches, however, is in formulating, expressing, and recording new ideas in technical work.

2.2 Sketching Materials. Only three objects are needed to make a freehand sketch: a pencil, an eraser, and a piece of paper. Sketches are often made on the backs of envelopes or on scraps of paper, but the draftsman or engineer usually has a sketching pad or several sheets fastened to a clip board. Cross-section paper is often used, Fig. 2–2. The ruled lines help to keep the sketch lines straight, and the squares (usually $\frac{1}{8}''$ or $\frac{1}{4}''$) can be used to sketch approximately to scale, if desired. However, most sketches need not be drawn to scale, but only in proportion, Sec. 2.7. It is desirable for the beginner to learn to sketch well as soon as possible without the aid of cross-section paper.

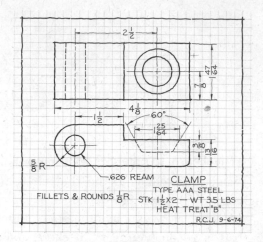

Fig. 2–2. Sketch on Cross-Section Paper.

For sketching, always use a soft pencil, such as F or HB. Two erasers are desirable, an Artgum and an ordinary pencil eraser, Fig. 3–24 (a) and (b).

2.3 Sharpening the Pencil. Sharpen the pencil to a conical point, Fig. 2–3. Three thicknesses of lines are used in sketching: *thin*, *medium*, and *thick*, as shown at (a), (b), and (c). To make these, the pencil should be *sharp*, *nearly sharp*, or *slightly dull*, respectively. All three should be clean-cut and *dark*. Avoid fuzzy, gray, or sloppy lines. Only construction lines should be *very light and gray*, (d).

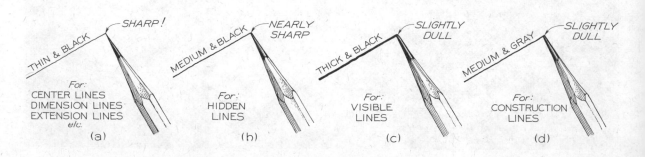

Fig. 2–3. Pencil Points.

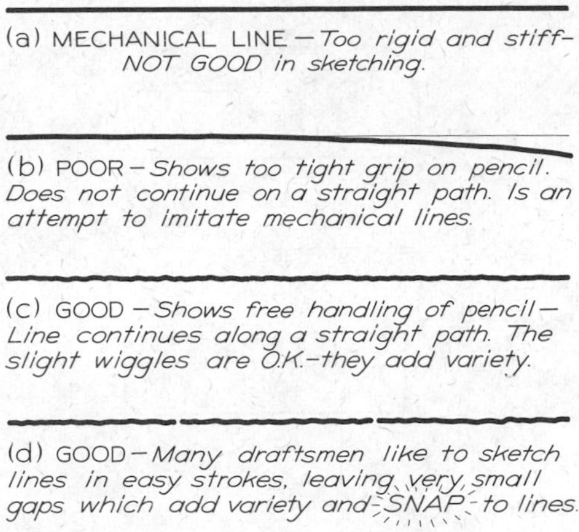

(a) MECHANICAL LINE — *Too rigid and stiff—NOT GOOD in sketching.*

(b) POOR — *Shows too tight grip on pencil. Does not continue on a straight path. Is an attempt to imitate mechanical lines.*

(c) GOOD — *Shows free handling of pencil— Line continues along a straight path. The slight wiggles are O.K.—they add variety.*

(d) GOOD — *Many draftsmen like to sketch lines in easy strokes, leaving very small gaps which add variety and "SNAP" to lines.*

Fig. 2–4. Character of Lines.

2.4 Character of Lines. A good freehand line should not be rigid and stiff like a mechanical line, Fig. 2–4 (a). The effectiveness of a mechanical line lies in its *exacting uniformity,* and no attempt should be made in sketching to imitate mechanical lines. The important thing is to sketch the line in the right direction, and not as shown at (b).

A good freehand line has the quality of *freedom* and *variety,* and continues in the correct path, as shown at (c) and (d). Long lines should not be drawn in a single stroke, but with several strokes end-to-end, the hand being shifted after each stroke. Small gaps (if *very small*) may be left between strokes if desired, (d). All final lines should be clean-cut and dark. They should never overlap or be sloppy and indefinite. Avoid uncertain intermingled lines, appropriately referred to as "hen-scratch lines."

2.5 Straight Lines. Hold the pencil naturally, about $1\frac{1}{2}$ inches from the point. To draw horizontal lines, Fig. 2–5 (a), first spot your beginning and end points, then swing the pencil

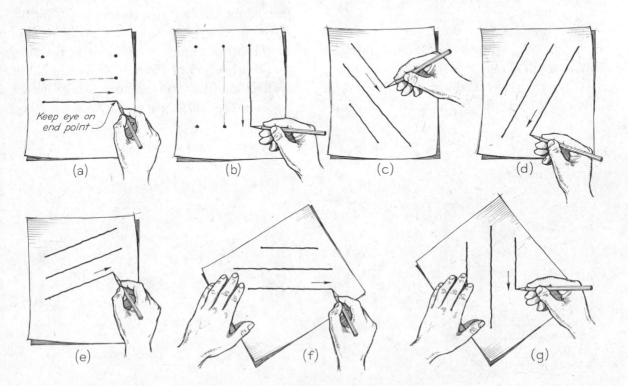

Keep eye on end point

(a)   (b)   (c)   (d)

(e)   (f)   (g)

Fig. 2–5. Sketching Straight Lines.

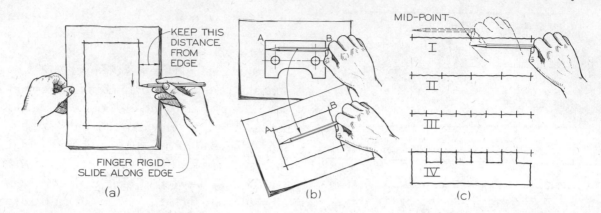

Fig. 2-6. Aids in Sketching.

back and forth between the points, barely touching the paper, until the direction is clearly established, and finally draw the line firmly with a free and easy wrist-and-arm motion, *keeping the eye on the point toward which you are drawing*, not the pencil point. This is something like golf, where you watch the ball, not the club.

Draw vertical lines downward, or toward your stomach, with a free finger-and-wrist motion, as shown at (b). If an inclined line is to be nearly vertical, draw it downward, as shown at (c) and (d). If it is to be nearly hori-

zontal, draw it to the right, (e). You can draw an inclined line as a horizontal or a vertical line merely by turning the paper to the desired position, (f) and (g).

2.6 Aids in Sketching. In sketching, any method you can use which requires only your pencil, paper, and your hands is permitted. If you are clever, you can use many "tricks" to aid in sketching. In Fig. 2-6 (a) is shown a useful method of blocking in horizontal or vertical lines by drawing lines parallel to the edges of the tablet or pad. At (b) is a method

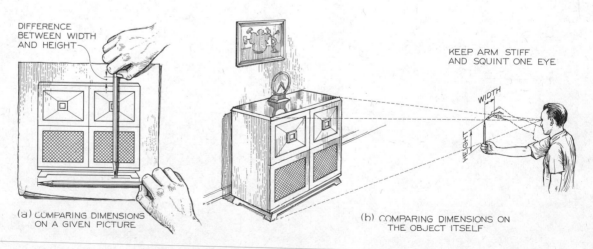

Fig. 2-7. Estimating Proportions of a Television Cabinet.

of transferring a distance by using a pencil as a measuring stick. At (c) is a method of dividing a line into a number of equal parts. At I, the pencil is used to estimate half the distance by trying an estimated distance on the left and then on the right. At II and III, the divisions are further subdivided by eye. The final drawing is shown at IV.

2.7 Estimating Proportions.

You have heard the advice, "Be the labor great or small, do it well or not at all." This rule applies particularly to *proportions* in sketching. The proportions "make or break" a sketch! A drawing is *in proportion* if all areas of it are the correct sizes as compared to all other areas. The larger the size, the more important it is to get it in correct proportion.

For example, if you want to sketch the front view of a television cabinet, you must first get the width and height correctly proportioned. If you are working from a given picture, Fig. 2–7 (a), you can compare measurements with your pencil, as shown. If you are sketching directly from the object itself, you can compare measurements by sighting dimensions on the object and noting how long they appear on your pencil, as shown at (b). For these comparisons, always hold your pencil at arm's length.

Another method, if you are working from a picture, is to mark off by eye convenient units on the picture, as shown in Fig. 2–8 (a). Or use a strip of paper upon whose edge you have marked off one unit, and see how many units high and how many units wide the draw-

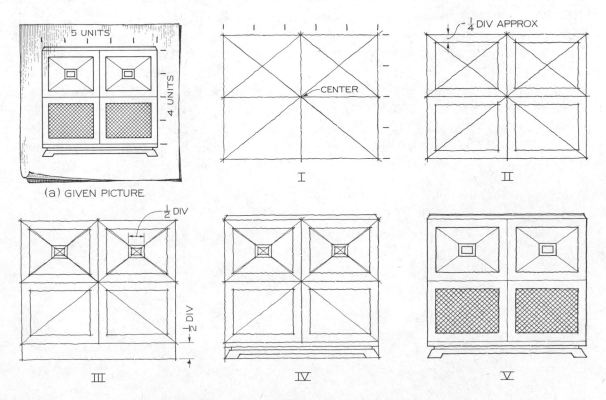

Fig. 2–8. Steps in Sketching Television Cabinet.

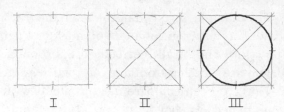

Fig. 2–9. Starting with a Square.

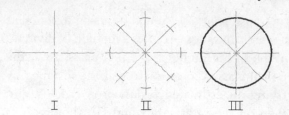

Fig. 2–10. Starting with Center Lines.

ing is. In this case, it is convenient to mark off five units for the width, and it is found that about four of these units are required for the height of the main cabinet portion. Start the sketch, I, by sketching the large rectangle four units high and five units wide, carefully comparing each unit by eye with every other unit. Then draw light diagonals to locate the center, and through the center draw a horizontal line and a vertical line to divide the rectangle into four equal panels. Next, II, sketch the frames of the panels, making them $\frac{1}{4}$ unit wide. Then add the remaining details, as shown at III and IV. All lines should be very light up to this point.

Finally, dim all the lines with the Artgum until they can hardly be seen, and then "punch in" the final lines, V, making them clean-cut and dark. The main outline should be slightly heavier than the interior lines.

Remember, *the secret of sketching is to draw the large areas first in correct proportion, and then to add the smaller features in their correct relative sizes.*

2.8 Circles and Arcs. You can easily sketch a small circle, Fig. 2–9 (I), by first lightly sketching the enclosing square and marking the midpoints of the sides. Then, II, draw light diagonals and mark off the estimated radius-distance on each, and finally draw the circle through the eight points, III.

Another method is to start with the center lines, Fig. 2–10 (I). Add light radial lines, or "spokes," in between these as at II, and sketch small arcs at the radius-distance from the center on each. Finally, III, sketch the full circle.

In both methods, keep the construction lines very light. If necessary, dim all lines with Artgum before heavying-in the final circle.

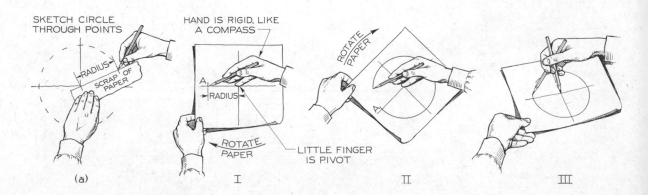

Fig. 2–11. Sketching Large Circles.

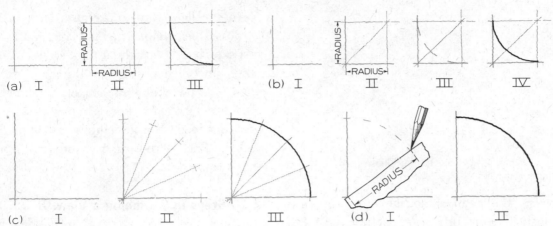

Fig. 2–12. Sketching Arcs.

Another excellent method, Fig. 2–11 (a), especially for large circles or arcs, is to set off the radius on a scrap of paper and to use it to set off from the center as many points on the circle as desired. The circle is then sketched through these points.

After a little practice, you can make excellent large circles by using your pencil and hand as a compass. Place the little finger at the center to serve as a pivot, and set the pencil point at the radius-distance from the center, Fig. 2–11 (I). While the hand is held rigidly in this position, rotate the paper slowly, as shown at II, while the pencil marks the circle. You can use the same procedure while holding two pencils rigidly in position, III.

In sketching arcs, Fig. 2–12, use the same general methods as in sketching circles. Where the construction lines are too noticeable, dim the lines with the Artgum before heavying-in the final arcs.

2.9 Sketching Ellipses. If you look straight at circles, they appear as true circles, Fig. 2–13 (a). If you view circles at an angle, they appear as *ellipses*, (b), and if you view them

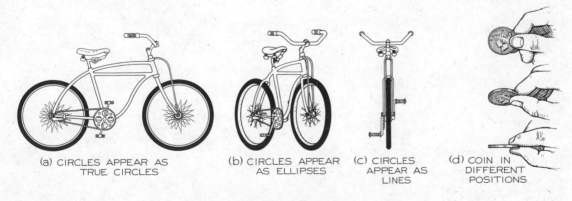

(a) CIRCLES APPEAR AS TRUE CIRCLES

(b) CIRCLES APPEAR AS ELLIPSES

(c) CIRCLES APPEAR AS LINES

(d) COIN IN DIFFERENT POSITIONS

Fig. 2–13. Circles and Ellipses.

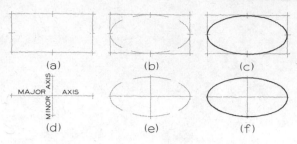

Fig. 2–14. Sketching Ellipses.

An excellent way to sketch a large ellipse is to use the *trammel method*, Fig. 5–21. You first sketch the two axes at right angles to each other, and then prepare a "trammel" on the edge of a piece of scrap paper, as shown in the figure. As the trammel is moved to different positions, points on the ellipse can be marked and the final ellipse can be sketched through these points.

edgewise, they appear as lines, (c). You can demonstrate this principle another way by viewing a coin in different positions, as shown at (d). The long axis of an ellipse is called the *major axis*, and the short axis the *minor axis*.

To sketch an ellipse, Fig. 2–14 (a), sketch lightly the enclosing rectangle and mark the approximate mid-points of the sides. Then, (b), sketch light tangent arcs at the mid-points, and complete the ellipse, (c). Before heavying-in the final ellipse, dim all construction with the Artgum.

A second method is to start with the major and minor axes, and sketch the ellipse as shown at (d) to (f).

2.10 Steps in Sketching a View. If we view the Lock Plate, Fig. 2–15 (a), in the direction of the arrow, we see a "view" of the object. The steps in sketching this view are shown in the figure. The sketch is not made to scale, but is carefully proportioned.

I. Sketch in the large main areas lightly. This is the most important part of the sketch. No sketch can be satisfactory if these main areas are incorrectly proportioned.

II. Block in construction for areas and circles, lightly.

III. Dim all lines with the Artgum, and then heavy-in all final lines, making them clean-cut and dark.

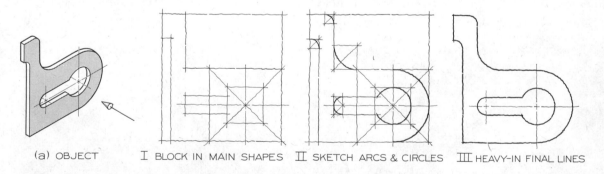

(a) OBJECT   I BLOCK IN MAIN SHAPES   II SKETCH ARCS & CIRCLES   III HEAVY-IN FINAL LINES

Fig. 2–15. Steps in Sketching Lock Plate.

2.11 Sketching Problems. A number of one-view sketching problems are given in Figs. 2–16 to 2–25. For each problem, use an $8\frac{1}{2}'' \times 11''$ sheet of cross-section paper, plain bond typewriter paper, or drawing paper. Sketch a border, leaving a margin of about $\frac{1}{2}''$ on all sides; then sketch the assigned problem carefully in proportion, fitting it on the sheet approximately as shown. Write the date and your name at the bottom of each sheet.

For problems 2–24 and 2–25, you may sketch the plans as shown; or, with your instructor's approval, you may sketch from your own home.

Additional one-view sketching problems are available in Figs. 3–44 to 3–65. Sketching from actual objects similar to those shown below will be most helpful. Select some object and obtain your instructor's approval before you start work.

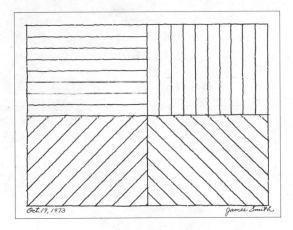

Fig. 2–16. Straight Lines.

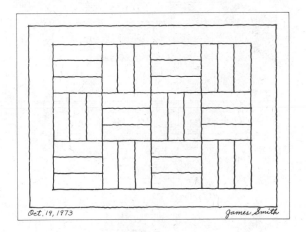

Fig. 2–17. Parquet Floor.

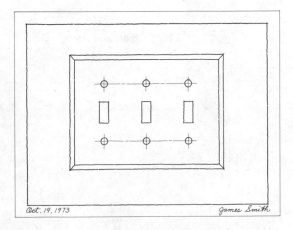

Fig. 2–18. Switch Cover.

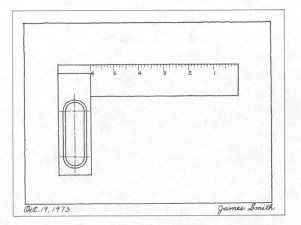

Fig. 2–19. Try Square.

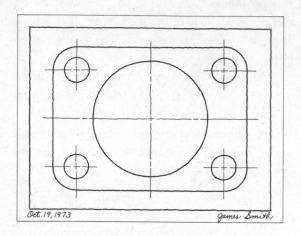

Fig. 2–20. Cover Gasket.

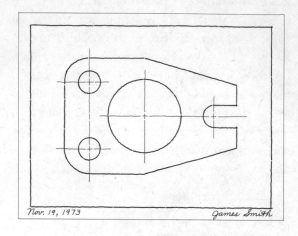

Fig. 2–21. Stamping.

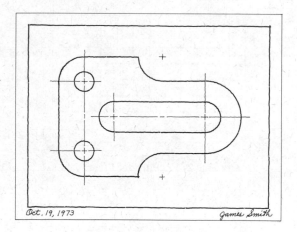

Fig. 2–22. Stamping.

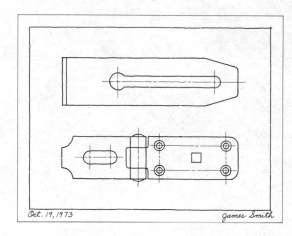

Fig. 2–23. Plane Blade and Hasp.

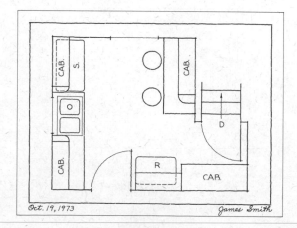

Fig. 2–24. Kitchen Plan.

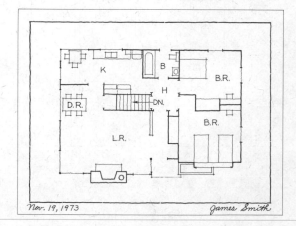

Fig. 2–25. House Plan.

CHAPTER 3

MECHANICAL DRAWING

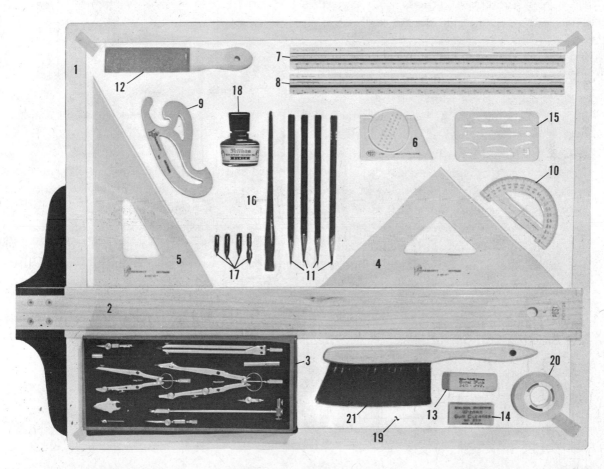

1. Drawing board (approx. 20″ × 24″)
2. T-square (24″ transparent edge)
3. Set of drawing instruments
4. 45° triangle (8″ sides)
5. 30° × 60° triangle (10″ long side)
6. Ames Lettering Instrument (or lettering triangle)
7. Engineers scale
8. Architects scale
9. Irregular curve
10. Protractor

11. Drawing pencils (2H, 4H, F, HB)
12. Sandpaper pad (or file)
13. Weldon Roberts Coral Pink (India) or Eberhard Faber Ruby pencil eraser
14. Weldon Roberts Gum Cleaner or Artgum eraser
15. Erasing shield
16. Pen staff
17. Gillott's 303 and 404 pen points; No. 12 Henry Tank Pen
18. Black drawing ink

19. Drawing paper
20. Drafting tape (or thumbtacks)
21. Dusting brush (or dust cloth)

Optional equipment (not illustrated above):

Pencil lead pointer
Mechanical pencils with leads
Cleaning powder (or pad)

Fig. 3–1. Equipment Used in Technical Drawing.

3.1 What Is Mechanical Drawing? Most people who think they could never learn to draw just "think" they couldn't. Their usual apology is: "I never could learn to draw—I can't even draw a straight line." This is in a sense true; no one can draw a really straight line without a guiding edge.

A clear distinction should be made between *mechanical drawing* and *freehand drawing* or *sketching,* both of which are used to express the graphic language. As a matter of fact, Fig. 2–4, a good freehand sketch should not be drawn with rigidly straight lines; the lines should have a certain freedom and variety, unlike mechanically drawn lines. Mechanical drawings, however, with which we will concern ourselves in this chapter, are made with precision drawing instruments and require no artistic or special ability. Any intelligent person can learn to execute good mechanical drawings rapidly and skillfully.

3.2 Drawing Equipment. The principal items of drawing equipment used by the draftsman

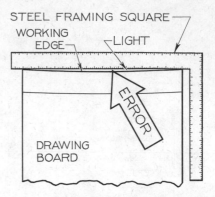

Fig. 3–2. Testing Working Edge of Board.

are shown in Fig. 3–1, and their uses are described on the following pages. Since good drawing instruments (item 3) are fairly expensive, and since it is difficult for beginners to tell inferior instruments from those of high quality, you should consult your instructor before purchasing.

3.3 Drawing Boards. At least one end of the drawing board must be true. Test both ends of the board with a framing square, Fig. 3–2,

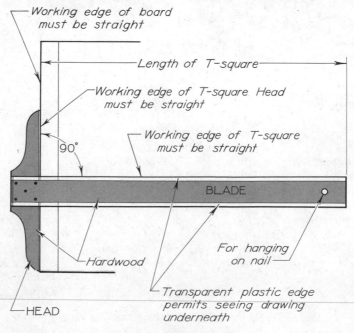

Fig. 3–3. T-Square.

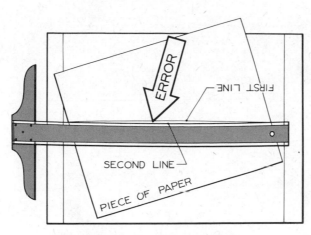

Fig. 3–4. Testing T-Square Blade.

or with a T-square whose blade is known to be true. If neither edge of the board is true, use a hand plane or a jointer to produce at least one straight edge. Mark this edge "working edge." Always use your T-square head against this edge.

3.4 T-Squares. The T-square is composed of two parts, the *head* and the *blade,* Fig. 3–3. They must be rigidly fastened at right angles to each other, and their *working edges* must be straight. Test the working edge of the head with a framing square, a triangle, or any true straightedge. You can easily test the working edge of the blade, Fig. 3–4, by drawing an accurate sharp line along the working edge on a piece of paper, then turning the paper around until the other side of the line is against the T-square, and drawing a second line. If the two lines do not coincide, the space between them represents *double the error* of the blade.

A new but faulty T-square should be returned to the dealer. Slight errors in the working edges can be corrected with a fine sandpaper block, but to avoid ruining the T-square, you should obtain your instructor's assistance.

Do not use the T-square to drive tacks into the board, or for any rough purpose. Never cut paper along its working edge, as this will produce nicks which will ruin the T-square.

3.5 Drawing Paper. White, cream, and light green drawing papers are used, with greatest preference for the cream or "buff." White paper is used for display drawings, but seldom for working drawings because it soils so easily. The light green paper is used by some to lessen eyestrain.

Most industrial drawings are made today in pencil directly on tracing paper, vellum, pencil tracing cloth, or tracing film so that blueprints can be made from them immediately without "tracing," Sec. 7.1. The use of ink has decreased considerably in recent years. When ink is used, it is generally applied to tracing cloth. For inking, see Secs. 8.1 to 8.6.

Two systems of paper sizes are in general use and are now USA Standard. One is based on $9'' \times 12''$ (trimmed size), followed by the multiples $12'' \times 18''$, $18'' \times 24''$, etc. The other is based on $8\frac{1}{2}'' \times 11''$, as follows: $8\frac{1}{2}'' \times 11''$, $11'' \times 17''$, $17'' \times 22''$, $22'' \times 34''$, and $34'' \times 44''$. These sheets when folded fit into standard letter files.

For sheet layouts for problems, see Appendix pages 474–475.

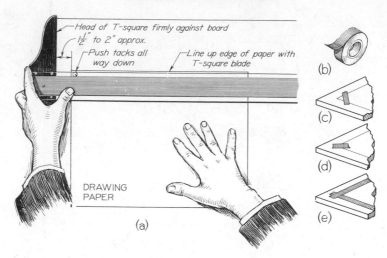

Fig. 3–5. Fastening Paper to Drawing Board.

3.6 Fastening Paper to the Drawing Board. Place the paper fairly close to the working edge of the board, Fig. 3–5 (a), to decrease the error from the slight swing or "give" of the T-square blade, and about equally far from the top and bottom of the board. Insert one thumbtack in the upper left corner, pushing it down firmly. Adjust the paper with the right hand about this tack as a pivot until the top edge lines up with the T-square; then insert the second tack in the lower right-hand corner, followed by tacks in the other two corners. Smooth the paper from the center toward each new tack before inserting it. For larger sheets use additional tacks as necessary. For small flat sheets two tacks in the upper corners may be sufficient.

Left-handers: Place the working edge of the drawing board and the head of the T-square on your right and insert the first tack in the upper right corner.

In recent years, drafting tape has become

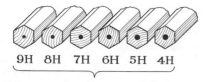

9H 8H 7H 6H 5H 4H

HARD

The harder pencils in this group (left) are used where extreme accuracy is required, as on graphical computations, charts and diagrams. The softer pencils in this group (right) are used by some for line work on engineering drawings, but their use is restricted because the lines are apt to be too light.

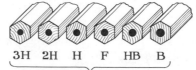

3H 2H H F HB B

MEDIUM

These grades are for general-purpose work in technical drawing. The softer grades (right) are used for technical sketching, for lettering, arrowheads, and other freehand work on mechanical drawings. The harder pencils (left) are used for line work on machine and architectural drawings. The H and 2H pencils are used on pencil tracings for blueprinting.

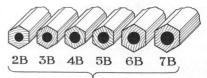

2B 3B 4B 5B 6B 7B

SOFT

These pencils are too soft to be useful in mechanical drafting. Their use for such work results in smudged, rough lines which are hard to erase, and the pencil must be sharpened continually. These grades are used for art work of various kinds, and for full-size details in architectural drawing.

Fig. 3–6. Pencil Grade Chart.

Fig. 3–7. Mechanical Pencil.

popular as a paper fastener. It is available in rolls, Fig. 3–5 (b), and is applied in a variety of ways, as shown at (c), (d), and (e). The advantage of tape is that it does not damage the board as thumbtacks do and may be used on hard surfaces, such as masonite, steel, or glass. Some prefer transparent cellulose tape, which does not tend to roll up, but is more difficult to remove.

3.7 Drawing Pencils. High-quality *drawing pencils* should be used in industrial drafting—never ordinary writing pencils. Your drawing pencils are your most important tools. The leads are made of graphite with clay added in varying amounts to make eighteen grades from 9H (the hardest) down to 7B (the softest), Fig. 3–6. The general uses of the different grades are explained in the figure.

Short pencil stubs should not be used (less than 3″), as they are too short to be handled properly. Stubs can be used if inserted in a *pencil holder,* which extends their length. Many makes of mechanical pencils are also

available, together with refill drafting leads in all grades, Fig. 3–7. These are preferred by the professional draftsman, and are recommended for students who can afford the higher cost.

3.8 Choice of Pencil. The grade mark, Fig. 3–8 (b), is supposed to indicate the exact hardness of lead, but actually it is only approximate. Since you cannot depend entirely on the grade mark, you must learn to use judgment in selecting pencils for the kind of lines required.

For light *construction lines,* Sec. 3.10, guide lines for lettering, and for constructions where great accuracy is necessary, use a hard pencil, such as 4H, 5H, or 6H. For general line work and lettering, all lines should be dark, so they will show up clearly. Select a pencil that is soft enough to produce a black line but does

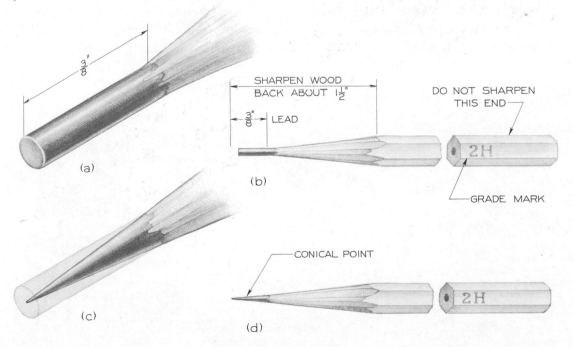

Fig. 3–8. Pencil Points.

not smudge too easily and require too-frequent sharpening, such as an F or H. The texture of the paper must be considered; for example, a hard paper will take a harder pencil than will a soft paper. Also, the weather must be considered; for example, in humid weather, all papers tend to soften and require softer pencils. The draftsman must learn to select the pencil that produces the quality of line he needs.

3.9 Sharpening the Pencil. *Keep your pencil sharp!* This is certainly the instruction most frequently needed by the beginning student. A dull pencil produces fuzzy, indefinite, sloppy lines, which typify the listless and careless student. Only a *sharp* pencil can produce accurate, clean-cut, dark lines that sparkle with clarity and are characteristic of the alert and skillful draftsman.

First, Fig. 3–8 (a) and (b), with a sharp pocket knife (never a razor blade or other crude makeshift), cut the wood away from the *unmarked end* of the pencil a full $1\frac{1}{2}''$ from the point, leaving about $\frac{3}{8}''$ of lead extending uncut beyond the wood. Pencil sharpeners,

with special draftsman's cutters, are available for cutting away the wood in this manner.

Second, Fig. 3–8 (c) and (d), to produce the conical point (for all line work and lettering), dress the lead down to a long, sharp, symmetrical cone by rubbing it while rotating it on a sandpaper pad or file, Fig. 3–9 (a). *Keep the pencil almost flat on the file or pad.* Many draftsmen prefer to finish the point by "burnishing" on a scrap of rough paper, such as drawing paper, (b).

Be careful not to get loose graphite on your drawing, your tools, or your hands. Never sharpen your pencil over your drawing or over any of your equipment, but to one side, where the loose graphite will fall on the floor, Fig. 3–9 (c). After sharpening, wipe the point on a clean cloth (not your handkerchief). Keep your sandpaper pad or file in an envelope to prevent it from soiling your equipment. See Fig. 3–25 (5).

If a mechanical pencil is used, a conical point may be produced with the sandpaper pad or file as described before, or by using a lead pointer, Fig. 3–10. The pencil lead pointer is especially designed for pointing leads in standard drafting mechanical pencils and

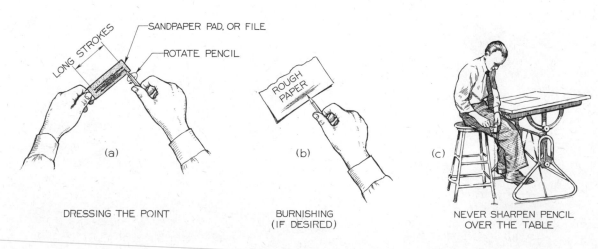

DRESSING THE POINT

BURNISHING
(IF DESIRED)

NEVER SHARPEN PENCIL
OVER THE TABLE

Fig. 3–9. Dressing the Lead.

Courtesy Tru Point Products, Inc.

Fig. 3–10. Pencil Lead Pointer.

wooden pencils after the wood has been cut back. The pencil lead is inserted in the receptacle in the top of the pointer and rotated to shape the point.

3.10 Alphabet of Lines.

The lines used in drafting are often referred to as the Alphabet of Lines, Fig. 3–11, because they have the same relation to drawings that letters do to words. Pencil lines are shown on the left, and ink lines on the right. The heavier types of ink lines are thicker than the corresponding pencil lines, while the thinner lines are about the same in pencil or in ink.

There are three distinct thicknesses of lines, as follows: (1) *thick* (border lines, visible lines, cutting-plane lines, and short-break lines), (2) *medium* (hidden lines), and (3) *thin* (long-break lines, section lines, center lines, dimension lines, extension lines, and phantom lines).

For pencil drawing, all lines except the construction line should be dense black—never gray, fuzzy, or indefinite—so that they will reproduce clearly by blueprinting or otherwise, Secs. 7.3 to 7.6. The thin lines, such as the center line, should be just as dark as the visible or hidden lines—the contrast being in the thickness and not in the degree of blackness. A fairly soft pencil, such as F or H, should be used for the thicker lines, and a slightly harder pencil, such as the 2H, should be used for the thinner lines.

Construction lines should always be extremely light—so light that they can be barely seen when viewed at arm's length—and made with a hard pencil, such as the 4H. Construction lines are used for "blocking in," or constructing a drawing before the lines are made heavy, Fig. 3–45 (V). They are also used for guide lines for lettering, Sec. 4.7.

Applications of the various lines are illustrated in the center column of Fig. 3–11.

On a drawing the visible lines should be the outstanding feature.

3.11 Horizontal Lines.

To draw a horizontal line, Fig. 3–12 (a): With the left hand press the T-square head firmly against the working edge of the board, and with the same hand smooth the blade to the right, pressing the blade firmly against the paper. Then draw the line from left to right, with the little finger gliding lightly along the T-square blade. At the same time, in order to produce lines of uniform width, rotate the pencil slowly by pressing the thumb forward so as to roll the pencil. This prevents the pencil from wearing down in one place, which would increase the thickness of the line and make the point lopsided.

Lean the pencil at an angle of about 60° with the paper in the direction of the line, as shown by the triangle in Fig. 3–12 (b), while keeping the pencil in a vertical plane. In this

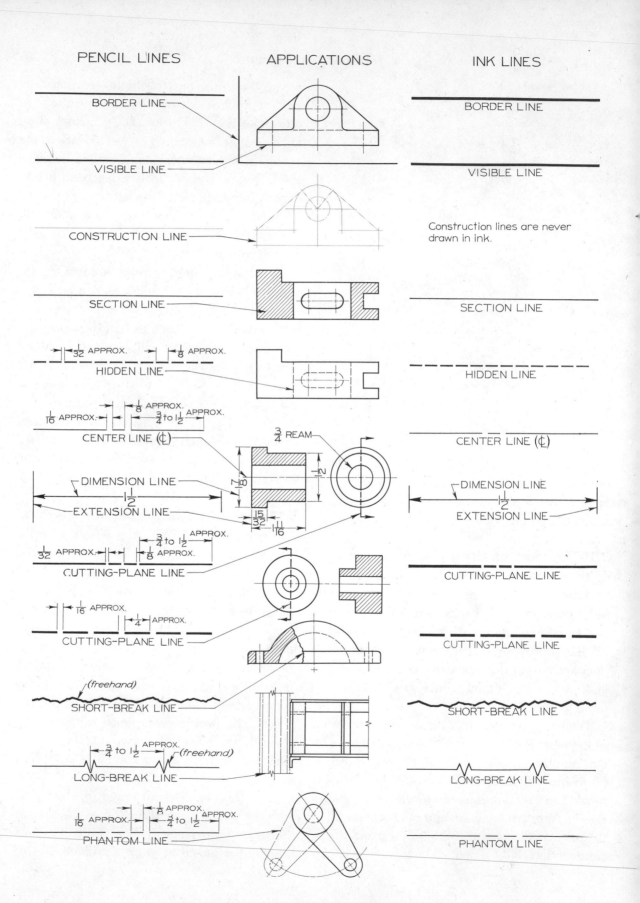

PENCIL LINES   APPLICATIONS   INK LINES

BORDER LINE —   BORDER LINE

VISIBLE LINE —   VISIBLE LINE

CONSTRUCTION LINE —   Construction lines are never drawn in ink.

SECTION LINE —   SECTION LINE

$\frac{1}{32}$ APPROX. $\frac{1}{8}$ APPROX.
HIDDEN LINE —   HIDDEN LINE

$\frac{1}{16}$ APPROX. $\frac{1}{8}$ APPROX. $\frac{3}{4}$ to $1\frac{1}{2}$ APPROX.
CENTER LINE (℄) —   CENTER LINE (℄)

$\frac{3}{4}$ REAM

—DIMENSION LINE
$1\frac{1}{2}$ —DIMENSION LINE
—EXTENSION LINE— $1\frac{1}{2}$
    EXTENSION LINE—

$\frac{7}{8}$ $\frac{1}{2}$ $\frac{15}{32}$ $\frac{11}{16}$

$\frac{3}{4}$ to $1\frac{1}{2}$ APPROX.
$\frac{1}{32}$ APPROX. $\frac{1}{8}$ APPROX.
CUTTING-PLANE LINE —   CUTTING-PLANE LINE

$\frac{1}{16}$ APPROX. $\frac{1}{4}$ APPROX.
CUTTING-PLANE LINE —   CUTTING-PLANE LINE

(freehand)
SHORT-BREAK LINE —   SHORT-BREAK LINE

$\frac{3}{4}$ to $1\frac{1}{2}$ APPROX.   (freehand)
LONG-BREAK LINE —   LONG-BREAK LINE

$\frac{1}{16}$ APPROX. $\frac{1}{8}$ APPROX. $\frac{3}{4}$ to $1\frac{1}{2}$ APPROX.
PHANTOM LINE —   PHANTOM LINE

Fig. 3–11. Alphabet of Lines.

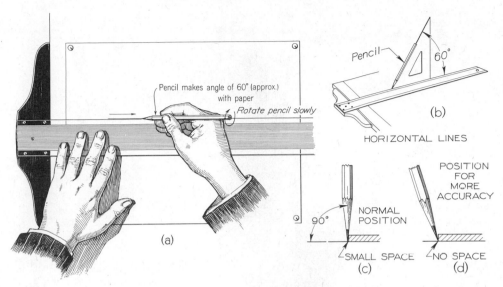

Fig. 3–12. Drawing a Horizontal Line.

position, the point will be slightly away from the T-square, (c). The line will not be straight if the pencil is leaned away from this position while a line is being drawn. Where great accuracy is required, the pencil may be "toed in," (d), to assure the straightness of the line.

Never draw lines along the lower edge of the T-square.

Left-handers: If you are left-handed (and many excellent draftsmen are), press the T-square head against the right edge of the board, and draw the line from right to left.

3.12 Vertical Lines. Either of the two triangles may be used to draw vertical lines, Fig. 3–13. Notice that the *vertical side of the triangle is on the left,* from which direction the light comes. Press the head of the T-square against the board, then slide the left hand to the position shown in Fig. 3–13 (a), where it holds both the T-square and the triangle together firmly in position. With the pencil leaning at 60° to the paper, (b), *draw the line upward,* rotating the pencil slowly throughout the progress of the line.

Left-handers: Since your T-square head will be on the right, your triangle will have its vertical edge on the right. Lean the pencil as shown, and draw the line *upward.*

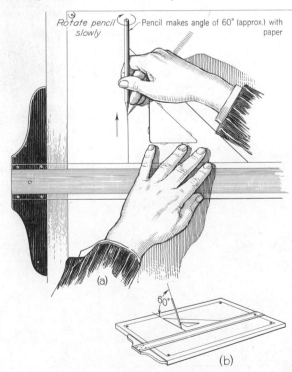

Fig. 3–13. Drawing a Vertical Line.

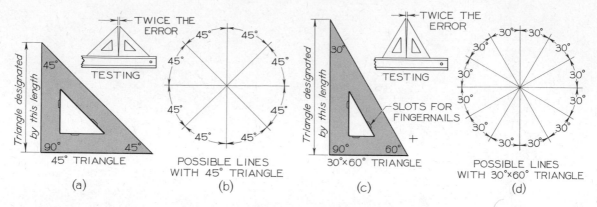

Fig. 3–14. The Triangles.

3.13 Triangles. All vertical lines and most inclined lines are drawn with the 45° triangle or the 30° × 60° triangle, Fig. 3–14. As shown at (b) the complete 360° can be divided into eight 45° sectors with the 45° triangle, and as shown at (d) twelve 30° sectors can be made with the 30° × 60° triangle.

3.14 Inclined Lines. The method of drawing lines at 45° with horizontal is shown in Fig. 3–15. When the inclined lines to be drawn are long, use the positions at (a) or (b). The position at (c) is more rigid and accurate and is recommended for general use. Arrows indicate the directions in which lines are drawn.

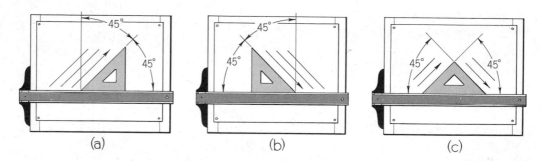

Fig. 3–15. Drawing Lines with 45° Triangle.

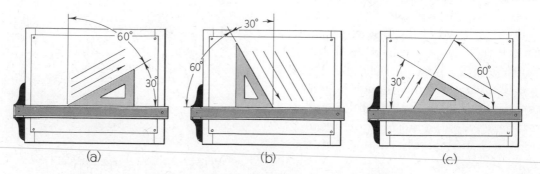

Fig. 3–16. Drawing Lines with 30° × 60° Triangle.

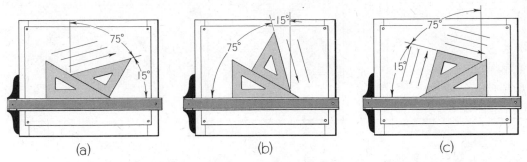

Fig. 3–17. Drawing Lines with Triangles in Combination.

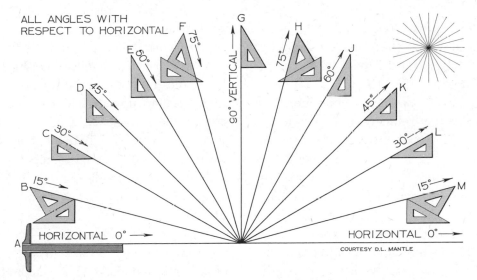

Fig. 3–18. Triangle Summary.

The method of drawing lines at 30° with horizontal is shown in Fig. 3–16 (a), and at 60° with horizontal at (b). These are recommended where the lines to be drawn are long, but the method of (c) is best for general use because of the greater stability and accuracy.

If the two triangles are combined, lines may be drawn at 15° with horizontal, Fig. 3–17 (a), or 75° with horizontal, (b). If the top triangle is arranged to rest upon its hypotenuse, more accurate lines at either 15° or 75° with horizontal may be drawn, (c).

The entire 360° can be divided into 24 sectors of 15° each with the T-square and the triangles either singly or in combination, Fig.

3–18. The arrows indicate the directions in which lines should be drawn.

Angles other than those shown in Fig. 3–18 are measured with the protractor, Fig. 3–19. The drafting machine, Fig. 3–42, is ideal for drawing inclined lines, as well as for many other uses.

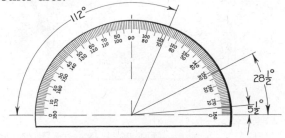

Fig. 3–19. Protractor.

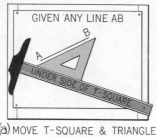
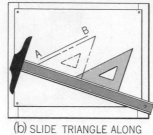
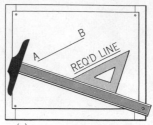

(a) MOVE T-SQUARE & TRIANGLE TO LINE UP WITH AB

(b) SLIDE TRIANGLE ALONG T-SQUARE

(c) DRAW REQUIRED LINE PARALLEL TO AB

Fig. 3–20. Drawing Parallel Lines.

3.15 Parallel Lines. You can easily draw parallel lines at any of the standard angles by sliding the triangles and repeating the lines, as shown in Figs. 3–15, 3–16, and 3–17.

To draw a line parallel to *any* line, Fig. 3–20 (a), move the T-square and triangle together until the hypotenuse of the triangle lines up with the given line; then, with the T-square held firmly in position, slide the triangle along the T-square away from the line, (b), and draw the required line, (c). The 30° × 60° triangle could be used equally well for this, and any side of either triangle could be used in contact with the T-square. Instead of the T-square as the supporting member, another triangle may be used.

3.16 Perpendicular Lines. In Figs. 3–15, 3–16, and 3–17, the lines at (b) are perpendicu-

lar to those shown at (a). At (c) the lines are perpendicular to each other. To draw a line perpendicular to *any* given line, Fig. 3–21 (a), move the T-square and triangle (either triangle, resting upon its hypotenuse), until a side of the triangle lines up with the given line; then, with the T-square held firmly in position, slide the triangle across the line, (b), and draw the perpendicular, (c). Instead of the T-square as the supporting member, another triangle may be used.

A perpendicular to a given line may also be drawn by the *revolved-triangle method*, Figs. 3–22 and 3–23. This method is used when the relatively long hypotenuse is needed for the required perpendicular.

Again, instead of the T-square as the supporting member, another triangle may be used if desired.

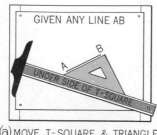
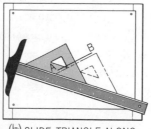
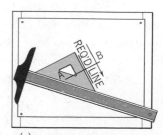

(a) MOVE T-SQUARE & TRIANGLE TO LINE UP WITH AB

(b) SLIDE TRIANGLE ALONG T-SQUARE

(c) DRAW REQUIRED LINE PERPENDICULAR TO AB

Fig. 3–21. Drawing Perpendicular Lines—Adjacent-Sides Method.

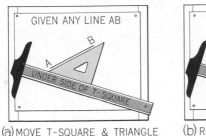
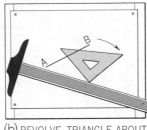
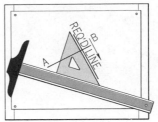

(a) MOVE T-SQUARE & TRIANGLE TO LINE UP WITH AB (b) REVOLVE TRIANGLE ABOUT 90° CORNER (c) DRAW REQUIRED LINE PERPENDICULAR TO AB

Fig. 3–22. Drawing Perpendicular Lines—Revolved-Triangle Method (45° Triangle).

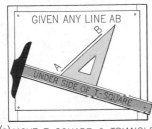
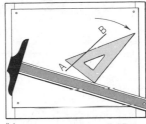
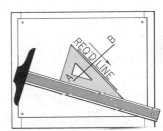

(a) MOVE T-SQUARE & TRIANGLE TO LINE UP WITH AB (b) REVOLVE TRIANGLE ABOUT 90° CORNER (c) DRAW REQUIRED LINE PERPENDICULAR TO AB

Fig. 3–23. Drawing Perpendicular Lines—Revolved-Triangle Method (30° × 60° Triangle).

3.17 Erasing. Erasers are made because we all make mistakes. You can avoid most mistakes by following the rule: *Do not draw a line until you know what you are doing!* When you do have to erase, exercise care to avoid spoiling the drawing altogether.

The Artgum, Fig. 3–24 (a), is used for cleaning large areas, and not for general erasing. Avoid the practice of making a careless, dirty pencil drawing, and then scrubbing it with Artgum before heavying-in the final lines of the drawing.

For general use, the medium-hard Weldon Roberts Coral Pink (India), (b), or an Eberhard Faber Ruby eraser, is recommended for both pencil and ink erasing. Never use gritty ink erasers, and never use a razor blade or a knife for erasing purposes.

The erasing shield, (c) to (c), is used to protect lines near the line being erased. Electric *erasing machines,* (e), are available in industrial drafting rooms to save valuable time.

3.18 Neatness. Neatness is a personal habit.

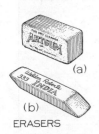

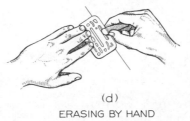
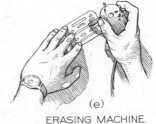

(a)

(b)

ERASERS

(c)

ERASING SHIELD

(d)

ERASING BY HAND

(e)

ERASING MACHINE

Fig. 3–24. Erasing.

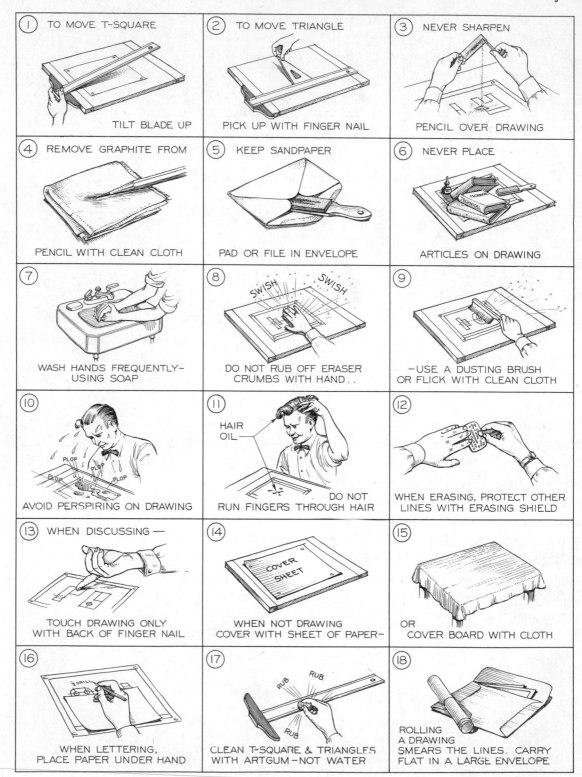

Fig. 3–25. Neatness Is Not Accidental.

There is no more reason for a sloppy, dirty drawing than there is for the draftsman himself being slouchy and unclean. Most of the "dirt" that tends to get on drawings is really graphite from the pencil and is directly the result of carelessness in letting particles of graphite drop on the drawing or in smearing the graphite of the lines already drawn. When a line is drawn, a trail of loose graphite particles is left along the line. These should be blown off with the breath at frequent intervals. Do not allow the drawing to accumulate dust or graphite, *but keep the drawing clean from the beginning.* Keep your hands and your equipment clean, and do everything you can to protect your drawing from smearing. For perspiring hands, wash frequently with soap and water, and apply talcum powder. Some of the precautions are illustrated in Fig. 3–25.

3.19 Drawing to Scale. Since drawings are made on certain standard sizes of paper, Sec. 3.5, and objects to be drawn may be relatively small (such as a machine part) or large (such as a building), it is necessary to consider what size to make the drawing. Whenever possible, an object should be drawn full size, but if necessary it may be drawn $\frac{1}{2}$ size, $\frac{1}{4}$ size, or smaller. A building may have to be drawn $\frac{1}{48}$ size or a map $\frac{1}{1200}$ size.

The conventional unit of measurement used on engineering drawings in the United States is the inch. In some industries all dimensions on engineering drawings are expressed in inches; in others both feet and inches are used. The standard symbols are $''$ for inch and $'$ for foot.

Small objects, such as machine parts, sheet-metal parts, or furniture, are drawn in terms of inches. Since this is understood, *all inch marks are omitted* from dimensions on such drawings. Thus, $2\frac{1}{2}$ *by itself* means $2\frac{1}{2}$ *inches.*

Large structures, such as buildings, bridges, ships, dams, and the terrain (as on maps), are drawn in feet and inches. It is common practice on such drawings to show the symbol for feet and to omit the symbol for inches. Dimensions in feet and inches together would therefore be given as: $5'-0$, $3'-2\frac{1}{2}$, $23'-0\frac{1}{4}$, etc.

The scale to which a drawing is made should be indicated on the drawing, either in the title box or in another appropriate space. For example, if a machine drawing is made using the architects scale, the scale would be indicated as:

SCALE: FULL SIZE or 1.00 = 1.00 or 1 = 1
   or 1/1
SCALE: HALF SIZE or .50 = 1.00 or $\frac{1}{2}$ = 1 or
   1/2

The scale of drawings made quarter or eighth size would be indicated in a similar manner. If an enlarged drawing is made, the scale would be shown as:

SCALE: TWICE SIZE or 2.00 = 1.00 or 2 = 1
   or 2/1

For architectural and construction drawings in which the dimensions are expressed in terms of feet and inches, the scale would be indicated as $3'' = 1'-0$, $1\frac{1}{2}'' = 1'-0$, $\frac{1}{4}'' = 1'-0$, etc. For drawings consisting of maps, diagrams, or other graphical constructions for which the engineers scale is used, the scale of the drawing might be indicated to scales such as $1'' = 400'$, $1'' = 200$ lbs., $1'' = 30°$ Fahrenheit, etc.

3.20 Types of Scales.* The *architects scale,* Fig. 3–26 (a), is an all-round scale for many uses. It has a full-size scale of inches divided

---

* Never call a scale a "rule" or a "ruler."

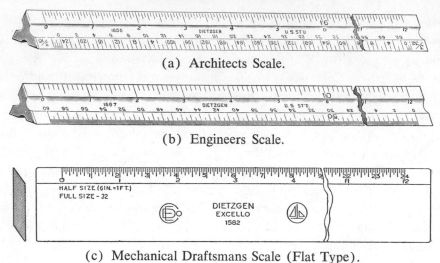

(a) Architects Scale.

(b) Engineers Scale.

(c) Mechanical Draftsmans Scale (Flat Type).

Courtesy Eugene Dietzgen Co.

Fig. 3–26. Types of Scales.

into sixteenths, and a number of reduced-size scales in which inches or fractions of an inch represent *feet*. Drawings can be made full size, $\frac{1}{2}$ size, $\frac{1}{4}$ size, and so on down to $\frac{1}{128}$ size.

The *engineers scale*, Fig. 3–26 (b), has a series of scales in which inches are divided into 10, 20, 30, 40, 50, 60, or 80 parts. These scales are used in many ways. For example, on a map the "50-scale" can be used so that $1'' = 50'$, or $1'' = 500'$, or $1'' = 5$ miles. Or, using the "10-scale" on a machine drawing, you can lay off a decimal dimension of, say, 3.652 by measuring $3''$ and then adding sixtenths and then slightly more than half a tenth.

The engineers scale is often called the civil engineers scale because it was originally used primarily in civil engineering.

The mechanical draftsmans scale, Fig. 3–26 (c), may be obtained with scales of full size and $\frac{1}{2}''$, $\frac{1}{4}''$, and $\frac{1}{8}''$ representing $1''$. For example, if a drawing is to be made $\frac{1}{2}$ size, the scale designated $\frac{1}{2}$ size would be selected, which would indicate that each $\frac{1}{2}''$ represents $1''$.

Scales are made either in triangular form,

Fig. 3–26 (a) and (b), or in flat form, (c). The triangular form is the more economical since it shows many scales on one stick, whereas several flat scales would be required to show the same number. The flat-type scale shown is very popular with professional draftsmen because of its simplicity.

Fully divided scales have all of the main units subdivided throughout the entire length of the scale. Fully divided scales have the advantage that several values can be read from the same origin without resetting, Fig. 3–26 (a) and (b). Engineers fully divided scales are often called chain scales. Open divided scales have only the main units graduated, but an extra unit is fully subdivided in the opposite direction from the zero point.

3.21 Use of Architects Scale. The standard scales for machine drawing are the full size, half size, quarter size, and eighth size. Fig. 3–27 shows an object $4\frac{13}{16}''$ long drawn to each of the following four scales:

(a) *Full Size* ($12'' = 1'-0''$). Simply set off $4''$ from zero, and add to this $\frac{13}{16}''$, as shown.

Note that to set off $\frac{1}{32}''$ it would be necessary to estimate half of $\frac{1}{16}''$.

(b) *Half Size* $(6'' = 1'-0'')$. Use the full-size scale, but regard each $\frac{1}{2}''$ on the scale as $1''$ on the drawing. (Do not use the $\frac{1}{2}''$ scale for half size because this is an architectural scale in which $\frac{1}{2}'' = 1'-0''$, or $\frac{1}{24}$ size.) To set

off $4\frac{13}{16}''$ to half size, set off $2''$ (half of $4''$), and add $6\frac{1}{2}$ sixteenths (half of $\frac{13}{16}''$).

(c) *Quarter Size* $(3'' = 1'-0'')$. To the right of zero on the scale is a *foot* reduced to $3''$, or quarter size. This foot is subdivided into inches and fractions of an inch. To set off $4\frac{13}{16}''$, set off $4''$ toward the right from zero

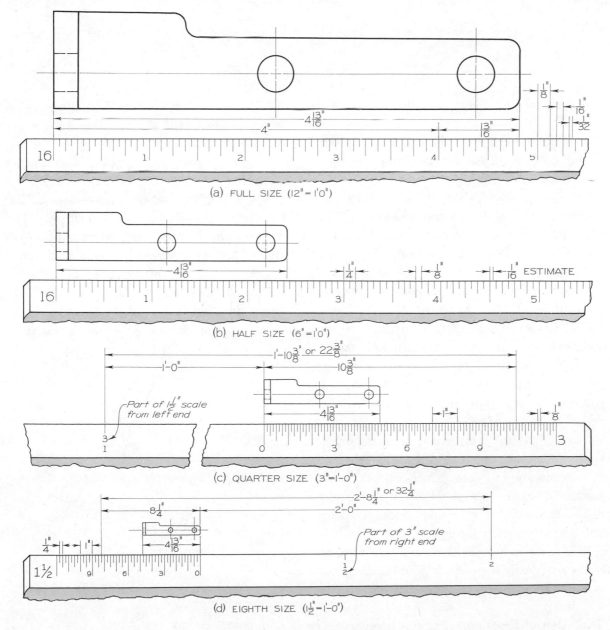

Fig. 3–27. Use of Architects Scale.

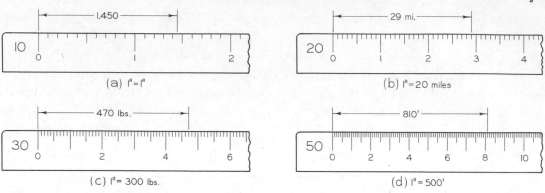

Fig. 3–28. Use of Engineers Scale.

and then add to this $6\frac{1}{2}$ small divisions (each division represents $\frac{1}{8}''$ or $\frac{2}{16}''$). To set off $1'-10\frac{3}{8}''$, take the foot from the first main division to the left of zero, and the $10\frac{3}{8}''$ from the right of zero.

(d) *Eighth Size* ($1\frac{1}{2}'' = 1'-0''$). To the left of zero on the scale is a foot reduced to $1\frac{1}{2}''$, or eighth size. This foot is subdivided into inches and fractions of an inch. To set off $4\frac{13}{16}''$, set off $4''$ toward the left from zero and then add to this $3\frac{1}{4}$ small divisions (each small division $= \frac{1}{4}''$ or $\frac{4}{16}''$). To set off $2'-8\frac{1}{4}''$, take the $2'$ from the second main division to the right of zero, and the $8\frac{1}{4}''$ from the left of zero.

All the remaining scales on the architects scale are reduced scales *in which inches or fractions of an inch stand for feet*. The reduced scales marked $1''$, $\frac{3}{4}''$, $\frac{1}{2}''$, $\frac{3}{8}''$, $\frac{1}{4}''$, $\frac{3}{16}''$, $\frac{1}{8}''$, and $\frac{3}{32}''$ each have main divisions representing feet, and one foot at the end is divided into inches and fractions. Architects frequently use the $\frac{3}{4}''$ scale for large details of doors, windows, etc., and the $\frac{1}{4}''$ and $\frac{1}{8}''$ scales for plans and elevations of buildings.

If a machine part is very small, it is drawn oversize, usually *double size*. The full-size scale is used, $1'' = \frac{1}{2}''$ on the object.

3.22 Use of Engineers Scale. The engineers scale is frequently used in drawing maps, charts, and diagrams and whenever decimal dimensions are encountered graphically. A linear measurement of $1''$ on any of the scales may be used to represent distances such as inches, feet, miles, etc., or to represent other quantities such as time, weight, force, etc.

For example, to set off a distance of $1.450''$ full size, Fig. 3–28 (a), the 10-scale would be used by simply measuring one main division and adding $4\frac{1}{2}$ subdivisions. At (b) a distance of 29 miles is set off by using the 20-scale ($1'' = 20$ miles) and measuring two main divisions plus nine subdivisions. Other examples of use of the engineers scale are shown at (c) and (d).

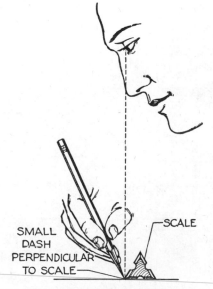

Fig. 3–29. Accuracy in Use of Scale.

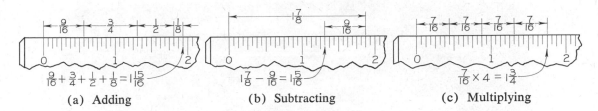

(a) Adding	(b) Subtracting	(c) Multiplying
$\frac{9}{16}+\frac{3}{4}+\frac{1}{2}+\frac{1}{8}=1\frac{15}{16}$	$1\frac{7}{8}-\frac{9}{16}=1\frac{5}{16}$	$\frac{7}{16}\times 4=1\frac{3}{4}$

Fig. 3–30. Scale Arithmetic.

3.23 Measuring with the Scale. The habit of accuracy is important in drafting, and is essential to success in any technical work. Accurate drafting depends largely upon the correct use of the scale. Measurements should not be taken directly off the scale with compass or dividers, as this will mar the subdivisions and ruin the scale. Always place the scale along the line to be measured (if no line is present, draw a construction line), and with the eye directly above the graduation mark on the scale, Fig. 3–29, make a short dash (use a sharp pencil) at right angles to the scale. After setting off a dimension, always double-check with the scale to make sure the distance is accurate. For horizontal measurements, place the scale on the paper so that the scale

in use is at the top; for vertical measurements, place the scale so that the scale in use is on the left.

Avoid cumulative errors in the use of the scale. If a series of measurements is to be set off end to end, such as $\frac{9}{16}$, $\frac{3}{4}$, $\frac{1}{2}$, and $\frac{1}{8}$, do not set off the first dimension and then move the scale to start the next dimension from zero, etc. All of these should be set off *without moving the scale*, as shown in Fig. 3–30 (a). The scale is thus a simple adding machine. The scale may also be used for subtraction and multiplication, as shown at (b) and (c).

3.24 Drawing Instruments. A set of drawing instruments suitable for school or professional use is illustrated in Fig. 3–31. This set contains

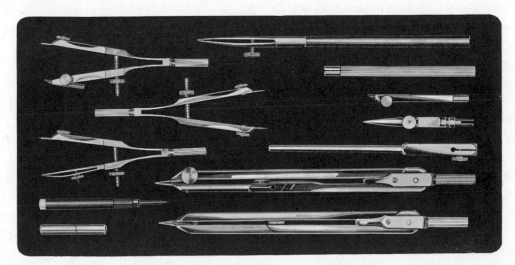

Courtesy Gramercy Guild Group, Inc.

Fig. 3–31. Conventional Three-Bow Set.

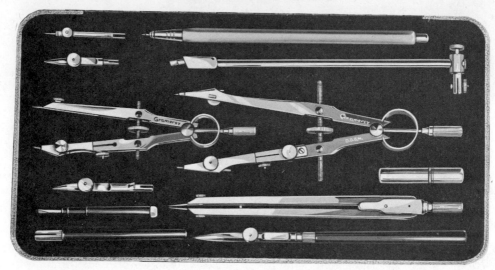

Fig. 3–32. Giant Bow Set.

the following instruments: compass, dividers, bow pencil, bow pen, bow dividers, ruling pen, and various accessories.

In order to do highly skilled work you must have good tools, and this is particularly true in mechanical drawing. First-class work is impossible with cheap or defective instruments. However, the qualities of high-grade instruments are likely not to be recognized by the beginner, who is apt to be attracted by sets which have beautiful cases and whose instruments have a cheap shine. Therefore, you should obtain the advice of your instructor, of an experienced draftsman, or of a reliable dealer.

3.25 The Giant Bow Set. Since, in modern drafting, reproductions are so often made directly from pencil drawings or tracings, the lines must be dense black, and a very rigid compass is needed so that the draftsman can "bear down" in drawing arcs and circles. For this reason, the *giant bow compass* is coming into increasing use. This is simply a large bow compass that has a maximum radius about equal to the conventional large compass, but

has the rigidity of the small bow instrument. A typical giant bow set is shown in Fig. 3–32. This set contains a large bow compass, a small bow compass, dividers, a ruling pen, and various accessories.

3.26 The Compass. The compass, with attachments, is used to draw circles from about 1″ radius to about 6″ radius, Fig. 3–33. If the lengthening bar is used, Fig. 8–4 (c), circles up to about 12″ radius can be drawn. A special *beam compass* must be used for larger sizes, (d).

When the compass is lifted from the case, the legs are squeezed apart with the thumb and forefinger, the needle point set at the center of the circle, and the compass adjusted to the required radius previously set off with a scale on the center line as in Fig. 3–33 (a). Start drawing the circle on the left-hand side, (b), lean the compass slightly forward, and draw the circle in a clockwise direction with the handle rotating between the thumb and forefinger, (c).

Left-handers: Hold the compass with the left hand and draw circles *counterclockwise*.

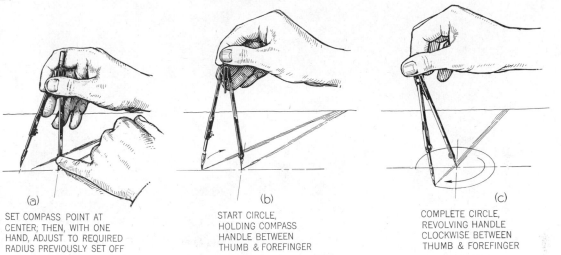

(a)

SET COMPASS POINT AT
CENTER; THEN, WITH ONE
HAND, ADJUST TO REQUIRED
RADIUS PREVIOUSLY SET OFF

(b)

START CIRCLE,
HOLDING COMPASS
HANDLE BETWEEN
THUMB & FOREFINGER

(c)

COMPLETE CIRCLE,
REVOLVING HANDLE
CLOCKWISE BETWEEN
THUMB & FOREFINGER

Fig. 3–33. Using the Compass.

As shown in Fig. 3–34, the scale should be used to set off the radius along a center line. Do not place the compass directly on the scale to obtain the radius, as this will eventually damage the subdivisions on the scale. After the circle is drawn, III, check the diameter with the scale, IV, to make sure an error in radius has not resulted in a double error in diameter. The circle should be drawn lightly and not made heavy until the correctness is established. A good method is to draw a trial circle on a piece of scrap paper and check it with the scale before drawing the circle on the drawing.

In drawing circles, you usually know the diameter and must determine the radius by mentally dividing the diameter in half. If a diameter of $3\frac{1}{2}''$ is given, the radius is figured as follows: Half of $3'' = 1\frac{1}{2}''$. Half of $\frac{1}{2}'' = \frac{1}{4}''$. The radius $= 1\frac{1}{2}'' + \frac{1}{4}'' = 1\frac{3}{4}''$. If the drawing is to half scale, it is necessary to divide $1\frac{3}{4}''$ again in the same manner.

3.27 Sharpening the Compass Lead. For construction arcs or circles, use a hard lead, such as 4H, 5H, or 6H. For general work, use a softer lead, which will produce dark lines without smudging too easily, such as an F or H. Since you cannot exert as much pressure on a compass as on a drawing pencil, it may be necessary to use a compass lead about one grade softer than on the straight-line work.

Compass leads come with your drawing set

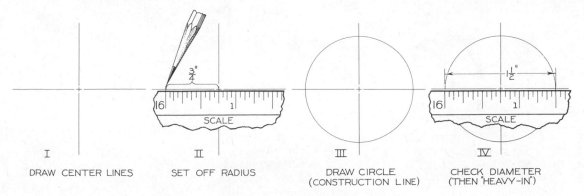

I

DRAW CENTER LINES

II

SET OFF RADIUS

III

DRAW CIRCLE
(CONSTRUCTION LINE)

IV

CHECK DIAMETER
(THEN "HEAVY-IN")

Fig. 3–34. Use of Scale.

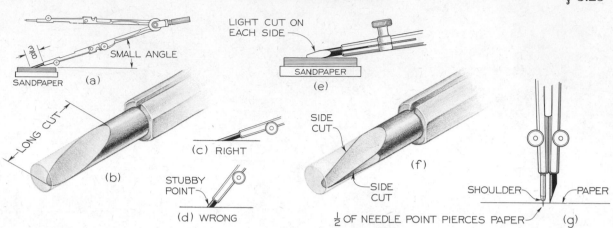

Fig. 3–35. Sharpening Compass Lead.

and can also be purchased separately in all grades. If you do not have the right lead, cut off a piece of your drawing pencil, remove the wood, and use the lead. Used pencil stubs should be saved for this purpose.

Adjust the lead so that it extends about $\frac{3}{8}''$ from the compass and rub the lead on the sandpaper pad or file to produce a long inclined cut on the lead, Fig. 3–35 (a) to (c). Always keep a liberal length of lead in the compass so that a long cut can be made. Avoid a stubby point as shown at (d).

Some draftsmen prefer to dress the point still further by making two additional very light side cuts, as shown at (e) and (f). Adjust the needle point so that it extends slightly

longer than the lead, as shown at (g). Use the "shoulder" end of the needle point, not the plain end.

3.28 Dividers. The *dividers*, Fig. 3–36, are used for subdividing distances into equal spaces or for transferring distances in which the spacing between the points is approximately $1''$ or over. To divide a line AB into equal parts, say three, lift the dividers from the case, squeeze the legs apart between the thumb and forefinger until the points are separated an estimated one-third of the total distance, and set the point at one end of the line as in Fig. 3–36 (a). Then turn the dividers between the thumb and forefinger to the sec-

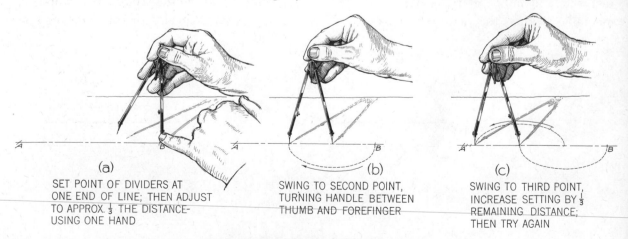

(a) SET POINT OF DIVIDERS AT ONE END OF LINE; THEN ADJUST TO APPROX. $\frac{1}{3}$ THE DISTANCE USING ONE HAND

(b) SWING TO SECOND POINT, TURNING HANDLE BETWEEN THUMB AND FOREFINGER

(c) SWING TO THIRD POINT, INCREASE SETTING BY $\frac{1}{3}$ REMAINING DISTANCE; THEN TRY AGAIN

Fig. 3–36. Dividing a Line into Equal Parts.

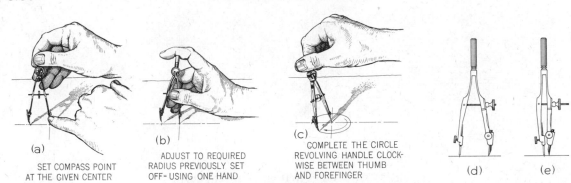

Fig. 3–37. Using the Bow Pencil.

ond point, as shown at (b), and finally to the third point, as at (c). Here the spacing of the divider points is too short, and it is necessary to increase the space by an estimated one-third of the remaining distance and to try again. If this is carefully done, the correct spacing can be obtained in two or three trials.

3.29 Bow Instruments. The *bow pencil, bow pen,* and *bow dividers* are known collectively as the *bow instruments,* Fig. 3–31. They are used in a manner similar to the larger instruments, but they are smaller and more rigid and should always be used when spacings or radii are under 1″, which is the approximate capacity of these instruments. The proper use

of the bow pencil is shown in Fig. 3–37. Some draftsmen prefer the "center-wheel" instruments shown in this figure, (a) to (c); others prefer the "side-wheel" instruments, (d) and (e). One is as good as the other. Notice that bow instruments, like all drawing instruments, are operated with one hand.

Sharpen the compass lead as described in Sec. 3.27. For average use, turn the slanted side of the lead to the outside, Fig. 3–37 (d). For small radii, turn the slanted side to the inside, as shown above at (e).

3.30 Irregular Curves. *Irregular,* or *French, curves* are made of amber or clear plastic, Fig. 3–38 (a) to (f), and are used to draw ir-

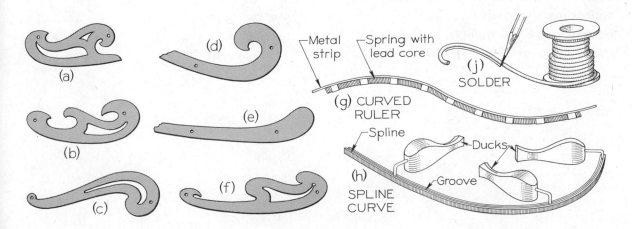

Fig. 3–38. Irregular Curves.

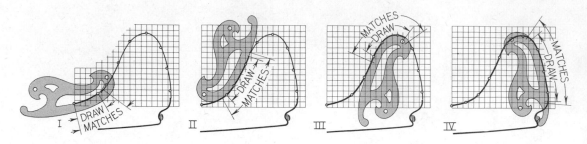

Fig. 3–39. Settings to Draw a Smooth Curve.

regular curves. The irregular curve is not used to establish the original curve, but only to make the final curve smooth, Fig. 3–39. First, plot enough points to establish the curve accurately. Second, sketch *very lightly*, by eye, a smooth curve through the points. Third, match the irregular curve to the sketched curve, the sketched line determining the direction or flow of the curve.

The student usually tries to make his curve fit too many points in one setting in order to reduce the number of settings. Watch the original curve carefully and make sure that each setting *flows smoothly* from the previous one. You do this by never drawing the curve the full length of the portion which appears to match the sketched curve, but by overlapping each successive setting, Fig. 3–39.

For ink work, the ruling pen should be held nearly vertical, and the blades should be kept approximately parallel to the edge of the irregular curve. A triangle placed under the curve will keep ink from running under. Leave

small gaps between the inked segments, and fill these in by hand. See Secs. 8.1 and 8.4.

Adjustable curves are shown in Fig. 3–38 (g) and (h). Ordinary solder wire, (j), can be bent easily to almost any desired curve and is a very satisfactory substitute for commercial adjustable curves.

If the curve is symmetrical, such as an ellipse, it is well to use the same segment of the irregular curve in two or more opposite places. For example, Fig. 3–40 (a), the irregular curve has been fitted and the line drawn from 1 to 2. Light pencil dashes are drawn directly on the irregular curve at these points (the curve will take pencil marks better if it is "frosted" by light sandpapering with #00 sandpaper). At (b) the irregular curve is turned over and matched so that the line may be drawn from 2 to 1. In similar manner, the same segment is used again at (c) and (d). To complete the ellipse, the gaps at the ends may be filled in by means of the irregular curve or, if desired, with the compass.

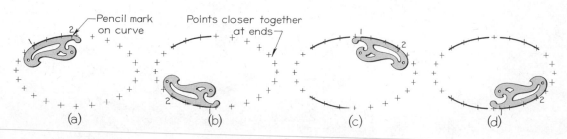

Fig. 3–40. Symmetrical Figures.

Fig. 3–41. Parallel Ruling Straightedge.

3.33 Sheet Layouts. Standard sheet layouts for all problems are given in the Appendix of this book where they can be readily found when needed. These conform to the USA Standard sheet sizes, based on $8\frac{1}{2}'' \times 11''$ and multiples thereof, Sec. 3.5. The $8\frac{1}{2}'' \times 11''$ basic size has been adopted by the majority of industries largely because the $8\frac{1}{2}'' \times 11''$ size, and the folded multiples, will fit in standard envelopes and can be filed in standard letter files. Three sizes of sheets are shown, $8\frac{1}{2}'' \times 11''$, $11'' \times 17''$, and $17'' \times 22''$, with title blocks or strips for each layout.

3.31 Parallel Ruling Straightedge. For large drawings, the T-square becomes unwieldy because of the length of the blade, and considerable inaccuracy may result from the "give," or swing, of the blade. In such cases the *parallel ruling straightedge*, Fig. 3–41, is recommended. The ends of the straightedge are controlled by a system of cords and pulleys, permitting it to be moved up or down on the board, always maintaining a true horizontal position.

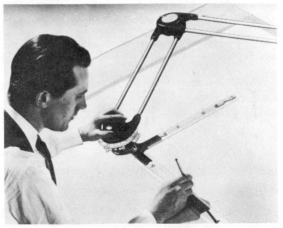

Fig. 3–42. Drafting Machine.

3.32 Drafting Machine. The *drafting machine*, Fig. 3–42, has long been regarded as a practical and effective time-saver in drafting. The removable scales are attached securely to the head where a knob is provided to turn the scales to any desired angle as indicated on the protractor around the knob. The scales may then be moved anywhere on the board without changing the angle. The scales are used both for measuring and for drawing lines. The drafting machine thus combines in convenient form all the functions of the T-square, triangles, scales, and protractor.

3.34 To Lay Out a Sheet. After the sheet has been attached to the board as described in Sec. 3.6, lay out the sheet as shown in Fig. 3–43. Construction lines, drawn lightly with a sharp hard pencil, are used in the first five steps, and clean black border lines, Fig. 3–11, are used in step VI. Note that in steps II and III all measurements are set off at once for the title strip, including guide lines for lettering.

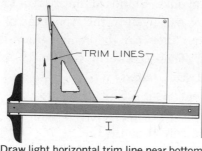

TRIM LINES

I

Draw light horizontal trim line near bottom of paper; then draw light vertical trim line near left edge.

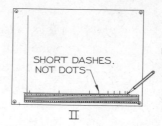

SHORT DASHES. NOT DOTS

II

Set off all width dimensions along lower trim line. The full-size scale should be up.

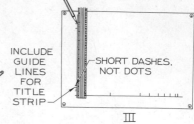

INCLUDE GUIDE LINES FOR TITLE STRIP

SHORT DASHES. NOT DOTS

III

Set off all height dimensions along left trim line. The full-size scale is to the left.

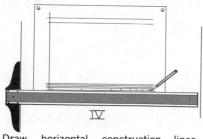

IV

Draw horizontal construction lines through marks at left side of sheet.

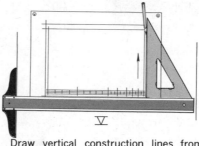

V

Draw vertical construction lines from bottom upward, through marks at bottom of sheet.

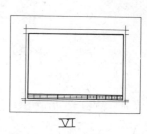

VI

Retrace border and title strip to make heavier. Use softer pencil.

Fig. 3–43. **Laying Out Sheet** (Layout A, Appendix, page 474).

3.35 Mechanical Drawing Problems. The problems which follow for this chapter are all drawn on Layout A, as shown on page 474 in the Appendix. If the instructor wishes to use 9″ × 12″ sheets, this can be easily done by the lengthening of the two left-hand spaces in the title strip of Layout A. The problems may be dimensioned, if desired by the instructor. In such a case, the student should study Sec. 4.11 or 4.13, on the lettering of numerals, and Secs. 9.1 to 9.12, on dimensioning conventions.

The problems for this chapter are all one-view problems, which afford practice in using the T-square, triangles, drawing instruments, and other equipment.

In Fig. 3–45, the steps in making a one-view mechanical drawing of Fig. 3–44, involv-

ing only horizontal and vertical lines, are given. It is suggested that this problem be drawn first, and then assignments may be made from Figs. 3–46 to 3–51.

Next, the student should draw the problem in Figs. 3–52 and 3–53, involving inclined

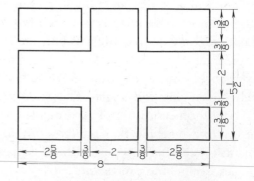

Fig. 3–44. **Inlaid Linoleum Design.**

lines. Further assignments can be made from Figs. 3–54 and 3–55.

A problem involving circular arcs and circles is given in Figs. 3–56 and 3–57. Further assignments can be made from Figs. 3–58 to 3–65.

The steps in drawing the Inlaid Linoleum Design, Fig. 3–44, are shown in Fig. 3–45.

First, lay out the sheet, as shown in Fig. 3–43 (Layout A, Appendix, page 474). The working space inside the border is $10\frac{1}{2}''$ wide and $7\frac{3}{8}''$ high, and the drawing is to be centered in this space. Make spacing calculations on a scrap of paper or on your sheet outside the trim line. Follow the steps below and apply them to all of the problems that follow.

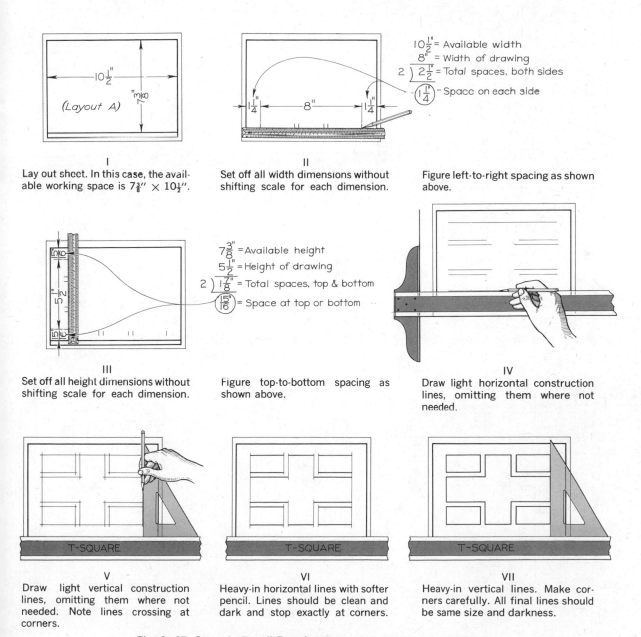

I
Lay out sheet. In this case, the available working space is $7\frac{3}{8}'' \times 10\frac{1}{2}''$.

II
Set off all width dimensions without shifting scale for each dimension.

Figure left-to-right spacing as shown above.

$$10\frac{1}{2}'' = \text{Available width}$$
$$8'' = \text{Width of drawing}$$
$$2\,\overline{)\,2\frac{1}{2}''} = \text{Total spaces, both sides}$$
$$\left(1\frac{1}{4}\right) = \text{Space on each side}$$

III
Set off all height dimensions without shifting scale for each dimension.

IV
Draw light horizontal construction lines, omitting them where not needed.

$$7\frac{3}{8}'' = \text{Available height}$$
$$5\frac{1}{2}'' = \text{Height of drawing}$$
$$2\,\overline{)\,1\frac{7}{8}''} = \text{Total spaces, top \& bottom}$$
$$\left(\frac{15}{16}''\right) = \text{Space at top or bottom}$$

Figure top-to-bottom spacing as shown above.

V
Draw light vertical construction lines, omitting them where not needed. Note lines crossing at corners.

VI
Heavy-in horizontal lines with softer pencil. Lines should be clean and dark and stop exactly at corners.

VII
Heavy-in vertical lines. Make corners carefully. All final lines should be same size and darkness.

Fig. 3–45. Steps in Pencil Drawing (Layout A, Appendix, page 474).

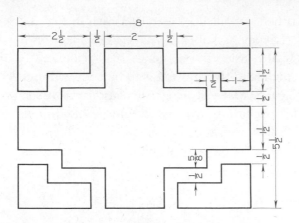

Fig. 3–46. Inlaid Linoleum Center Design. Draw full size.*

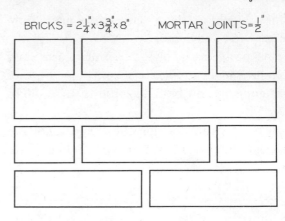

Fig. 3–47. Brick Wall. Draw half size.* (Small rectangles are ends of bricks.)

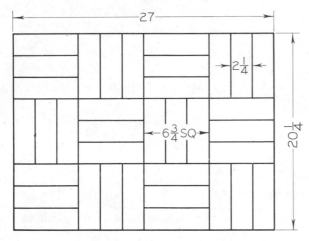

Fig. 3–48. Oak Floor Pattern. Draw quarter size.*

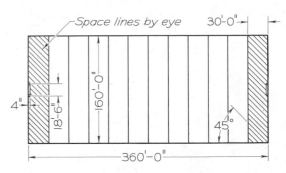

Fig. 3–49. Football Gridiron. Draw to scale 1″ = 40′–0″.* Use architects or engineers scale.

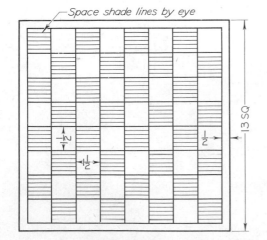

Fig. 3–50. Inlaid Wood Checker Board. Draw half size.*

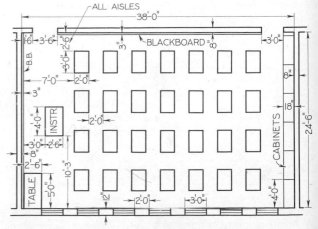

Fig. 3–51. Drafting Room Layout. Draw to scale ¼″ = 1′–0″.*

* Use Layout A (Appendix, page 474).

The steps in drawing the Base Plate, Fig. 3–52, are shown in Fig. 3–53. First, lay out the sheet as shown in Fig. 3–43 (Layout A, Appendix, page 474). The working space inside the border is 10½″ wide × 7⅜″ high, and the drawing is to be centered in this space. One method of spacing is shown in Fig. 3–45. Another is shown in Fig. 3–53, in which you locate the center of the working space by drawing diagonals and then making all measurements from this point.

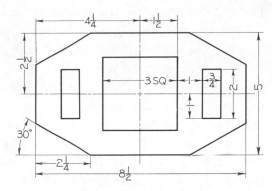

Fig. 3–52. Base Plate.

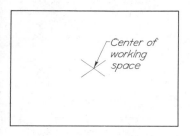

I. Locate center by drawing light diagonals.

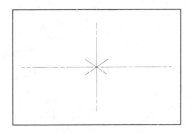

II. Draw center lines.

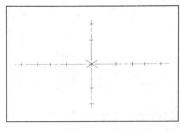

III. Set off vertical and horizontal distances.

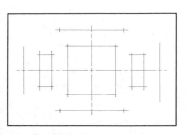

IV. Draw horizontal and vertical construction lines.

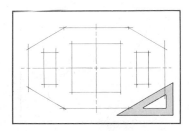

V. Draw angles (note lines crossing at corners).

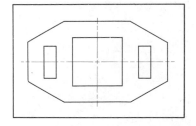

VI. Heavy-in final lines.

Fig. 3–53. Steps in Drawing Base Plate (Layout A, Appendix, page 474).

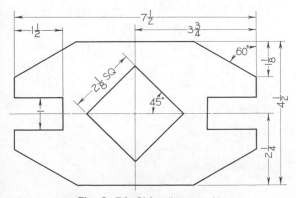

Fig. 3–54. Shim (Layout A).

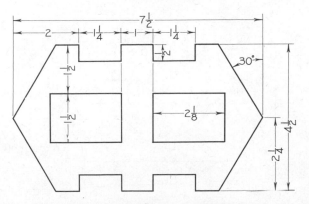

Fig. 3–55. Sheet Metal Stamping (Layout A).

The steps in drawing the Adjusting Arm, Fig. 3–56, are shown in Fig. 3–57. First, lay out the sheet, using Layout A (Appendix, page 474). The working space inside the border is $10\frac{1}{2}''$ wide $\times$ $7\frac{3}{8}''$ high, and the drawing is to be centered in this space. As shown in step I, draw the horizontal center line at mid-height on the sheet ($7\frac{3}{8}'' \div 2 = 3\frac{11}{16}''$). Since the object is $8\frac{1}{4}''$ long overall, and the space is $10\frac{1}{2}''$ wide, the space on each side is $1\frac{1}{8}''$, as shown.

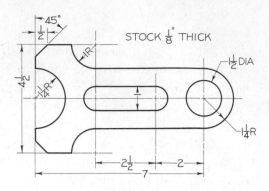

Fig. 3–56. Adjusting Arm.

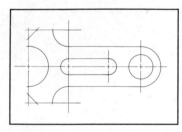

I. Draw horizontal center line and set off horizontal distances with scale.

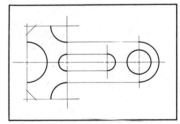

II. Draw vertical lines and construction as shown.

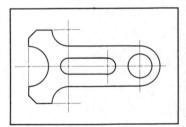

III. Draw construction arcs.

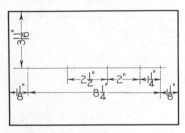

IV. Draw connecting straight lines.

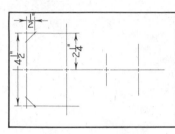

V. Heavy-in arcs and circles.

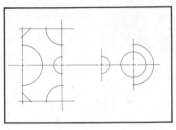

VI. Heavy-in straight lines to complete drawing.

Fig. 3–57. Steps in Drawing Adjusting Arm (Layout A, Appendix, page 474).

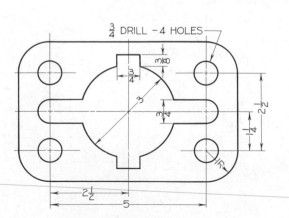

Fig. 3–58. Key Plate (Layout A).

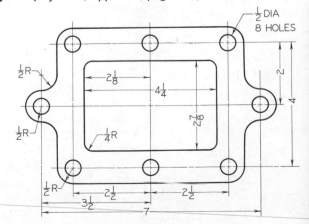

Fig. 3–59. Gasket (Layout A).

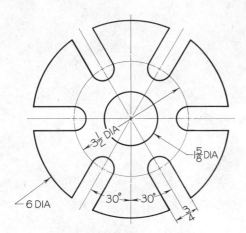

Fig. 3–60. Slotted Cam.*

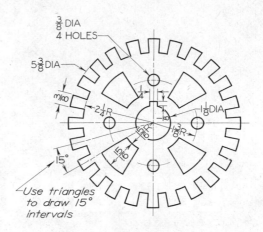

Fig. 3–61. Armature Lamination.* Use triangles to draw 15° sectors, Fig. 3–18.

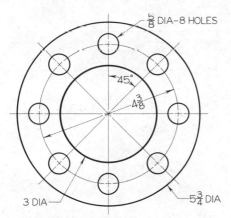

Fig. 3–62. Gasket.* Use 45° triangle to draw 45° intervals, Fig. 3–14 (b).

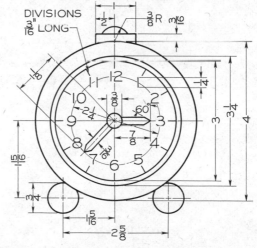

Fig. 3–63. Clock.* Use 30° × 60° triangle to draw 30° intervals, Fig. 3–14 (d).

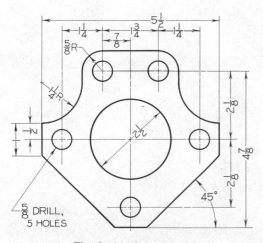

Fig. 3–64. Template.*

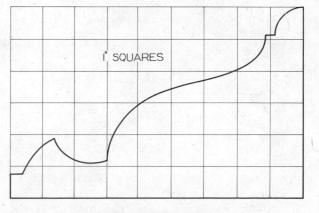

Fig. 3–65. Table Leaf Support Wing.* Draw full size. Use French curve.

* Use Layout A (Appendix, page 474).

CHAPTER 4
LETTERING

4.1 Origin of Letters.* The letters of our alphabet† are *symbols* for sounds. When letters are placed in combinations, *words* result that have meanings. All writing or lettering started in some form of *pictograph* (picture writing), but it took prehistoric man thousands of years to develop letters first from the earliest cave drawings, Fig. 4–1, then from pictographs. Among the earliest forms of

Fig. 4–1. Cave Drawings.‡

pictographs were the ancient Egyptian hieroglyphics (pronounced high-a-row-glif ′iks), Fig. 4–2. Present-day Chinese characters are a form of picture writing, but one must go back to earlier forms to see the resemblance to real

Fig. 4–2. Egyptian Hieroglyphics (means "Cleopatra").

things, Fig. 4–3. In order to understand how ideas can be expressed by pictures, try a pictograph yourself, such as Fig. 4–4.

	SUN	MOON	MOUNTAIN
PRESENT FORMS	日	月	山
OLD FORMS	 ☉	☽	 ⌒⌒

Fig. 4–3. Chinese Characters.‡

The story of the alphabet is an interesting subject that would require a book in itself. It is enough for our purposes to know that the early Phoenicians developed an alphabet of twenty-two characters from earlier picture writing and that these were later developed by the Greeks and then the Romans. As a result of Roman conquests the Latin alphabet was adopted throughout most of the then civilized world. The Old Roman alphabet, Fig. 4–34, is the direct parent of all our present letter forms.

Fig. 4–4. "I Saw Many People Walking."‡

---

* "Printing" is done from type. *Letters* are made by hand. Children "print"; draftsmen *letter*.

† From the Greek letters α (alpha) and β (beta).

‡ From *The Story of Writing* (American Council on Education).

51

Gothic

Gothic ABCDEFGH abcdefgh

All letters composed of uniform width elements are classified with the Gothics
Before this simplified classification was adopted, Text letters were known as Gothic

Roman
Roman ABCDEFGJ abcdefghi

All letters composed of thick and thin elements are called Roman

Text
Text ABCDEFGH abcdefghijkl

Includes all styles of Old English Text, Cloister Text, Church Text and Black Text, German Text, Gordon or Bradley Texts and many others.

Fig. 4–5. The Three Basic Groups of Letters.*

4.2 Modern Letter Forms. The three basic types of letters used in the western world are the Roman, Gothic, and Text letters, Fig. 4–5. These may be capital letters, or lowercase letters (called *lowercase* because they were found by compositors in the lower cases of type). If the letters stand upright, they are *vertical letters*. If they are inclined, they are called *inclined letters* or *italic letters*. The letters in Fig. 4–5 were made with Speedball pens, the Gothic letters with the Style B Speedball, and the Roman and Text letters with the Style C Speedball, Fig. 4–32. Alphabets of

* Courtesy Ross F. George, *Speedball Text Book,* published by the Hunt Pen Co., Camden, N. J.

various styles of letters are given in Figs. 4–29 to 4–31 and 4–34 to 4–36.

4.3 Single-Stroke Gothic Letters. Early industrial drawings were lettered with what we would regard as "fancy" letters conforming to the historical styles, usually Roman. In those days it was the fashion for houses, furniture, and other manufactured products to be over-decorated with meaningless "curlicues." As industry advanced, everything became more streamlined or functional, and fancy lettering frills were abandoned. About seventy years ago C. W. Reinhardt developed alphabets based upon the Gothic letters that could be

ABCDEFGHIJKLMNOP
QRSTUVWXYZ&
1234567890

abcdefghijklmnopqrstuvwxyz

Fig. 4–6. USA Standard Vertical Letters. (Adapted from USAS Y14.2—1965 Tentative.)

ABCDEFGHIJKLMNOP
QRSTUVWXYZ&
1234567890

abcdefghijklmnopqrstuvwxyz

Fig. 4–7. USA Standard Inclined Letters. (Adapted from USAS Y14.2—1965 Tentative.)

easily made with single strokes of an ordinary pencil or pen. These were called single-stroke Gothic letters, and are very similar to our present letters.

4.4 USA Standard Letters. In 1935 the letters were further standardized when the United States of America Standards Institute adopted standard alphabets of vertical and inclined letters, and with slight revisions in 1936, 1956, and 1965,* these have become accepted generally as the "last word" for use in drafting, Figs. 4–6 and 4–7. These letters are used on all drawings in this book.

According to the USA Standard: "Lettering on drawings must be legible and suitable for easy and rapid execution." In some industrial drafting rooms, vertical letters are required, while in others inclined letters are used. The student should learn both before he completes his training, starting with the vertical.

4.5 Uniformity in Lettering. One of the main requirements for good lettering is uniformity, Fig. 4–8 (a). Never mix capitals and lowercase letters as at (b). Use guide lines to prevent irregularities as at (c) to (f). Avoid thick-and-thin strokes as at (g) and (h). The background *areas* between letters should appear equal as

* *USA Standard Drafting Practices*, USAS Y14.2—1965 (Tentative). Published by The American Society of Mechanical Engineers, 345 E. 47 St., New York, N.Y. 10017.

at (a) and not as at (j). The spaces between words should not be too small or too large as at (k), but should be equal. Spacing of letters and words is further discussed in Sec. 4.10.

4.6 Pencil Technique. Anyone can learn to letter well if he (1) learns the *shapes*, (2) learns the *strokes*, (3) learns the rules of *spacing*, and (4) *practices* with a real determination to improve. "Practice" without real *effort* to improve is useless. Basically lettering is a form of freehand sketching.

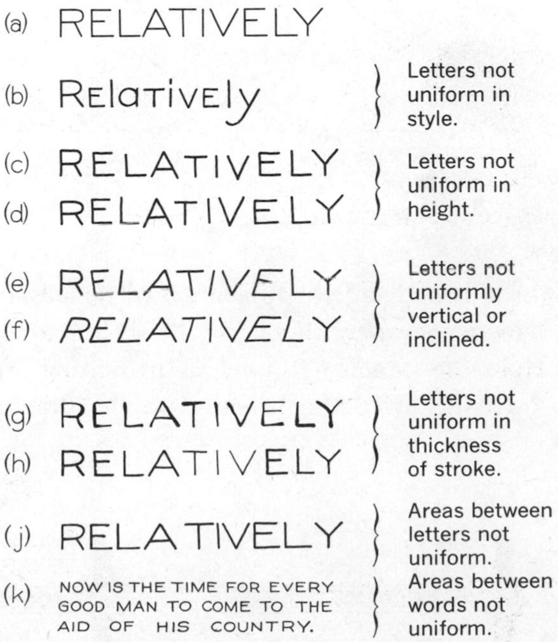

(a) RELATIVELY

(b) Relatively   } Letters not uniform in style.

(c) RELATIVELY
(d) RELATIVELY   } Letters not uniform in height.

(e) RELATIVELY
(f) *RELATIVELY*   } Letters not uniformly vertical or inclined.

(g) RELATIVELY
(h) RELATIVELY   } Letters not uniform in thickness of stroke.

(j) RELATIVELY   } Areas between letters not uniform.

(k) NOW IS THE TIME FOR EVERY GOOD MAN TO COME TO THE AID OF HIS COUNTRY.   } Areas between words not uniform.

Fig. 4–8. Uniformity in Lettering.

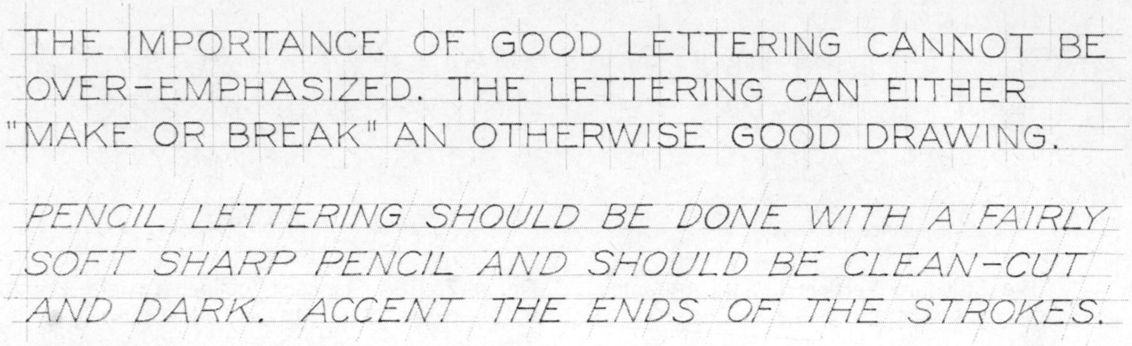

THE IMPORTANCE OF GOOD LETTERING CANNOT BE OVER-EMPHASIZED. THE LETTERING CAN EITHER "MAKE OR BREAK" AN OTHERWISE GOOD DRAWING.

PENCIL LETTERING SHOULD BE DONE WITH A FAIRLY SOFT SHARP PENCIL AND SHOULD BE CLEAN-CUT AND DARK. ACCENT THE ENDS OF THE STROKES.

Fig. 4–9. Example of Pencil Lettering (Full Size).

Since most drawings today are reproduced by blueprinting or otherwise directly from pencil tracings, it is most important to learn to do good pencil lettering, Fig. 4–9. Above all, the letters must be uniform and legible. Use a fairly soft pencil, such as HB, F, or H, sharpened to a sharp conical point, Fig. 4–10

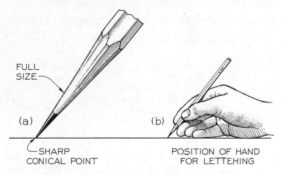

FULL SIZE

(a)   (b)

SHARP CONICAL POINT   POSITION OF HAND FOR LETTERING

Fig. 4–10. Lettering Pencil.

(a). For a detailed explanation of pencils and how to sharpen them, see Secs. 3.7 to 3.9. Hold the pencil naturally, as in writing, Fig. 4–10 (b), with the forearm on the drawing board. *Never letter with the forearm off the board.* Make the letters clean-cut and dark. If the fingers tend to cramp, stop and rest for a few seconds.

Make the beginnings and ends of all strokes *definite by accenting* or "bearing down" on the pencil. Shift the position of the pencil frequently to prevent the lead from wearing down in one place and producing dull lettering.

For ink lettering, see Sec. 8.5.

4.7 Guide Lines. For vertical letters, draw both horizontal and vertical guide lines, Fig. 4–11 (a) and (b). The horizontal guide lines are necessary to keep the letters exactly the same height. The vertical guide lines are spaced at random, and are not used to space the letters but only to keep them vertical. Use a "needle-sharp" hard pencil (4H to 6H), and draw the lines so lightly that they can barely be seen when viewed at arm's length.

For inclined letters, the inclined guide lines

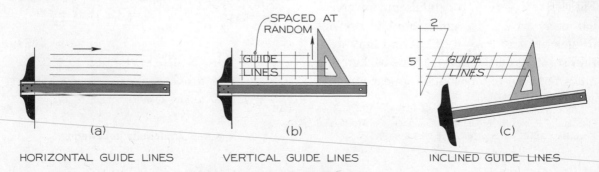

HORIZONTAL GUIDE LINES   VERTICAL GUIDE LINES   INCLINED GUIDE LINES

Fig. 4–11. Guide Lines.

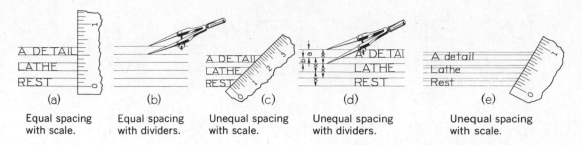

(a)	(b)	(c)	(d)	(e)
Equal spacing with scale.	Equal spacing with dividers.	Unequal spacing with scale.	Unequal spacing with dividers.	Unequal spacing with scale.

Fig. 4–12. Spacing of Guide Lines.

are drawn at a slant of about 68° with horizontal, or parallel to the hypotenuse of a right triangle whose sides are 2 and 5 units, respectively, Fig. 4–11 (c).

The most common spacing for horizontal guide lines is to make the letters $\frac{1}{8}''$ high and the space between lines of lettering $\frac{1}{8}''$. Place the scale in a vertical position, and mark light dashes at right angles to the scale, Fig. 4–12 (a). Another method is to space the guide lines with the bow dividers, (b).

Slightly better appearance results if the space between lines of lettering is less than the height of the letters (but never less than half the height of the letters). Place the scale diagonally, (c), and set off, for example, 4 units for the height of the letters and 3 units for the height of the spaces. Another method

is to use the bow dividers as at (d), making spaces x equal to $a + b$. Guide lines for lower-case letters may be spaced as shown at (e).

The Braddock-Rowe Triangle is convenient for drawing guide lines and also as a regular 45° triangle, Fig. 4–13 (a). A sharp-pointed hard pencil is inserted in the holes, and the triangle is moved along the T-square, (b). The holes at the left of the inclined slot produce guide lines for dimension figures $\frac{1}{8}''$ high and fractions $\frac{1}{4}''$ high.

The *Ames Lettering Instrument*, Fig. 4–13 (c), is highly recommended. Heights are varied as the central disc is turned to one of the numbers at the bottom of the disc. These numbers indicate the heights of letters in 32nds of an inch. Thus, to draw guide lines for letters $\frac{3}{16}''$ high, set the disc at number 6, since $\frac{3}{16}'' = \frac{6}{32}''$.

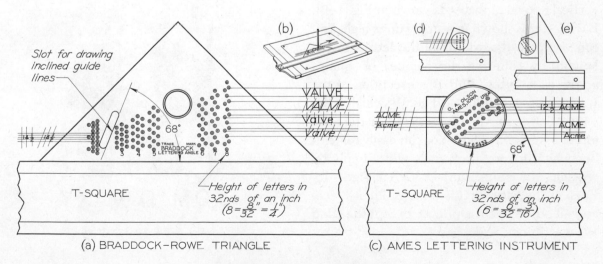

(a) BRADDOCK–ROWE TRIANGLE   (c) AMES LETTERING INSTRUMENT

Fig. 4–13. Guide-Line Devices.

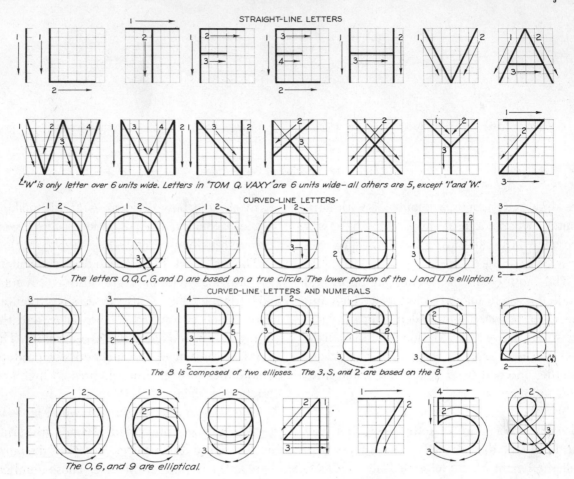

STRAIGHT-LINE LETTERS

"W" is only letter over 6 units wide. Letters in "TOM Q. VAXY" are 6 units wide – all others are 5, except "I" and "W".

CURVED-LINE LETTERS·

The letters O, Q, C, G, and D are based on a true circle. The lower portion of the J and U is elliptical.

CURVED-LINE LETTERS AND NUMERALS

The 8 is composed of two ellipses. The 3, S, and 2 are based on the 8.

The 0, 6, and 9 are elliptical.

Fig. 4–14. Vertical Capital Letters and Numerals.

4.8 Vertical Capital Letters. The standard vertical capital alphabet is shown in Fig. 4–14. Each letter is shown in a grid 6 units high. You can compare the widths of the letters to the heights by counting the number of squares across each letter. With the exception of the I, which has no width, and the W, which is the widest letter of the alphabet, all letters are either 5 or 6 units wide. You can easily remember which letters are 6 units wide simply by recalling the name TOM Q. VAXY, Fig. 4–15. Each of the letters in his name is 6 units wide, and all others in the alphabet, except the I and W, are 5 units wide.

Fig. 4–15. Tom Q. Vaxy.

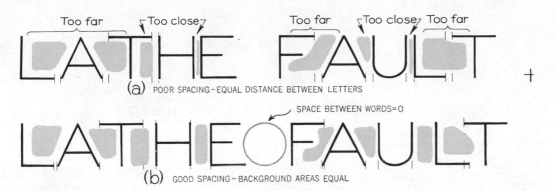

Fig. 4–16. Spacing Letters and Words.

In addition to the widths of the letters, you must also learn the order of strokes, as indicated in Fig. 4–14. Note that horizontal strokes are made from left to right, and vertical strokes are drawn downward. The best way to learn the proportions and strokes of the letters is to practice sketching them large on cross-section paper, making the letters 6 squares high.

4.9 Left-Handers. Are you left-handed? If so, you can learn to letter as well as anyone else. Many of the very best draftsmen are left-handed. The most important step is to learn the correct shapes and proportions, and you can do this as well as any right-hander. The habits of left-handers vary so much that no standard system of strokes can be used by all. Therefore, instead of using the strokes indicated in Fig. 4–14, work out a system of strokes of your own. However, do not adopt any strokes in which the pencil tends to dig into the paper, because later on when you do ink lettering such strokes cannot be used.

4.10 Spacing of Letters and Words. Do not space letters *equal distances* apart, Fig. 4–16 (a), but space them by eye so that the background areas between letters *appear* approximately equal, (b). Certain letters which have straight sides, such as the H and E, must be spaced apart to avoid small areas between them as shown at (a). Other letters, such as the L and T, are of such shape that they may actually be overlapped to produce good spacing. The lower stroke of the L may be slightly shortened when an A or J follows.

Most beginners space letters too far apart and words too close together, Fig. 4–17. Letters within a word should be evenly but compactly spaced, and words should be separated from each other by a space at least equal to the letter O. Make each word stand out as a separate unit, like the printed words in this book. An example of good spacing is shown in Fig. 4–9.

For a discussion of titles on drawings, see Sec. 15.3.

Space between words = letter O

(a) CORRECT ➤ I SPACE LETTERS CLOSELY & WORDS OPENLY

(b) *Wrong* ➤ I SPACE LETTERS OPENLY & WORDS CLOSELY

Fig. 4–17. Spacing Letters and Words.

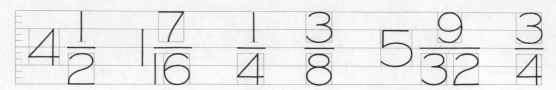

Fig. 4–18. Vertical Numerals and Fractions.

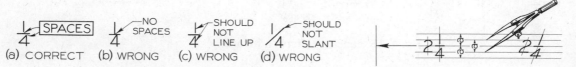

Fig. 4–19. Fractions (Full Size).   **Fig. 4–20. Spacing with Bow Dividers (Full Size).**

STRAIGHT-LINE LETTERS

"W" is only letter over 6 units wide. Letters in "TOM Q. VAXY" are 6 units wide – all others are 5, except "I" and "W".

CURVED-LINE LETTERS

The letters O, Q, C, G, and D are based on a true ellipse.   The lower portion of the J and U is elliptical.

CURVED-LINE LETTERS & NUMERALS

The 8 is composed of two ellipses. The 3, S, and 2 are based on the 8.

The 0, 6, and 9 are elliptical.

Fig. 4–21. Inclined Capital Letters and Numerals.

4.11 Vertical Numerals. All numerical values on drawings are expressed by various combinations of the ten basic numerals in Fig. 4–14. Therefore, it is worth your time to study carefully the proportions and strokes of these characters. All numerals, except the 1, are five units wide. The 3 and the S are both based on the shape of the 8, which is made up of a small ellipse centered over a larger ellipse. The 6 and 9 are alike except reversed, and both fit into the elliptical 0 (zero).

Whole numbers and fractions are shown in Fig. 4–18. It is common practice in dimensioning to make whole numbers $\frac{1}{8}''$ high, and fractions double this, or $\frac{1}{4}''$ high. The numerator and denominator are each slightly less than $\frac{1}{8}''$ high. A clear space must be left between them and the fraction bar, as shown in Fig. 4–19.

Horizontal guide lines for numerals and fractions can be spaced with the bow dividers, Fig. 4–20, with spaces *a* made equal, or they can be drawn with guide-line devices, Fig. 4–13. Vertical guide lines should also be used, as shown.

4.12 Inclined Capital Letters. The standard inclined capital alphabet is shown in Fig. 4–21. The grids and letters are the same as for the vertical capitals, except that they are leaned to the right. The proportions and the strokes correspond to the vertical letters. In addition to horizontal guide lines, draw inclined guide lines at random with the triangle as shown in Fig. 4–11 (c), or with guide-line devices as in Fig. 4–13. The correct and incorrect ways of making certain letters that usually give difficulty are illustrated in Fig. 4–22. The letters O, C, Q, G, and D are elliptical in shape, and the ellipses should definitely *lean to the right*, as shown for the O and G. The letters V, A, W, X, and Y are balanced equally about an imaginary inclined center line, as shown for the A in the figure.

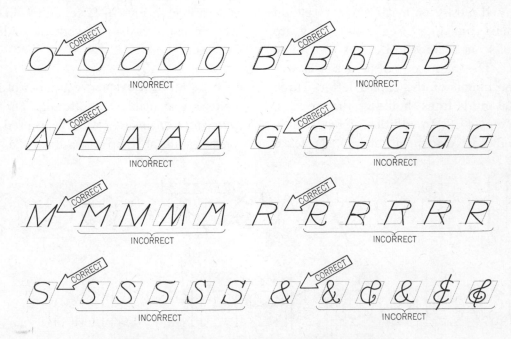

Fig. 4–22. Correct and Incorrect Inclined Capitals.

Fig. 4–23. Inclined Numerals and Fractions.

4.13 Inclined Numerals. Inclined letters and numerals are shown in Fig. 4–21. The numerals are like the vertical numerals in Fig. 4–14, except that they are leaned to the right. Whole numbers and fractions are shown in Fig. 4–23. It is common practice to make whole numerals $\frac{1}{8}''$ high and fractions twice this, or $\frac{1}{4}''$ high. Horizontal guide lines are drawn in a similar manner to those for vertical numerals, as shown in Fig. 4–20. Inclined guide lines should be used, as shown.

4.14 Vertical Lowercase Letters. These letters are used mostly on maps and very seldom on machine drawings, Fig. 4–24. They are formed of straight lines and circles with a few variations. The main body of the letter is two-thirds the height of the capital letter. Three horizontal guide lines should be drawn, Figs. 4–12 (e) and 4–13. In addition, vertical guide lines should be drawn at random. To learn the proportions and strokes, practice these letters on cross-section paper.

4.15 Inclined Lowercase Letters. The inclined lowercase letters are shown in Fig. 4–25. The grids and letters are the same as for the vertical lowercase letters except that they are leaned to the right. The circles become ellipses, and these should lean definitely to the right. Be careful to make the letters v, y, w, and x equally balanced about an imaginary inclined center line. Guide lines are drawn in a similar manner to those for vertical lowercase letters, as described in Figs. 4–12 (e) and 4–13. However, inclined instead of vertical guide lines are used.

4.16 Lettering Devices. A number of instruments are available for lettering. The *Leroy Lettering Instrument*, Fig. 4–26, is used with a special guide or template to form the letters.

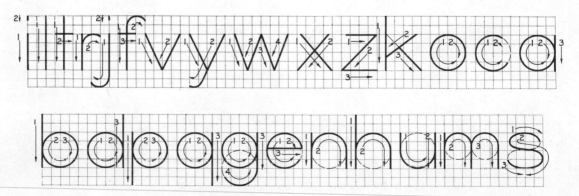

Fig. 4–24. Vertical Lowercase Letters.

Fig. 4–25. Inclined Lowercase Letters.

Courtesy Keuffel & Esser Co.

Fig. 4–26. Leroy Lettering Instrument.

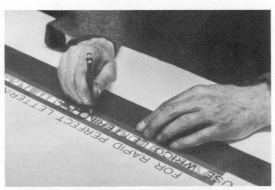

Courtesy Wood-Regan Instrument Co.

Fig. 4–27. Wrico Lettering Instrument.

Another device is the *Wrico*, Fig. 4–27. The Wrico has a guide with a series of perforations corresponding to parts of letters, and a special pen is used to trace these shapes. These instruments, especially the Leroy, are coming into increasing use in industry, particularly for lettering titles and drawing numbers. Other similar lettering instruments are the Varigraph, Tacro-Scriber, and Unitech Lettering Set.

4.17 Filled-in Letters. You can make large letters for titles, posters, and so forth by drawing the outlines, using the T-square, triangles, and compass, and some freehand work for curves. Then fill in the letters, Fig. 4–28. "Block letters," which can be easily drawn with the T-square and triangles, are shown in Fig. 4–29. Gothic and Roman alphabets are shown in Figs. 4–30 and 4–31. First, draw the grids lightly in pencil, or use the bow dividers to set off the unit distances, as shown; then draw the letters in pencil. Finally, ink them in, then erase the grid lines. The thicknesses of the

GOTHIC ROMAN

Use compass   Add spurs if desired   Leave in outline if desired

Fig. 4–28. Filled-in Letters.

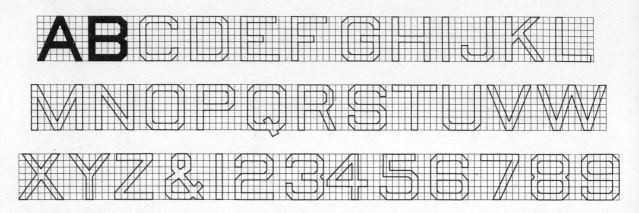

Fig. 4–29. Block Letters.

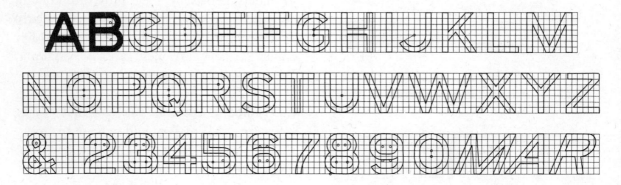

Fig. 4–30. Gothic Capital Letters.

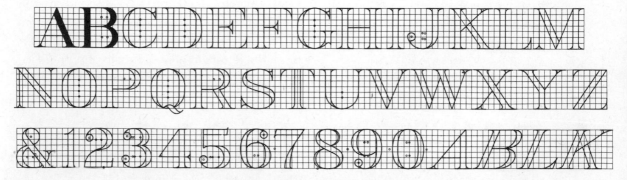

Fig. 4–31. Modern Roman Capitals.

stems may be varied from one-tenth to one-fifth the height of the letters, as desired; one-seventh has been adopted in Figs. 4–29 to 4–31.

4.18 Special Pens. Special "broad-nib" pens are available for making "single-stroke" large letters in the various styles shown in Fig. 4–5, as well as many variations of these. *Speedball* * pens, which are very popular, are made in four different styles of points, Fig. 4–32. Each style is available in different widths, from 0 (largest) to 5 or 6 (smallest). The alphabets in Figs. 4–34 to 4–36 are made with Speedball pens B, D, and C, respectively. Examples of posters and covers for sets of drawings are shown in Fig. 4–33.

In recent years a number of technical fountain pens have been developed for drafting, ruling, mechanical and freehand lettering,

* See Ross F. George, *Speedball Text Book*, published by the Hunt Pen Company, Camden, N.J.

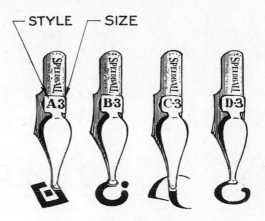

Fig. 4–32. Speedball Pens.

writing, and commercial art work. Among the more prominent ones are the Koh-I-Noor *Rapidograph*, *Gramo-Inker*, *Tacro-Graph*, *Unitech*, and *Mars* pens. The Koh-I-Noor *Rapidograph* set, shown in Fig. 4–37, consists of a single holder with seven interchangeable point sections in different line widths from No. 00 (extra extra fine) to No. 4 (extra broad). The outstanding feature of these sets

(a)

(b)

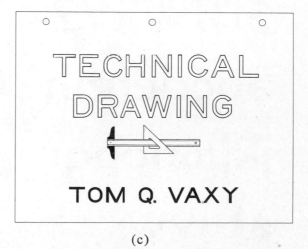

(c)

(a) and (b) courtesy Ross F. George, *Speedball Text Book*, published by Hunt Pen Co., Camden, N.J.

Fig. 4–33. Posters and Drawing Covers.

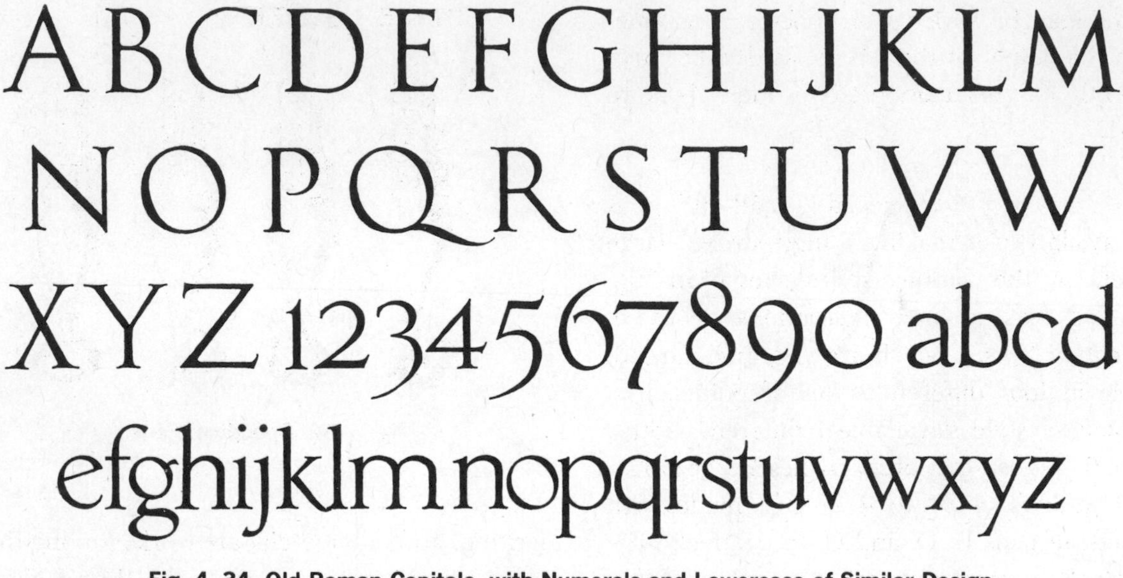
ABCDEFGHIJKLM
NOPQRSTUVW
XYZ 1234567890 abcd
efghijklmnopqrstuvwxyz

Fig. 4-34. Old Roman Capitals, with Numerals and Lowercase of Similar Design.

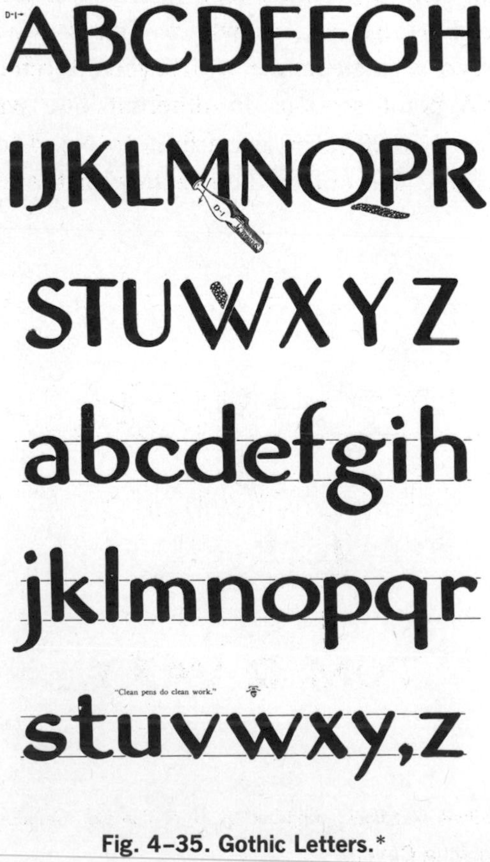
D-1
ABCDEFGH
IJKLMNOPR
STUWXYZ
abcdefgih
jklmnopqr
"Clean pens do clean work."
stuvwxy,z

Fig. 4-35. Gothic Letters.*

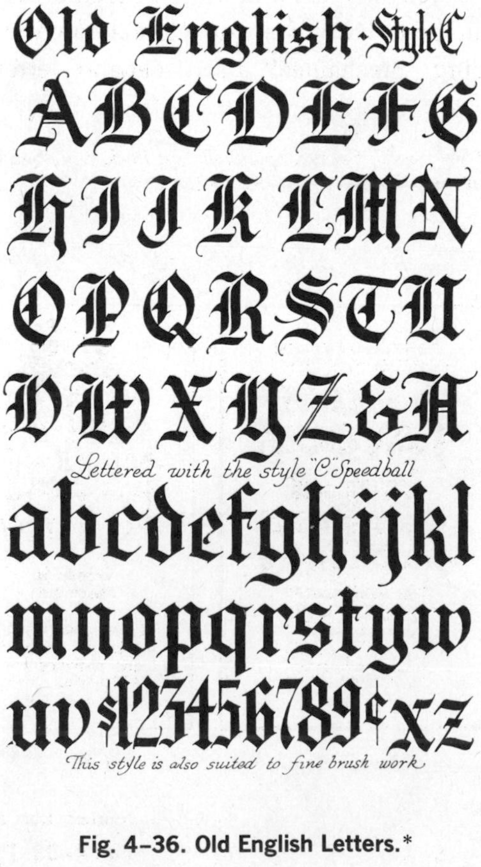
Old English·Style C
ABCDEFG
HIJKLMN
OPQRSTU
Lettered with the style "C" Speedball
abcdefghijkl
mnopqrstyw
uv $123456789¢ xz
This style is also suited to fine brush work

Fig. 4-36. Old English Letters.*

* Courtesy Ross F. George, *Speedball Text Book,* published by Hunt Pen Co., Camden, N.J.

Courtesy Koh-I-Noor, Inc.

Fig. 4–37. Koh-I-Noor Rapidograph Technical Fountain Pen Set.

is the assured constant line width of each point section. Interchange of points is accomplished easily and with complete cleanliness.

4.19 Lettering Practice. In order to learn the proportions and the strokes of the letters, it is suggested that you sketch them six squares high on cross-section paper, preferably paper with $\frac{1}{8}''$ divisions, Fig. 4–38. A variety of lettering pads or lettering practice booklets are available on the market. In these, the layouts are already printed and it is only necessary to fill in the lettering as indicated. These are excellent for the beginning student, since they do not require a complete layout before lettering.

For those classes where prepared lettering sheets are not available, the layouts in Figs. 4–39 to 4–46 are included here. For all of these, use Layout A, shown on page 474 in the Appendix. Draw light horizontal guide lines as shown, and *add light vertical or inclined guide lines,* the entire height of the sheet, using a sharp 4H pencil. Fill in all blank spaces with letters or words. Use an HB pencil for all $\frac{3}{8}''$ letters and an F pencil for the smaller letters. The last line on each sheet may be lettered first lightly in pencil and then in ink over the pencil if the instructor so assigns. Omit all spacing dimensions.

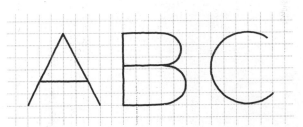

Fig. 4–38. Practice on Cross-Section Paper.

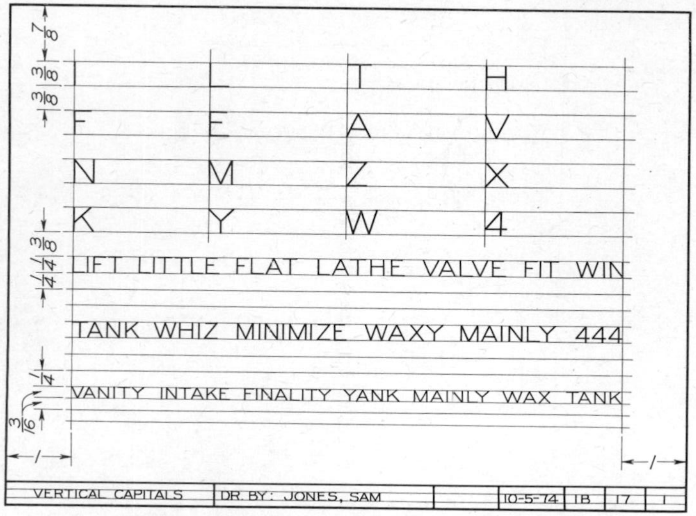

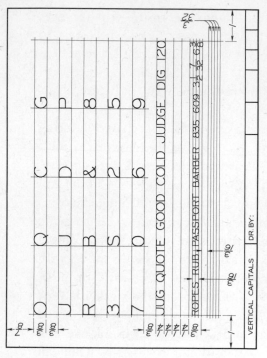

Fig. 4–40. Vertical Curved-Line Letters and Numerals.*

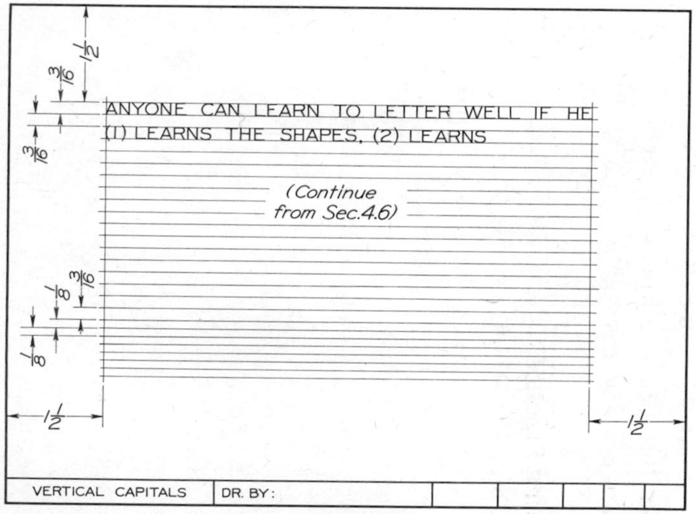

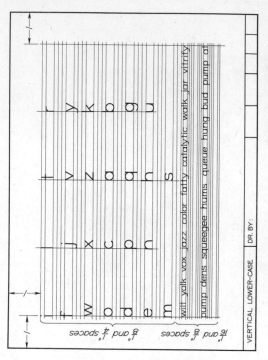

Fig. 4–42. Vertical Lowercase Lettering.*

Fig. 4–39. Vertical Straight-Line Letters.*

Fig. 4–41. Vertical Capital Lettering.*

* Use Layout A (Appendix, page 474). See instructions on page 65.

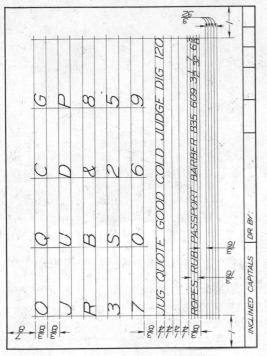

Fig. 4-44. Inclined Curved-Line Capitals. *

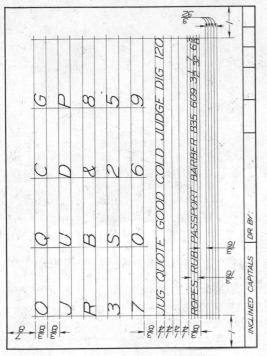

Fig. 4-43. Inclined Straight-Line Capitals. *

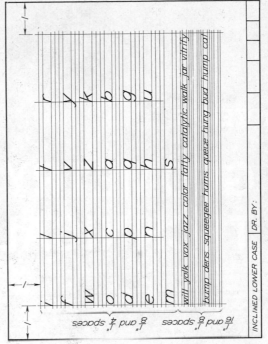

Fig. 4-46. Inclined Lowercase Lettering. *

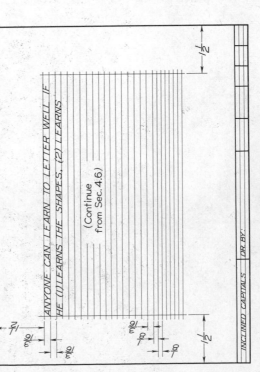

Fig. 4-45. Inclined Capital Lettering. *

* Use Layout A (Appendix, page 474). See instructions on page 65.

CHAPTER 5

GEOMETRY OF TECHNICAL DRAWING

5.1 Geometry in Drafting. In order to make drawings, and in some cases to solve problems by means of lines, certain geometrical constructions are frequently used. Some of these should be learned now in order for you to proceed with your work intelligently. Others are included here for reference when you need them later. This is not mathematician's geometry, but draftsman's geometry, and all of these constructions are easy to draw with the T-square, triangles, and drawing instruments. They are based mostly on *plane geometry,* a subject which you may study elsewhere from a mathematical point of view.

In drawing these constructions, accuracy is important. Use a sharp medium-hard lead (2H or 3H) in your pencil and compass. Draw construction lines lightly, and make all given and required lines thin but dark.

5.2 Geometric Shapes. The most common geometric shapes encountered in technical drawing are shown in Fig. 5–1, mainly for reference. Study each figure carefully and learn the terms used.

A common symbol for *angle* is ∠ (singular) or ∠s (plural). There are 360 degrees (360°) in a full circle. A degree is divided into 60 minutes (60′), and each minute is divided into 60 seconds (60″).

A *triangle* (symbols: △ singular, △s plural) is a plane (flat) figure, bounded by three sides, and the sum of the interior angles always equals 180°. The triangle is the basis for the study of *trigonometry,* a branch of mathematics of much value to the engineer and draftsman.

Quadrilaterals have four sides. *Regular polygons* have equal sides. *Regular solids* have faces that are regular polygons. *Prisms* have plane faces parallel to an imaginary axis. A prism having bases that are parallelograms is called a *parallelepiped. Pyramids* have plane triangular faces that intersect at a common point called the *vertex. Cylinders* and *cones* are said to be single-curved surfaces, and have straight-line elements. The *sphere* is like a round ball; the *torus* is like a "doughnut."

5.3 Geometric Constructions. The constructions given in Figs. 5–2 to 5–23 are the ones most often used by the draftsman. Purely geometric methods, which do not recognize the advantages in using drawing equipment, are omitted; the draftsman will and should use his instruments wherever convenient.

Two symbols that are commonly used in drafting are ₵, which means "center line," and ⊥, which means "perpendicular." See Figs. 5–13 and 5–14.

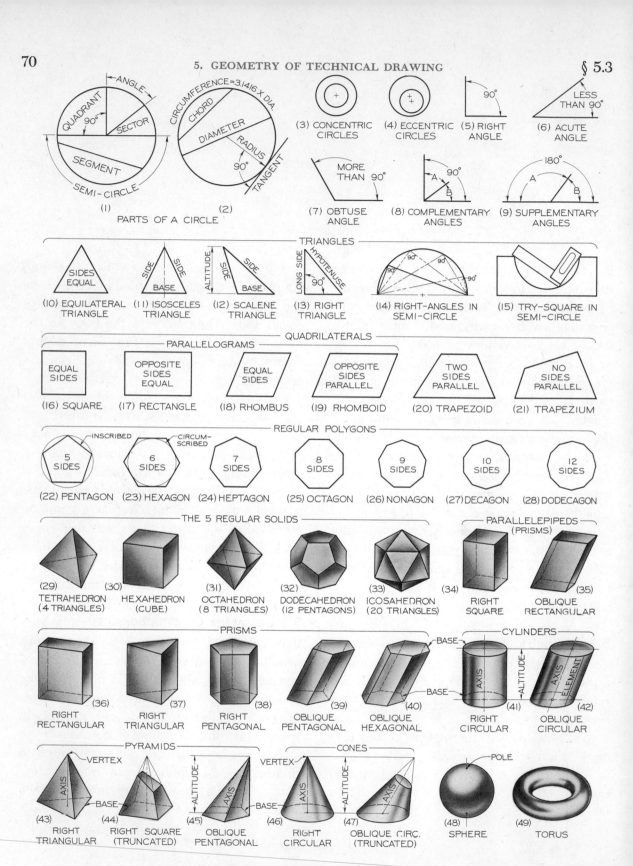

Fig. 5–1. Circles, Angles, Plane Figures, and Solids.

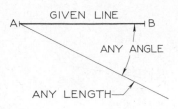

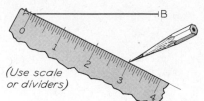

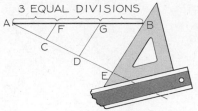

1. Draw light construction line at any convenient angle from either end of line.

2. With scale or dividers, set off the required number of equal divisions.

3. Join E to B, and draw DG and CF parallel to EB.

Fig. 5–2. To Divide Line into Equal Parts.

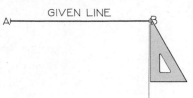

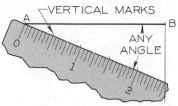

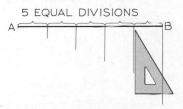

1. Draw vertical line at A or B of indefinite length.

2. Adjust scale with zero at A and 5th division on vertical line. Mark each division.

3. Draw vertical lines to AB to divide AB into 5 equal parts.

Fig. 5–3. To Divide Line into Equal Parts.

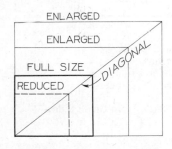

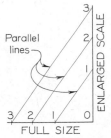

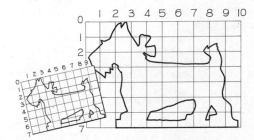

1. Enlarging or reducing a drawing sheet or any square or rectangle by use of diagonal.

2. Enlarging or reducing one or more measurements.

3. Draw grid of equal squares on drawing to be copied. Draw similar but larger grid on new sheet. Sketch lines on new grid through same grid intersections as on small drawing.

Fig. 5–4. Enlarging or Reducing.

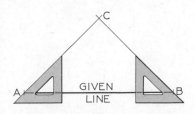

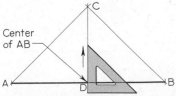

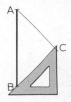

1. Draw 45° lines from ends A and B to locate C.

2. Draw vertical line through C to locate center D.

3. Draw 45° lines from A and B.

4. Draw horizontal line through C with T-square.

Fig. 5–5. To Bisect Horizontal and Vertical Lines.

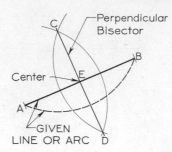

1. Draw equal arcs from end points A and B to intersect at points C and D, using radius more than half AB.
2. Draw perpendicular bisector through C and D.

Fig. 5–6. To Bisect a Line or an Arc.

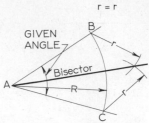

1. Strike arc R of any convenient radius, locating points B and C.
2. Strike equal arcs r to locate D.
3. Draw bisector through A and D.

Fig. 5–7. To Bisect an Angle.

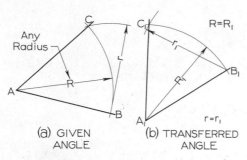

1. Draw AB in new location A_1B_1.
2. Strike arc R and make $R_1 = R$.
3. Strike arc r and make $r_1 = r$.
4. Draw AC in new location A_1C_1.

Fig. 5–8. To Copy Angle in New Location.

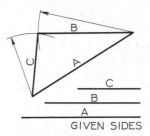

1. Draw one side, as A, in new location.
2. Strike arcs from ends of A, equal to B and C.

Fig. 5–9. Triangle from Given Sides.

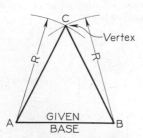

1. From ends A and B, construct any equal arcs R, to locate C.
2. Draw CA and CB.

Fig. 5–10. Isosceles Triangles.

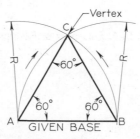

1. From ends A and B, construct arcs with radii equal to AB, to locate C.
2. Draw CA and CB.

Fig. 5–11. Equilateral Triangles.

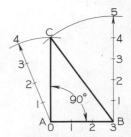

1. From ends of 3-unit base AB, strike 4-unit and 5-unit arcs to locate C.
2. Draw CA and CB.

Fig. 5–12. Right Triangle.

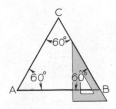
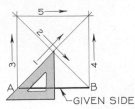
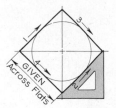
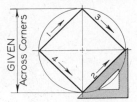

(a) EQUILATERAL TRIANGLE
Draw equal 60° angles, as shown.

(b) SQUARE
Draw lines in order shown.

(c) SQUARE
1. Draw ₵'s and circle.
2. Draw 45° tangents.

(d) SQUARE
1. Draw ₵'s and circle.
2. Draw 45° lines.

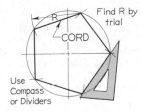
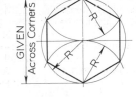
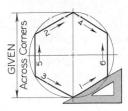
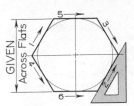

(e) PENTAGON
1. Divide circle into five equal parts.
2. Draw chords.

(f) HEXAGON
1. Draw circle and arcs R = radius.
2. Draw sides.

(g) HEXAGON
1. Draw ₵'s and circle.
2. Draw sides as shown.

(h) HEXAGON
1. Draw ₵'s and circle.
2. Draw tangent sides.

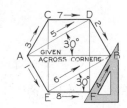
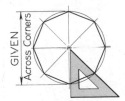
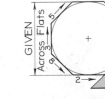

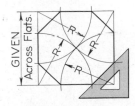

(j) HEXAGON
Draw lines in order shown.

(k) OCTAGON
1. Draw ₵'s and circle.
2. Draw 45° diagonals.
3. Connect intersections.

(m) OCTAGON
1. Draw circle.
2. Draw tangents.

(n) OCTAGON
1. Draw square.
2. Draw arcs and sides.

Fig. 5-13. To Draw Regular Polygons.

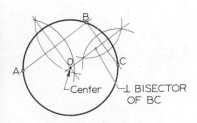
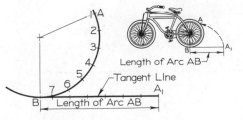
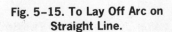
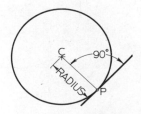

1. Connect A, B, and C.
2. Draw ⊥ bisectors.
3. Draw circle with center O, through A, B, and C.

1. From A on arc, set off equal divisions till near B.
2. Set off same divisions on line. (Use bow dividers.)

1. Erect ⊥ to line at P equal to radius desired.
2. Draw tangent circle, through P, using center C.

Fig. 5-14. Circle Through Three Given Points.

Fig. 5-15. To Lay Off Arc on Straight Line.

Fig. 5-16. To Draw Circle Tangent to Line.

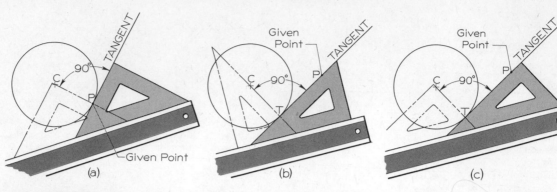

1. Move triangle until one side passes through C and tangent point P.
2. Slide triangle on T-square until other side passes through P.
3. Draw tangent line.

1. Move triangle until hypotenuse passes through P and just touches circle.
2. Revolve triangle about 90° corner, and mark T.
3. Revolve triangle back, and draw tangent line.

1. Move triangle until side passes through P and just touches circle.
2. Slide triangle until other side passes through C. Mark tangent point T.
3. Slide triangle back, and draw tangent line.

Fig. 5-17. To Draw Tangent to Circle.

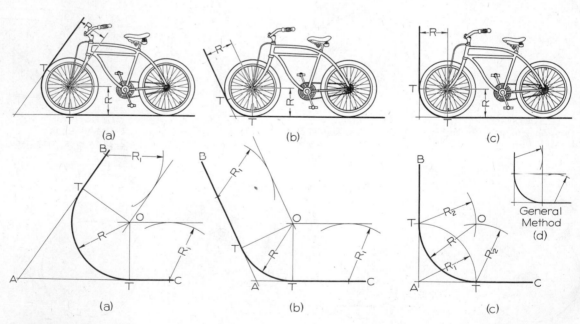

Instructions for (a) and (b):
1. Strike arcs R_1 equal to given radius R.
2. Draw construction lines parallel to given lines AB and AC, to intersect at O, center of arc.
3. Drop perpendiculars OT to get points of tangency T.
4. Draw tangent arc R from T to T, with center O.

Instructions for (c):
1. Strike arc R_1 = radius R.
2. Strike arcs R_2 = radius R, to locate center O of arc.
3. Draw tangent arc R from T to T, with center O.

Fig. 5-18. To Draw Arc Tangent to Two Straight Lines.

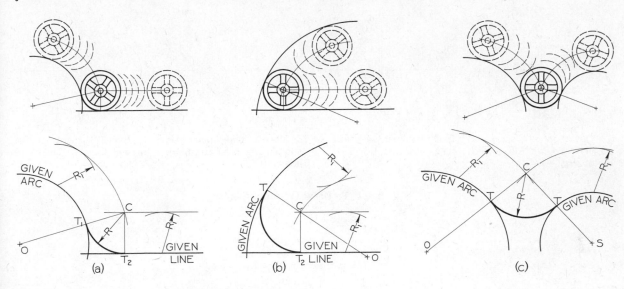

Instructions for (a) and (b):
1. Strike arcs R_1 = radius R (given radius).
2. Draw construction arc parallel to given arc, with center O.
3. Draw construction line parallel to given line.
4. From intersection C, draw CO to get tangent point T_1, and drop perpendicular to given line to get point of tangency T_2.
5. Draw tangent arc R from T_1 to T_2 with center C.

Instructions for (c):
1. Strike arcs R_1 = radius R.
2. Draw construction arcs parallel to given arcs, using centers O and S.
3. Join C to O and C to S to get tangent points T.
4. Draw tangent arc R from T to T, with center C.

Fig. 5–19. To Draw Arc Tangent to Straight Lines and Arcs.

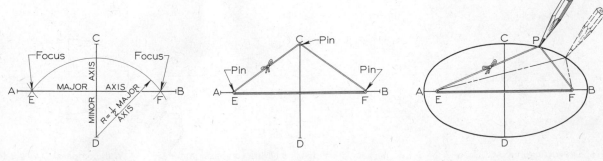

1. From C and D, strike arcs as shown, locating foci E and F.

2. Place pins at E, F, and C, and tie string around them without slack.

3. Remove pin C. Move pencil, keeping string without slack.

Fig. 5–20. To Draw "Pin and String" Ellipse.

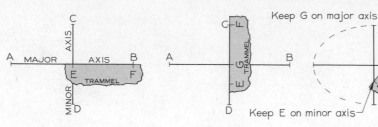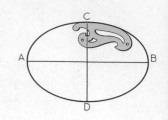

1. On trammel (scrap of paper) mark off EF = half major axis AB.

2. On trammel, set off GF = half minor axis CD.

3. Move E along minor axis and G along major axis. Mark points at F.

4. Sketch smooth curve through points. Then use French curve for final ellipse.

Fig. 5–21. To Draw Trammel Ellipse.

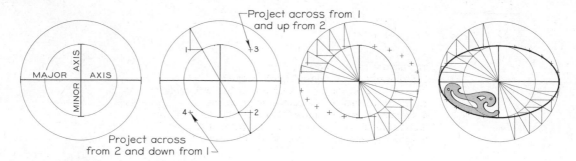

1. Draw light construction circles as shown, using the axes as diameters.

2. Draw any diagonal. At intersections with circles draw lines parallel to axes, as shown, to get 1, 2, 3, and 4.

3. Draw as many additional diagonals as needed, each diagonal producing 4 points.

4. Sketch smooth curve through points. Then use French curve for final ellipse.

Fig. 5–22. To Draw Concentric-Circle Ellipse.

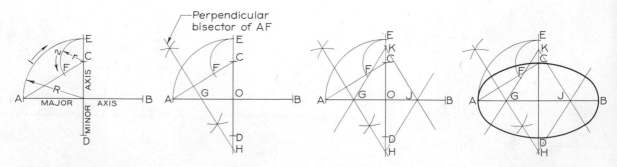

1. Draw diagonal AC. Draw arcs 1 and 2 to locate F.

2. Draw perpendicular bisector of AF to locate centers G and H.

3. With compass, make OJ = OG and OK = OH, and draw lines as shown.

4. With compass, draw small arcs from centers G and J and large arcs from centers K and H.

Fig. 5–23. To Draw Approximate Ellipse with Compass.

5.4 Drafting Geometry Problems. The problems that follow are to provide practice in using geometric constructions. Those in Figs. 5–24 to 5–26 cover the basic constructions, and those in Figs. 5–27 to 5–42 involve tangencies. Accuracy and clean drawing are the objectives. Dimensions may or may not be included, as assigned by the instructor. If they are assigned, the student should study Sec. 4.11 or 4.13 on lettering, and Secs. 9.1 to 9.12 on dimensioning conventions.

Use sheet Layout A (Appendix, page 474) and divide into four equal parts, as shown in Fig. 5–24. Other problems that may be assigned are given in Figs. 5–25 and 5–26, and all are printed one-third size. Apply your dividers directly to these problems, and step off

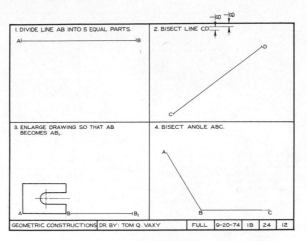

Fig. 5–24. Geometric Construction Problems (Layout A, Appendix, page 474).

all spacing measurements triple size. Lettering is $\frac{1}{8}$" high.

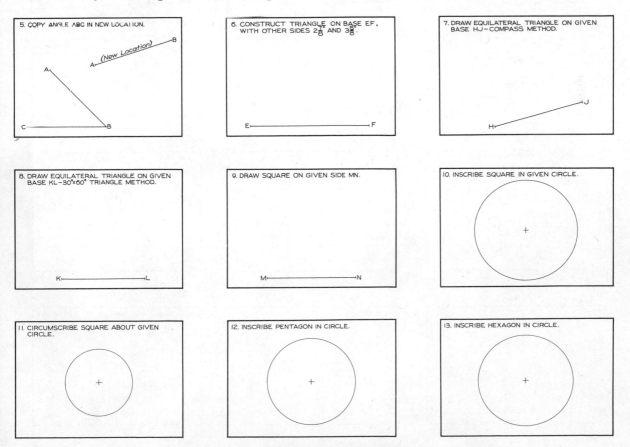

Fig. 5–25. Geometric Construction Problems.

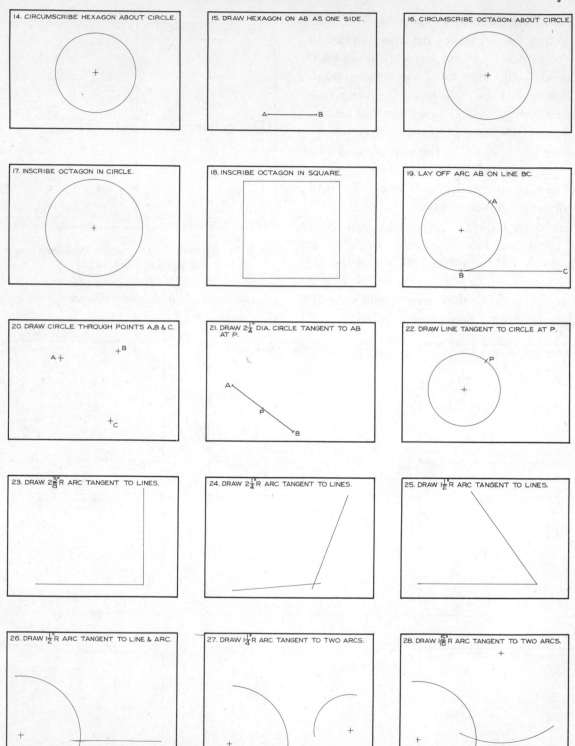

Fig. 5–26. Geometric Construction Problems. See instructions referring to Fig. 5–24.

The steps in drawing the Conveyor Link, Fig. 5–27, are shown in Fig. 5–28. First lay out the sheet, using Layout C (Appendix, page 474). Draw the main center lines, as shown in step I below; then draw the arcs, circles, and straight lines as shown in steps II to V. All lines should be light construction lines up through step V. Finally, heavy-in the final lines as shown in step VI.

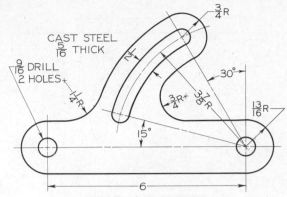

Fig. 5–27. Conveyor Link.

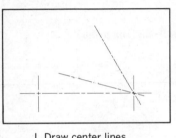

I. Draw center lines.

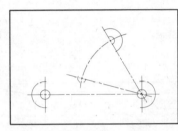

II. Draw circular arcs.

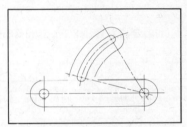

III. Draw connecting straight lines and arcs.

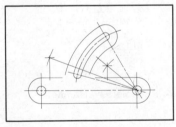

IV. Locate centers and points of tangency of tangent arcs.

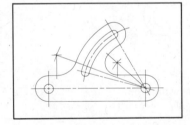

V. Draw tangent arcs.

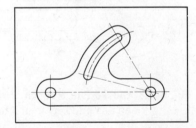

VI. Heavy-in all visible object lines.

Fig. 5–28. Steps in Drawing Conveyor Link.

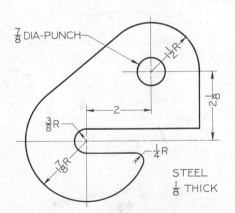

Fig. 5–29. Cover Plate (Layout C, Appendix, page 474).

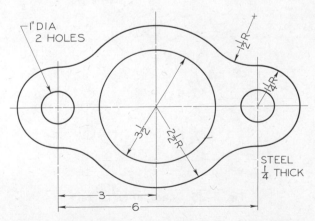

Fig. 5–30. Gasket (Layout C, Appendix, page 474).

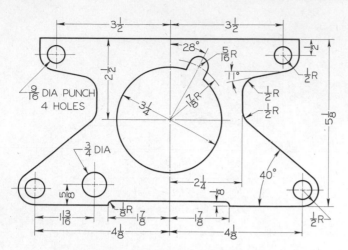

Fig. 5–31. Buick Transmission Gasket.*

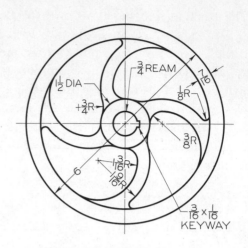

Fig. 5–32. Handwheel.*

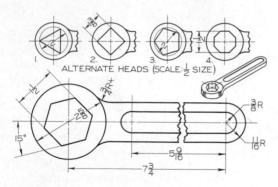

Fig. 5–33. Closed-End Wrench. (Draw with type head assigned by instructor.)*

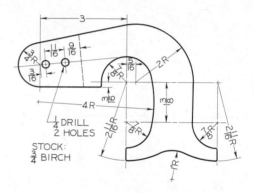

Fig. 5–34. Keyhole Saw Handle.*

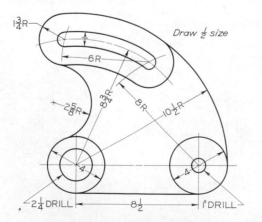

Fig. 5–35. Quadrant for Lathe.*

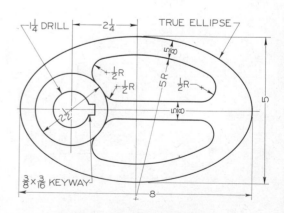

Fig. 5–36. Elliptical Cam.*

* Use Layout C (Appendix, page 474).

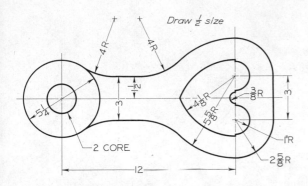

Fig. 5–37. Locomotive Truck Swing Link.*

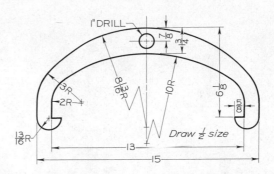

Fig. 5–38. Clamp for Laundry Machine.*

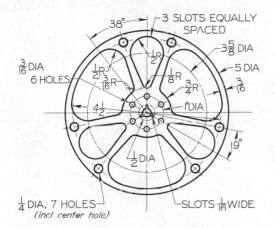

Fig. 5–39. Movie Film Reel.*

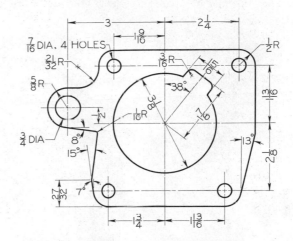

Fig. 5–40. Buick Transmission Gasket.*

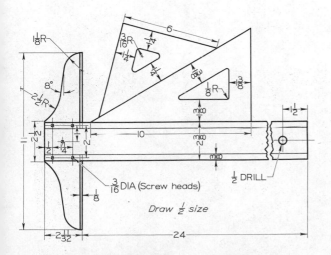

Fig. 5–41. T-Square and Triangles.*

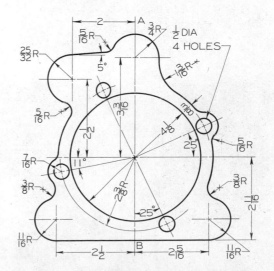

Fig. 5–42. Buick Rear Transmission Gasket.*

* Use Layout C (Appendix, page 474).

CHAPTER 6
VIEWS OF OBJECTS

6.1 Pictures and Views. An architect's sketch, made to show a client how his new home will look when built, is shown in Fig. 6–1. This drawing shows how the house will *appear* from one position, but provides no practical information as to the size of the house; the shapes, arrangement, and sizes of the rooms; or details of windows, doors, fireplace, kitchen cabinets, and so forth. The contractor will need all this information, and much more, to build the house. Further, the client himself will not agree to build until he can see exact drawings that show every detail unmistakably. In fact, the complete drawings (and specifications) are needed to figure costs. No one in his right mind would build a house until he and the contractor could agree on the price.

In order to describe the exterior of the house completely, a series of *views* must be drawn, each view showing the house from a different point of view. You can walk around the house and view it from all four sides, and if you fly over the house in a helicopter or airplane, you can look down and get a top view.

First, Fig. 6–2 (a), suppose you stand at a distance* in front of the house, and look directly toward the front. Imagine a glass plane between you and the house, and parallel to the house. The view on the plane is what you would see, and it is called a *front view* or *front elevation*. This view shows the true *width* and *height* of the house, but not the *depth*. More exactly, the view is obtained by dropping perpendicular *projectors* from all points on the house to the plane. Collectively, the piercing points of all these perpendiculars form the view.

---

* Theoretically at an infinite distance.

Fig. 6–1. Architect's Sketch of House.

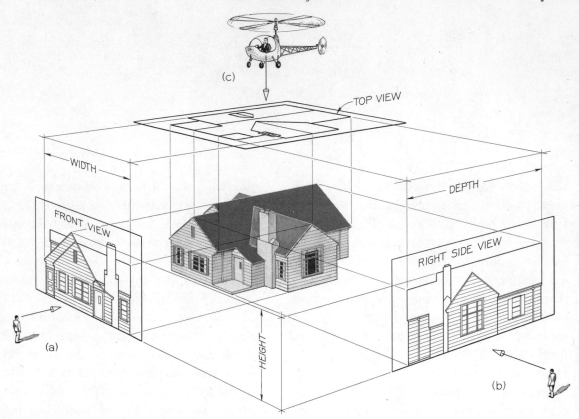

Fig. 6–2. Three Views of a House.

Next, take a position looking toward the right side of the house, (b). The view seen from here is the *right-side view* or *right-side elevation*. It shows the true *height* and *depth* of the house, but not the *width*.

Finally, take a position (or imagine it) above the house and looking down, (c). The resulting view is projected up to the plane, which is parallel to the ground, and is called the *top view* or *plan*. This view shows the true *width* and *depth* of the house, but not the *height*.

When these three views are accurately drawn to scale (usually $\frac{1}{4}'' = 1'-0''$), they can be dimensioned fully; thus they will provide detailed information to the client and to the contractor and his workers. Because of the large sizes of drawings of buildings, architects usually draw each view on a separate sheet, as shown in Fig. 6–3.

Actually, for an object as complicated as a house, additional views and sections would be necessary to provide complete information.

Fig. 6–3. Front View.

The three views here are sufficient to show how views are obtained.

6.2 Revolving the Object. In Fig. 6–2 the object to be drawn is very large and you can obtain views by *shifting your position* while the house remains stationary. If the object is small, such as a machine part, you can remain in your chair and *shift the object* to the desired position.

Suppose you wish to draw three views of the Holder for an Offset Press, Fig. 6–4 (a). The arrows indicate the directions in which you will view the object. Note the three principal dimensions: *width, height,* and *depth.*

Hold the object in your hand so that you look perpendicularly toward the front of the object, (b). *The front view shows the width and the height, but not the depth.*

To get the right-side view, revolve the object and bring the right side toward you, (c). *The right-side view shows the height and the depth, but not the width.*

To get the top view, revolve the object, bringing the top toward you, (d). *The top view shows the width and depth, but not the height.*

Note that each view shows two dimensions, but not the third dimension. Thus, a three-dimensional object requires at least two views to describe it,* and often three or more are needed, as will be shown later.

* Except as explained in Secs. 6.6 and 6.7.

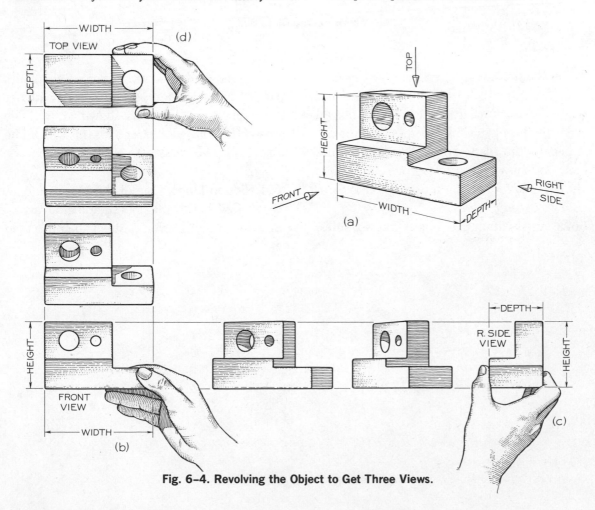

Fig. 6–4. Revolving the Object to Get Three Views.

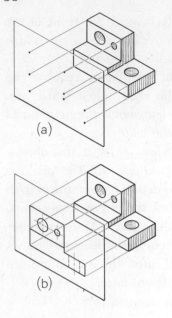

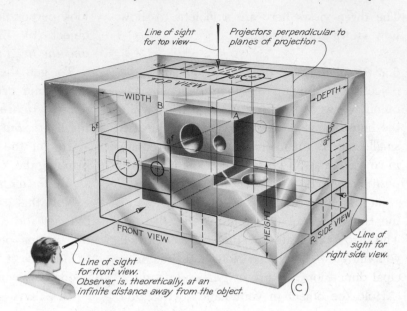

Fig. 6-5. The Glass Box.

6.3 The Glass Box. A more scientific explanation of how views are obtained is presented in Fig. 6-5. As shown at (a), the plane is placed parallel to the object, and perpendiculars are dropped to the plane from all points on the object. As shown at (b), the piercing points of these *projectors* are then connected to form the view, or *projection*.

Since any rectangular object has six sides,

we can obtain six views by placing six planes parallel to the six sides. Together, these planes form the "glass box," as shown at (c). The arrows indicate the directions in which the object is viewed to get the three regular views.

6.4 Hidden Lines. One of the big advantages that a view has over a photograph or an artist's drawing is that hidden parts of the object can

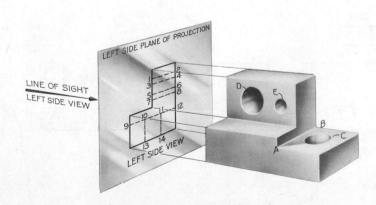

Fig. 6-6. Hidden Edges and Contours.

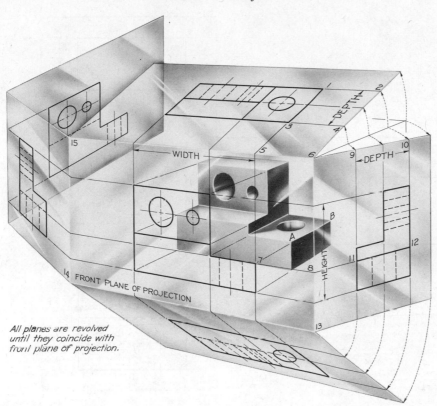

Fig. 6–7. Unfolding the Glass Box.

be shown by means of dashed lines, called *hidden lines*, Fig. 3–11. In Fig. 6–6, the edge AB on the object would not be visible as you look in the direction of the arrow. Therefore, its view 9–12 is represented by a hidden line. Likewise, the contours of hole C are invisible, and are projected as hidden lines 10–13 and 11–14. Hole E is projected as hidden lines 3–4 and 5–6. Note how other hidden lines are projected in this figure.

6.5 Unfolding the Glass Box. You perhaps wonder, "How can I get the views from the planes onto my flat sheet of drawing paper?" You do this, as shown in Fig. 6–7, by unfolding the glass box until all planes lie in the same plane as the front plane. The top, bottom, and both side planes are hinged to the front plane. The rear plane is hinged to the left-side plane.

Note, at the upper right of Fig. 6–7, that distances 4–2 and 9–10 are equal, which explains why the *depth* in the top view must always be the same as the *depth* in the side view. Observe also why the *width* must always be the same in the top, front, and bottom views, and the *height* must always be the same in the rear, left-side, front, and right-side views.

The six planes, after being revolved into one plane, are shown in Fig. 6–8. The top view is directly over the front view, the bottom view is directly under the front view, the right-side view is directly to the right of the front view, the left-side view is directly to the left of the front view, and the rear view is directly to the left of the left-side view.

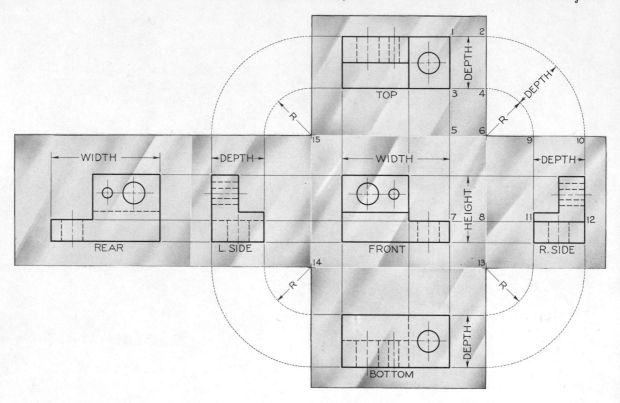

Fig. 6–8. The Glass Box Unfolded.

6.6 Elimination of Views. Clearly, all of the possible six views may not be needed to describe completely the shape of the object, as shown in Fig. 6–9. In practice, *only those views that are necessary to describe the shape of the object should be drawn.*

To choose the "necessary views," examine the object to see what shapes must be described. For example, in Fig. 6–9 (a), the object has two right-angled notches and three holes whose shapes must be described. These are all shown clearly in the six views, but with considerable duplication.

The front view shows the shapes of a right-angled notch and two holes. These are also shown in the rear view, but the rear view has a hidden line. The front view is preferred; the rear view is crossed out.

The top and bottom views both show the shape of the hole at the right end, but the top view is preferred because it has fewer hidden lines. The bottom view is, therefore, crossed out.

The left-side and right-side views both show the right-angled notch, but the right-side view is preferred because it has fewer hidden lines. The left-side view is, therefore, crossed out.

The remaining three views, the front, top, and right-side, are the necessary views of this object, as indicated by the border line around them in Fig. 6–9 (b). These are often referred to as the "three standard views," because they are the views most commonly required.

6.7 Choice of Views. Keep in mind that your job is to select views which are necessary to describe each contour or shape of the object. For example, in Fig. 6–10 (a), if the sheet-

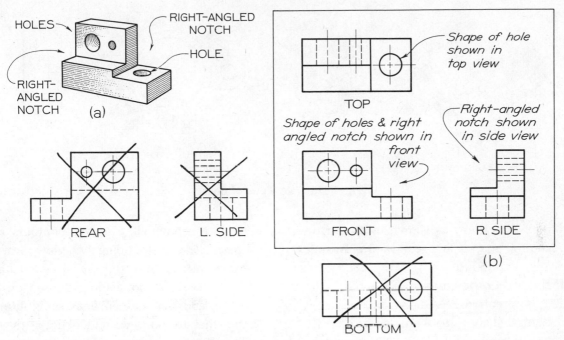

Fig. 6–9. Necessary Views.

metal part is viewed in the direction of the arrow, all the essential shapes are seen at once. Only the thickness is not seen. A one-view drawing is sufficient if the thickness is given in a note, or a second view showing the thickness may be added, producing a two-view drawing.

If the object shown at (b) is viewed in two

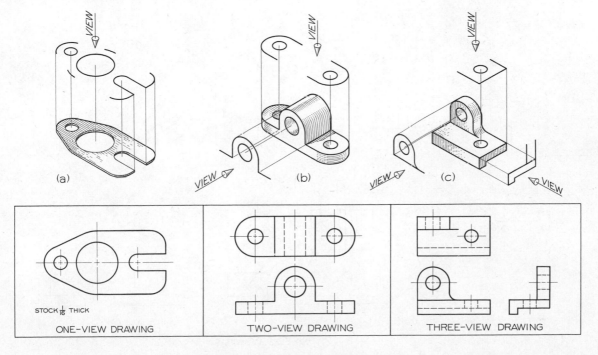

Fig. 6–10. Necessary Views.

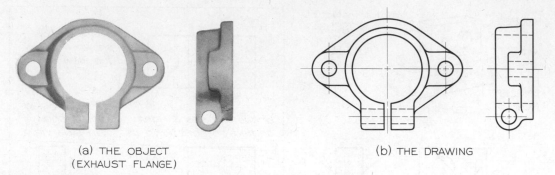

(a) THE OBJECT
(EXHAUST FLANGE)

(b) THE DRAWING

Fig. 6–11. Two Views of Collar.

different directions, as shown by the arrows, all essential shapes are seen, and a two-view drawing is required.

If the object shown at (c) is viewed in three different directions, as shown by the arrows, all essential shapes are seen, and a three-view drawing is required.

6.8 Two-View Drawings. In Fig. 6–11 (a) a machine part is shown in the front-view and right-side-view positions. These views are

enough to show all essential contours and shapes. The corresponding two-view drawing is shown at (b). Note the use of center lines. Also observe that no shading is used on the drawing and that hidden lines show interior shapes that are not seen by looking at the object itself at (a).

Some typical examples of objects which require only two views are shown in Fig. 6–12. At (a) the top view is omitted because it is a duplication of the front view. At (b) and (c)

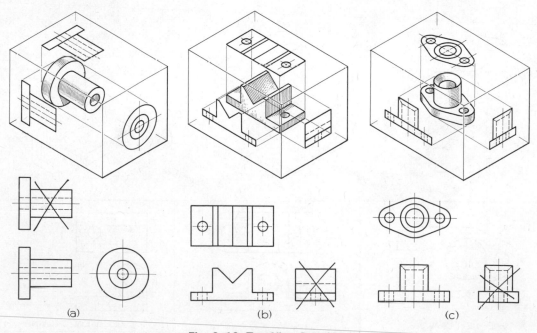

(a)   (b)   (c)

Fig. 6–12. Two-View Drawings.

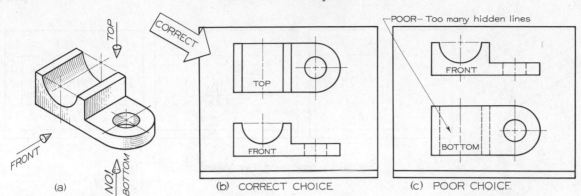

Fig. 6–13. Views with Fewest Hidden Lines.

the side views are not needed because they show no shapes not already given in the front and top views.

In choosing between two views that give the same information, such as between the top and bottom views of the object in Fig. 6–13 (a), choose the view that contains the least number of hidden lines, as shown at (b) and not as shown at (c). Note that the views at (c) could be called the front and top views, but still the lower view would not be desirable.

Sometimes, when there is little or nothing to choose between the top and side views, the views should be selected that will space best on the sheet. For example, in Fig. 6–14

(a), either a top or a side view could be used. If the side view is used, the result is a well-spaced drawing, as shown at (b), while if the top view is used, (c), the drawing will be poorly spaced.

6.9 Sketching Two Views. The best way to learn thoroughly the principles of shape description is to make freehand sketches of many objects, from simple shapes gradually up to more complicated shapes. Freehand sketching technique has been discussed in Chapter 2, to which you are referred for review.

To sketch two views of a simple object, such as the Clamp Block in Fig. 6–15 (a), proceed

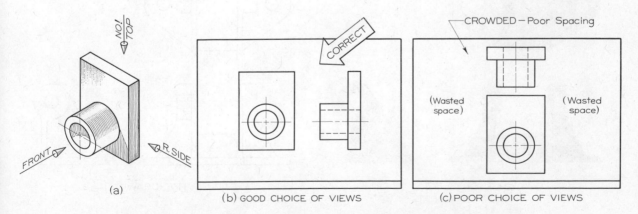

Fig. 6–14. Effect of Spacing on Choice of Views.

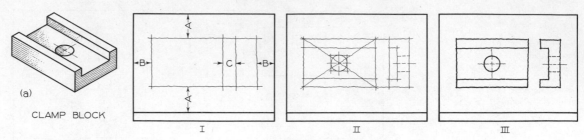

Fig. 6–15. Sketching Two Views.

as follows (use either plain paper or cross-section paper):

I. Sketch the enclosing rectangles for the two views, making spaces A about equal. Make spaces B about equal, and make space C about equal or slightly less than spaces B.

II. Sketch diagonals in front view to locate center. Construct circle and other details.

III. Dim all construction lines with the Artgum; then heavy-in all final lines, making them clean-cut and dark.

6.10 Three-View Drawings. In Fig. 6–16 (a) a Bracket is shown in the front-view, top-view, and right-side-view positions required to show all the essential shapes of the object. The cor-

responding three-view drawing is shown at (b). Note the use of center lines and hidden lines, and the absence of any shading.

As shown in Figs. 6–7 and 6–8, the relative positions of the views depend upon their locations when the glass box is unfolded, but errors in this are so common among beginners that this principle needs to be re-emphasized here.

One of the worst mistakes you can make in technical drawing is to draw a view out of place! For example, the object shown in Fig. 6–17 (a) requires front, top, and right-side views. As shown at (b), the right-side and top views must "line up" with the front view. Never draw the views out of alignment, as

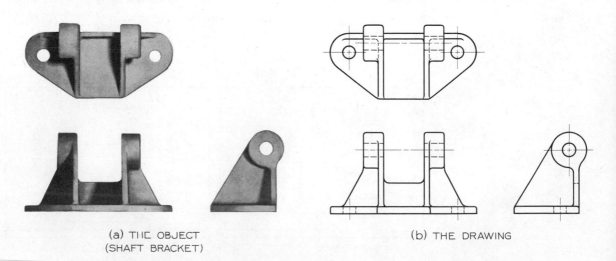

(a) THE OBJECT
(SHAFT BRACKET)

(b) THE DRAWING

Fig. 6–16. Three Views of Bracket.

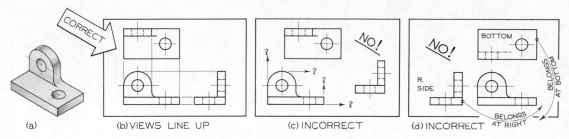

Fig. 6–17. Position of Views.

shown at (c). Also, never draw the views in reversed positions, as shown at (d), even though they do line up with the front view.

6.11 Sketching Three Views. To sketch the three views of the Stop Clamp, Fig. 6–18 (a), proceed as follows:

I. Sketch the enclosing rectangles for the three views, making spaces A about equal, and space B about equal or slightly less than either of spaces A. Make spaces C about equal, and space D about equal or slightly less than either of spaces C. A scrap of paper may be used, as shown, to transfer the depth from the top to the side view, and for transferring other equal distances. It is very important to draw these rectangles in correct proportion, Sec. 2.7, as the entire sketch depends upon your having the large main shapes of the object in correct relative proportion.

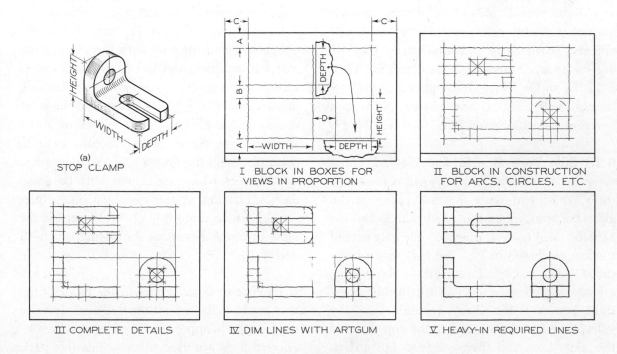

Fig. 6–18. Sketching Three Views.

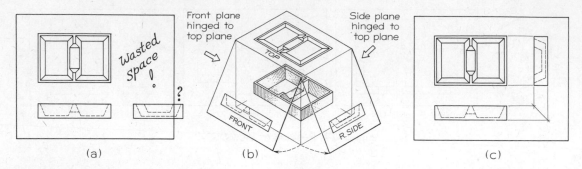

(a)   (b)   (c)

Fig. 6–19. Side View Beside Top View.

II. and III. Block in main shapes lightly; then add smaller details.

IV. Dim all lines with Artgum.

V. Heavy-in all final lines, clean-cut and dark.

6.12 Uprightness of Views. Objects that we are accustomed to seeing in a certain position, such as a house, Fig. 6–2, a telephone, a chair, or a light fixture, should be drawn in their normal upright positions. However, many objects, particularly machine parts, may be drawn in any convenient position. See Figs. 6–11 (b) and 6–16 (b). It is customary to draw screws, bolts, shafts, and similar parts in a horizontal position. See Figs. 15–6 and 15–11.

6.13 Side View Beside Top View. If three views of a wide flat object, such as a canasta tray, are drawn with the side plane of the glass box hinged to the front plane, like the glass box in Fig. 6–7, the side view may extend too far, as shown in Fig. 6–19 (a). Even if the three views can be drawn within the border, a large wasted space will be left in the upper-right corner of the sheet. In such cases, the side plane may be hinged to the top plane of the glass box, and then revolved upward, as shown at (b). This places the side view beside

the top view, as shown at (c), and the spacing is thereby improved.

6.14 Center Lines. A center line (symbol: ₵) is used to indicate an axis of a symmetrical part, Figs. 6–11 and 6–16. As such, the center line becomes a skeleton about which a view or a symmetrical feature is drawn. Center lines are necessary in dimensioning, for many important dimensions are given to them.

Center lines, Fig. 3–11, should be *thin* enough to contrast well with the visible lines and hidden lines, and the dashes and spaces should be carefully spaced by eye. The long dashes should be $\frac{3}{4}''$ to $1\frac{1}{2}''$ long, and the short dashes about $\frac{1}{8}''$ long, with spaces of about $\frac{1}{16}''$ between. Short dashes should occur in open places on the drawing. In circular views, crossed center lines are drawn, with the small dashes crossing at the center. Center lines should extend uniformly about $\frac{1}{4}''$ outside the view or circular part, as shown in Figs. 6–11 and 6–16.

6.15 Hidden Lines. Poorly made hidden lines can easily make an otherwise good drawing very sloppy in appearance and hard to "read." Hidden lines are used to show invisible parts that otherwise could not be shown at all, and

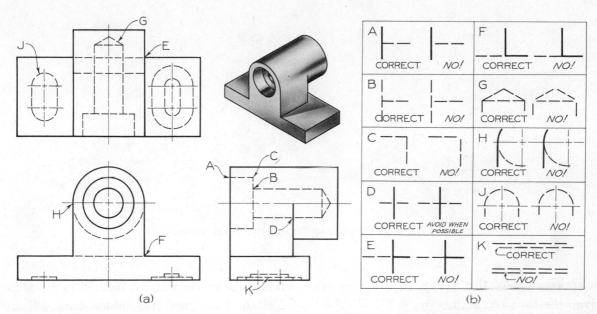

Fig. 6–20. Hidden Lines.

are just as important as visible lines. Make the dashes medium in thickness, and dark, about $\frac{1}{8}''$ long, with spaces about $\frac{1}{32}''$—all carefully estimated by eye. See Fig. 3–11.

The correct methods of drawing hidden-line dashes are illustrated in Fig. 6–20 (a). Special attention should be given to the places marked A, B, C, etc. The correct and incorrect methods for each of these, full size, are shown at (b). Study each example carefully. In drawings of complicated objects, hidden lines that are not necessary for clearness should be omitted. However, the beginner should draw all hidden lines until experience teaches him which lines are not necessary and may be omitted.

6.16 Lines That Coincide. When a visible line coincides with either a hidden line or a center line, the visible line is shown, Fig. 6–21 (a) and (b). When a hidden line coincides with a center line, (c), the hidden line is shown.

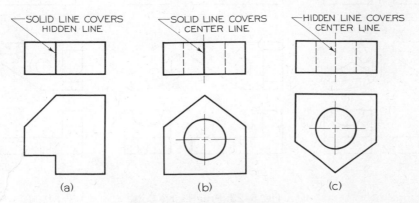

Fig. 6–21. Lines That Coincide.

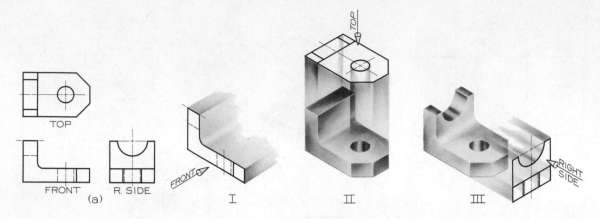

Fig. 6–22. Visualizing Object from Given Views.

6.17 Visualizing the Views. Suppose we are given the three-view drawing in Fig. 6–22 (a) to "read" or visualize. The front view tells us that the object is L-shaped as shown at I. Certain hidden lines and solid lines in the front view are still not clear, but these will be explained by the other views. The top view tells us that two corners are chamfered and that a hole is drilled through the base, as shown at II. Certain solid lines and hidden lines still are not clear. The right-side view tells us that the left end of the object has a semicircular cut, to complete the shape of the object.

6.18 Progressive Cuts. An excellent way to learn to read views and to understand the re-

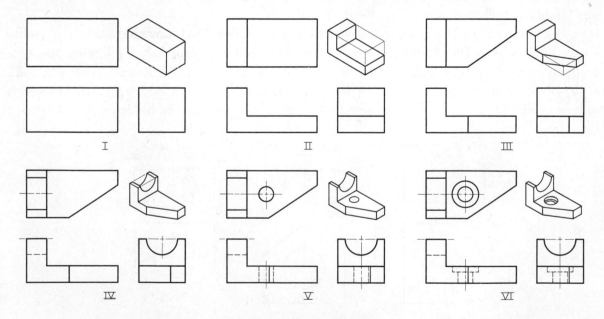

Fig. 6–23. Progressive Cuts.

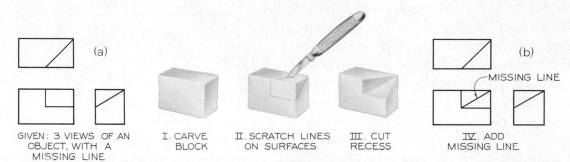

Fig. 6–24. Model Carving.

lationship between a picture of an object and the views of the object is to make a series of drawings showing the progressive cuts that a machinist might make on a piece of raw stock to complete some object, Fig. 6–23.

The first drawing, I, would show the three views of the piece of raw stock. In each succeeding drawing, the various cuts are shown on the views, as shown in II to VI. Instead of making six different drawings, all of these steps could be taken in a single drawing to produce the result shown at VI.

6.19 Model Carving. One of the best aids in reading drawings is to make an actual model of modeling clay, laundry soap, or wood. For example, suppose that three views are given in which one line is missing, Fig. 6–24 (a). First, form a block of the same overall dimensions (not necessarily to scale, but in proportion), as shown at I. Then, II, scratch lines to

represent the edges shown in the given views, and carve out the recess to agree with the lines, as shown at III. This reveals a diagonal line which does not appear on the given views. This line is then added to the sketch, as shown at IV.

6.20 Problems. In Figs. 6–26 to 6–29 are given a variety of objects to be sketched, either on cross-section paper or upon plain paper. Sketch Layout B (Appendix, page 474) freehand, and sketch two problems per sheet. Pictorial sketches, either in isometric or in oblique, may be drawn along with the sketches of views. These may be sketched on separate paper or in the upper right corner of each space on the sheet. Before making any pictorial sketches, study Secs. 16.3 to 16.5, 16.18, and 16.19.

Additional pictorial sketching problems are available at the end of Chapter 16.

Use $8\frac{1}{2}'' \times 11''$ cross-section paper or plain drawing paper. Sketch border and title strip of Layout B (Appendix, page 474). Sketch the necessary views of two assigned problems from this and on the following page. Each grid space on the isometrics is $\frac{1}{4}''$. Sketch the views full size. An example sketch is shown in Fig. 6–25.

If plain paper is used, count the isometric grid spaces to determine the proportions, but make the sketches by eye and not with the aid of the scale.

Make all required lines clean-cut and dark, using a soft pencil. Make center lines thin and hidden lines medium in thickness, but dark in both cases. For a discussion of freehand sketching techniques, see Chapter 2.

Additional sketching problems may be assigned from Figs. 7–27 to 7–29.

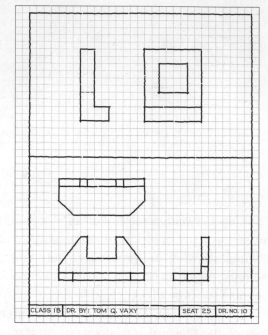

Fig. 6–25. Sketch on Cross-Section Paper.

Fig. 6–26. Sketching Problems.

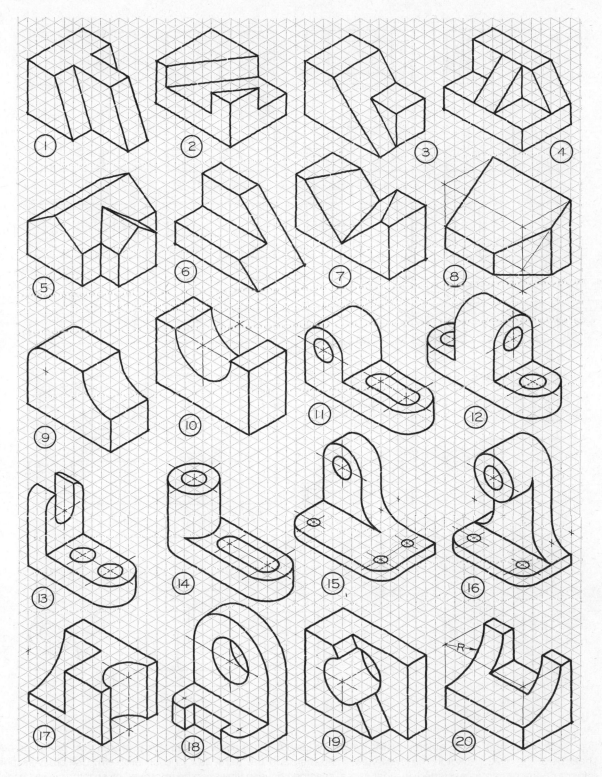

Fig. 6–27. Sketching Problems. See instructions referring to Fig. 6–25.

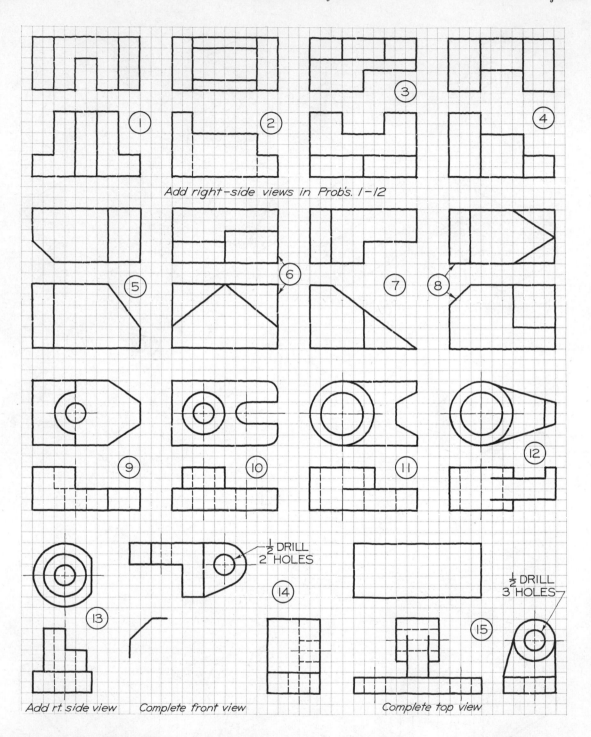

Add right-side views in Prob's. 1–12

Add rt. side view   Complete front view   Complete top view

Fig. 6–28. Missing-View Problems. Using freehand Layout B (Appendix, page 474), sketch two problems per sheet as in Fig. 6–25. Sketch the given views, and then add the third view in each problem. Use cross-section paper or plain paper, as assigned. In most cases, spacing between views can be improved by spacing views farther apart than shown here. Each grid space $= \frac{1}{4}''$.

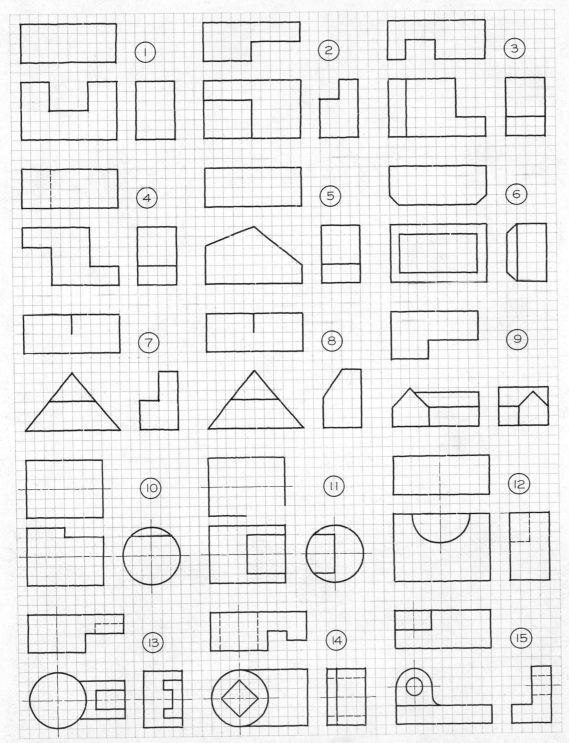

Fig. 6–29. Missing-Line Problems. Using Layout B (Appendix, page 474), divided as in Fig. 6–25, sketch the views of two assigned problems, adding all missing lines. Each grid space = $\frac{1}{4}''$. In most cases, spacing between views can be improved by spacing views farther apart.

CHAPTER 7
TECHNIQUES AND APPLICATIONS

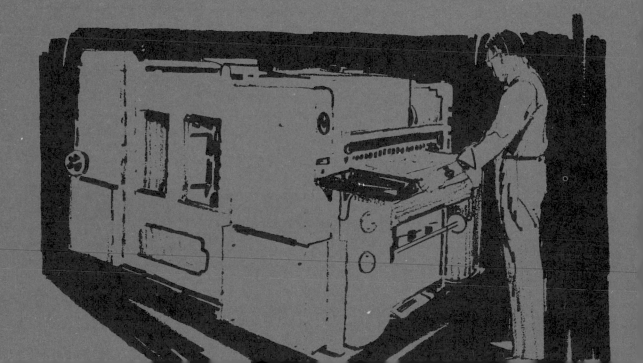

7.1 Pencil Drawing. Before blueprinting was introduced in this country at the Philadelphia Centennial Exposition in 1876, drawings were generally made upon imported handmade white paper, usually in ink and often in colors. Only the original drawing was available. This was often made by the owner of the shop, who handed it to a mechanic as his guide to production. Much was left to the imagination of the shop man, but in any case, the person who made the drawing was usually at hand to supervise the construction.

As industry developed, more people, often in different localities, needed the instructions on the drawing. It was at about this time that the blueprint process was introduced, by which duplicates could be made at low cost. It became the practice to make pencil drawings on detail paper and then to trace the drawings in ink on tracing cloth. As reproduction methods and transparent tracing papers were improved, much time was saved by making the drawings directly in pencil with black lines on the tracing paper and making blueprints therefrom. This did away with the preliminary pencil drawings.

Today most drawings are made directly in pencil, and ink drawings are usually made only when the drawings must be handled a great deal or when an exceptionally finished appearance is necessary. Skill in drafting is therefore concerned largely with the ability to make good pencil drawings.

7.2 Pencil Technique. A draftsman is said to have "good technique" if he knows how to use his tools correctly—especially the pencil. Lines and lettering are clean-cut and dense black, and the drawing is said to have "snap" or "punch." Good pencil technique consists of the following, Fig. 7–1:

1. *Light construction lines.* These should be so light that they can hardly be seen at arm's length. If so made, they will not show on the blueprint.

2. *Dense black lines.* A fairly soft pencil should be used, and enough pressure should be applied to produce black lines. See Secs. 3.7 to 3.10.

3. *Correct line thicknesses and dash lengths.* The different types of lines should contrast well. Visible lines should be thick enough to make the views stand out clearly. See the Alphabet of Lines, Fig. 3–11.

4. *Accented ends of dashes.* Bear down on the pencil at the beginning and end of each dash (hidden lines and center lines).

5. *Sharp, clean corners.* Be careful not to overrun corners or leave gaps. Make each corner square and sharp.

6. *Smooth tangencies.* Locate centers carefully, Figs. 5–17 to 5–19, and make sure your compass lead is sharpened properly, Fig. 3–35.

In Fig. 7–1 (a) a pencil drawing with good technique is shown, and below it at (b) is shown the blueprint from it. At (c) is shown a drawing with poor technique, and below it

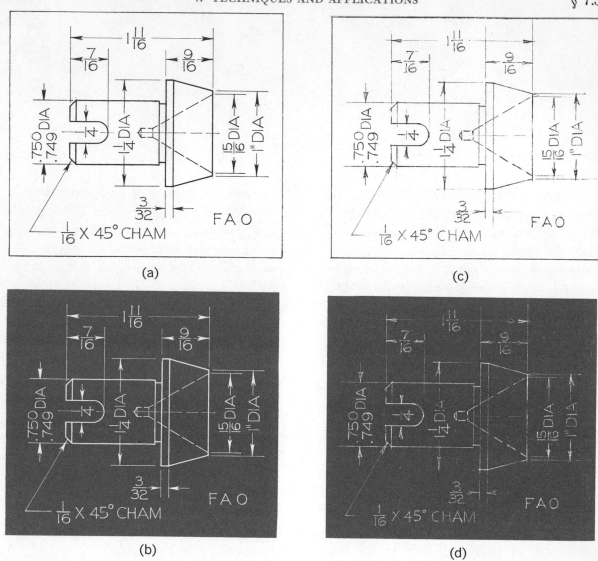

(a)

(b)

(c)

(d)

Fig. 7–1. Pencil Technique.

at (d) is shown the corresponding blueprint. In general, good technique consists in line work and lettering that will produce the clearest, most legible reproductions.

When a drawing is made directly on tracing paper or vellum, both of which are quite thin and transparent, a sheet of drawing paper should be placed underneath as a "backing sheet" to provide a smooth, firm surface. A white backing sheet is recommended so you can clearly see the density of the lines and lettering.

An excellent way to test the density or blackness of your lines and lettering is to hold the tracing up to the light, Fig. 7–2.

Steps in pencil drawing are illustrated in Figs. 7–9 and 7–12.

7.3 Reproduction Processes. An essential part of a designer's or draftsman's education is a thorough knowledge of reproduction techniques and processes, Fig. 7–3. Specifically, he should be familiar with the various processes available for the reproduction of drawings: *blueprint, diazo, microfilm,* etc. Equally important is a knowledge of industrial printing and duplicating methods: *letterpress, lithography, xerography,* and so on.

Each of these processes has very definite advantages and disadvantages. In some, such as blueprinting, the original must be transparent, while in others, such as xerography, opaque or translucent originals may be used. In addition to copies, some reproducing equipment can make enlargements or reductions of the original, which often are extremely

Fig. 7–2. Holding Tracing up to the Light.

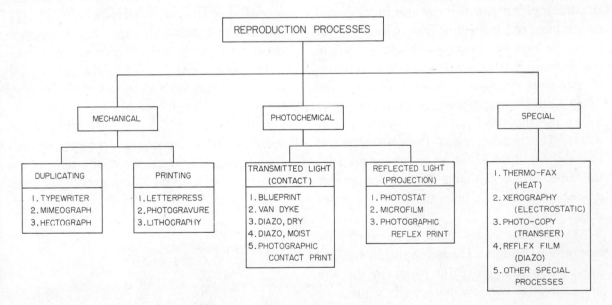

Fig. 7–3. Reproduction Processes.*

useful. A general familiarity with all of these reproduction processes is therefore absolutely necessary, since the reproduction method selected for a specific project will be dependent on the type of original used, number of copies required, size and appearance of copy desired, and the cost.

Since it is not possible to discuss here all of the reproduction processes listed in Fig. 7–3, we shall describe only the three most common processes in use today for reproduc-

* Adapted from table appearing in *Journal of Engineering Graphics,* Vol. 19, No. 2, published by Division of Engineering Graphics, American Society for Engineering Education.

ing drawings: blueprinting, diazo-dry, and microfilming. For an explanation of the other processes, we suggest that you consult your school or public library.

7.4 Blueprinting. The oldest of the several processes used today for reproducing drawings, *blueprinting*, is still extensively used, largely because of its low cost.

Blueprint paper is ordinary paper coated on one side with a chemical preparation sensitive to light. The coated surface is a pale green color. The tracing is placed flat against the coated surface of the blueprint paper, and light is allowed to pass through the tracing onto the sensitized paper. The light passes through the transparent background areas of the tracing, but not through the black pencil or ink lines of the tracing. Thus, the blueprint paper is *exposed* everywhere except where the lines are. After a few seconds or minutes of exposure to light (depending upon the "speed" of the paper and the intensity of the light), the blueprint paper is washed in clear water. This washes away the chemicals not affected by light—that is, those covered by the lines of the drawing. The background is a deep blue, and the lines are the white paper itself.

Anyone can easily make a blueprint by exposure in sunlight. He can do this by holding the tracing and blueprint paper in the sun flat against a window pane with a piece of cardboard, and timing the exposure for several minutes according to the brightness of the sunlight.

A simple *sun frame,* seen in Fig. 7-4, is merely a flat box, which you can easily make. It has a glass pane and a layer of felt and a wooden back that can be made to press tightly against the blueprint paper and tracing. The tracing is placed with its face against the glass,

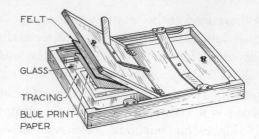

Fig. 7-4. Sun Frame.

and the blueprint paper is against the back of the tracing.

After exposure, the print should be washed in clear water for a minute or two. It will be improved if it is then coated (use a cloth or a brush) with a solution of potassium dichromate. This substance comes in orange-colored crystals. You can make a solution by placing a small quantity, say a tablespoonful, of crystals in a half-gallon of water. Exact measurements of crystals and water are unnecessary. A few seconds after coating the print, wash off the solution with clear water and hang up the wet print to dry in the shade.

If the blueprint is a weak light-blue color, it has not been exposed to light long enough. If the print is very dark and the lines of the drawing appear to be "eaten away," the print

Courtesy Charles Bruning Co.

Fig. 7-5. Automatic Blueprinting Machine.

is overexposed. A correct print has a bright clear blue background with clear white lines.

Modern blueprint machines are available in which cut sheets are fed through for exposure only, and then washed in a separate washer. Where large quantities of blueprints are made, a *continuous blueprint machine* is used, Fig. 7–5. This combines exposure on a continuous roll of blueprint paper with washing and drying in a single operation.

7.5 Diazo-Dry Process. One of the foremost diazo-dry processes in use today is the *Ozalid* process. Ozalid (pronounced Owes'a-lid) paper is coated with chemicals. When the paper is exposed to light through a tracing in the same manner as in blueprinting, and then developed by contact with ammonia vapors, these chemicals produce a white background with lines in black, blue, maroon, etc., depending upon the paper used.

Without special equipment, you can develop a roll of exposed Ozalid paper simply by placing it in a mailing tube, the lower end of which has been closed. Next, tie a string to a small open bottle of ammonia water, which can be purchased at any drugstore. Lower the bottle to the bottom of the mailing tube and close the top of the tube for a few seconds. You can then remove the developed print, dry and ready for use.

A wide range of machine models is available for both exposure and development. A popular machine for moderate quantities of prints is the Ozalid Streamliner, Fig. 7–6. You can make excellent prints quickly by simply feeding cut sheets through and waiting a few seconds for the finished dry print to come out. This process is becoming very popular and has already largely replaced blueprinting. A number of different companies make ammonia-process machines.

Courtesy Ozalid Div., General Aniline & Film Corp.

Fig. 7–6. Ozalid Streamliner.

7.6 Microfilming. The process known as *microfilming* reduces drawings and records to $\frac{1}{15}$–$\frac{1}{60}$ of their original size, and plays an important role in many phases of engineering and business. This process is now extensively used and permits storage of copies of drawings and records on 16 mm., 35 mm., 70 mm., and 105 mm. film, either in rolls or on individual frames. The 16 mm. and 35 mm. frames are usually mounted on aperture cards, Fig. 7–7, while the 70 mm. and 105 mm. frames are generally stored in envelopes. For quick refer-

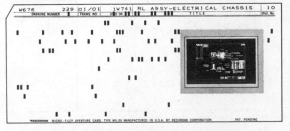

Courtesy Recordak Co., Business Systems Market Division of Eastman Kodak Co.

Fig. 7–7. Microfilm Aperture Card.

The obvious advantage of this process is the great reduction in the space required for, and therefore the cost of, storing engineering and business documents. The microfilm process, however, is much more versatile than mere physical reduction of documents. Information recorded on film produces reasonably permanent records that resist deterioration. Also, microfilm can be used with various specially designed automatic equipment (cameras, reader-printers, etc.) to provide an automatic information filing and retrieval system.

7.7 Mechanical Drawing of Two Views. The first problem in any mechanical drawing is spacing the views on the sheet, and no lines should be drawn until this has been done properly. To draw two views mechanically, Fig. 7–9, proceed as follows:

I. *Figure horizontal spacing.* We have a working space $10\frac{1}{4}''$ wide, and the width of the object is $6''$. Subtract $6''$ from $10\frac{1}{4}''$, leaving $4\frac{1}{4}''$. Divide $4\frac{1}{4}''$ by 2 to get the space on each

Courtesy Minnesota Mining and Manufacturing Company

Fig. 7–8. Microfilm Reader-Printer.

ence, the film can be viewed on special reader-printers, Fig. 7–8, which also furnish instant full-size copies of the drawing or record being projected.

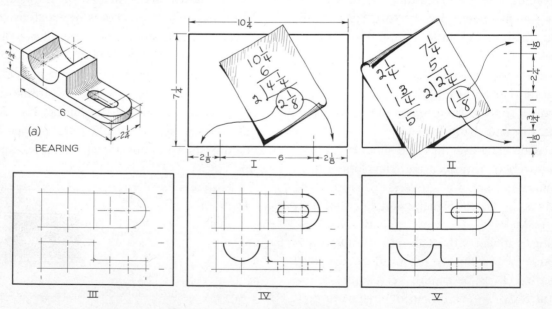

Fig. 7–9. Two-View Mechanical Drawing.

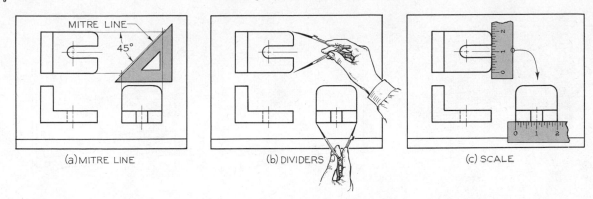

Fig. 7–10. Transferring Depth Dimension.

side. Set this off with the scale, making sure that the 6″ is accurate.

II. *Figure vertical spacing.* We have a working space $7\frac{1}{4}$″ high. The top view requires a vertical space of $2\frac{1}{4}$″, and the front view a height of $1\frac{3}{4}$″. Assume a space between views that will look well, say 1″, and then add $2\frac{1}{4}$″, 1″, and $1\frac{3}{4}$″, to get 5″. Subtract 5″ from $7\frac{1}{4}$″ to get $2\frac{1}{4}$″. Divide $2\frac{1}{4}$″ by 2, to get $1\frac{1}{8}$″, the space at the top and bottom, as shown.

III. *Block in the views* with light construction lines. Let construction lines cross at corners. *Do not draw construction lines between the views.* Use a hard pencil.

IV. *Draw arcs.* These arcs can be put in heavy at once, or if desired you can draw them lightly first. Project the edge of the large arc up to the top view. Project the edges of the small arcs down to the front view and draw the hidden lines heavy at once.

V. *Heavy-in all final lines.* Use a medium-soft pencil and make all lines dense black, but be sure to have three distinct thicknesses of lines: *heavy* for visible lines, *medium* for hidden lines, and *thin* for center lines. Heavy-in arcs and circles first.

NOTE: If dimensions are to be added, Secs. 9.1 to 9.21, allowances in spacing should be made in steps I and II.

7.8 Transferring Depth Dimension. Several methods are used in mechanical drawing to transfer the depth dimension from the top to the side view or the reverse, Fig. 7–10. At (a) is shown the use of the 45° *mitre line*, which is recommended for elementary work to emphasize the idea that the depths of the top and side views are the same.

Since there can be some cumulative error in projecting by means of the mitre line, the professional draftsman is more likely to use the dividers, (b), or the scale, (c), especially for complicated drawings where there are a great many transfers to be made between the top and side views.

7.9 Spacing Three Views. If three views of the Control Block in Fig. 7–11 (a) are to be drawn, the first problem is to determine the spacing of the three required views. If you use Layout C (Appendix, page 474), you will have a *working space* $10\frac{1}{2}$″ wide × $7\frac{1}{8}$″ high, Fig. 7–11 (b).

As shown at I, the front view will occupy a space 2″ high, and as shown at II, the top view will occupy a space 2″ high. Assume a space between the top and front views that will look well, say $1\frac{1}{8}$″, and then add 2″, $1\frac{1}{8}$″, and 2″, to get $5\frac{1}{8}$″. Subtract $5\frac{1}{8}$″ from $7\frac{1}{8}$″

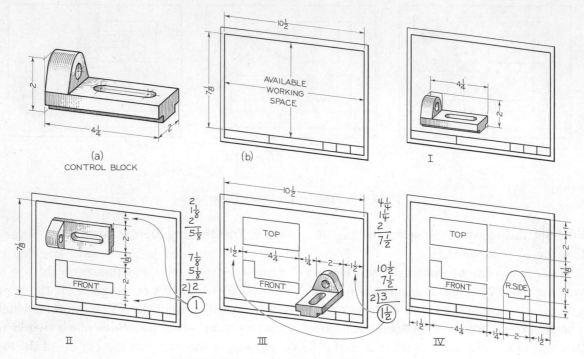

Fig. 7–11. Spacing Three Views on the Sheet.

to get 2″. Divide 2″ by 2 to get 1″, the space at the top and at the bottom.

As shown at III, the front and top views occupy spaces $4\frac{1}{4}$″ wide, and the right-side view requires a space 2″ wide. Assume a space between the front and right-side views that will look well, say $1\frac{1}{4}$″, and then add $4\frac{1}{4}$″, $1\frac{1}{4}$″, and 2″ to get $7\frac{1}{2}$″. Subtract $7\frac{1}{2}$″ from $10\frac{1}{2}$″ to get 3″. Divide 3″ by 2 to get $1\frac{1}{2}$″, the space on each side. Note that the distances between the top and front views and between the front and side views do not have to be equal.

7.10 Mechanical Drawing of Three Views.
To draw mechanically three views of the Control Block, Fig. 7–12 (a):

I. *Figure vertical spacing.* Place scale in vertical position with full-size scale to the left. Mark short dashes perpendicular to scale.

II. *Figure horizontal spacing.* Place scale horizontally with full-size scale up. Mark short dashes perpendicular to scale.

III. *Block in the views lightly,* using a hard pencil. Let construction lines cross at corners. Do not draw construction lines between the views. Construct all three views together, not one at a time. Draw construction for all points of tangency. Draw hidden lines in final weight.

IV. *Heavy-in arcs and circles.* Use medium-soft lead in compass.

V. *Heavy-in straight lines.* Use medium-soft pencil and make all lines dense black, but be sure to have three distinct thicknesses of lines: *heavy* for visible lines (it may be necessary to run the pencil back and forth over the lines to make them heavy enough), *medium* for hidden lines, and *thin* for center lines.

NOTE: If dimensions are to be added, Secs. 9.1 to 9.21, make allowances in spacing in steps I and II.

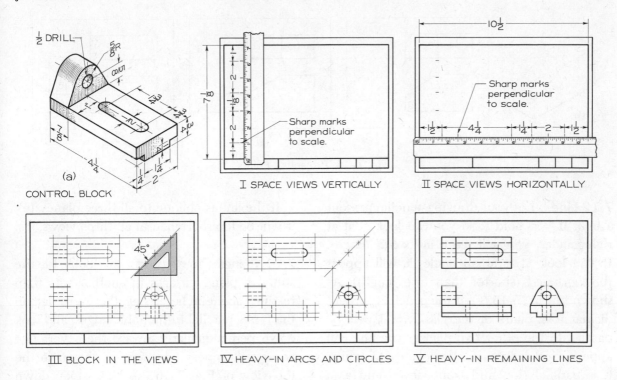

Fig. 7–12. Mechanical Drawing of Three Views.

7.11 Points. Any given point will show in all views. As shown in Fig. 7–13, each point has a front view, a top view, and a side view. If several points are lined up in a row perpendicular to the plane for a view, they will appear as one point, as shown for points A, B, and C in the front view.

If two views of an object are given, and a third view is required, it will be helpful to number the corners and to project the points one by one and connect them to get the required view. Suppose a right-side view of the object in Fig. 7–14 (a) is required, with the front and top views given, as shown at I. Number the four corners of surfaces A and B as shown at (a) and at I. Notice that if a point is visible in a given view, the number is placed *outside* the corner, but if the point is invisible, the number is placed *inside*. The four corners 1, 2, 3, 4 of surface A are projected as shown at II, and the four corners 1, 2, 5, 6 of surface B are projected to complete the view, as shown at III.

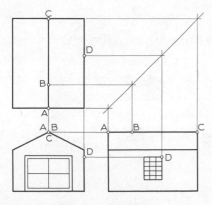

Fig. 7–13. Points.

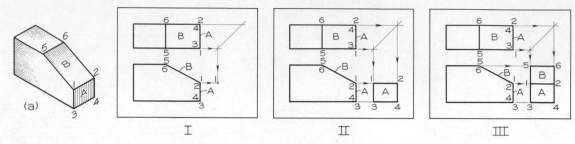

Fig. 7–14. Projecting Points.

7.12 Lines. Let your drawing pencil represent a line. If you hold it so you can look at it at right angles, you will see it in its *true length*. If you look at it at an angle, it will appear *foreshortened* (shorter than true length). As shown in Fig. 7–15 (a), if the pencil is perpendicular to a plane of the glass box, it will be parallel to the other two planes. The line will appear true length on the planes to which it is parallel (top and front views) and as a point on the plane to which it is perpendicular, (b).

If the line is parallel to one plane but inclined to the other two, (c), it will show true length on the plane to which it is parallel (top view) and foreshortened on the planes to which it is inclined, (d).

If the line is oblique to all three planes, (e), it will be foreshortened in all three views, (f).

7.13 Planes. You can use this book for the study of plane surfaces in another way than reading about it here. Lay the book flat on the table, parallel to the table edges, and look down perpendicularly toward the cover. The cover will appear true size, as shown in the top view of Fig. 7–16 (a). Now stoop down and look parallel to the cover to see the front view in which the surface appears as a line. The cover will also appear as a line in the side view, as shown.

Now lift the cover so that it makes an angle, as shown at (b). The edge view (right side) of the cover shows one dimension of the cover

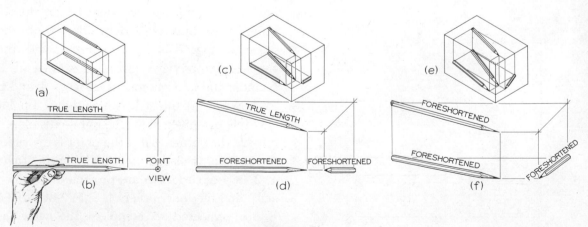

Fig. 7–15. Lines.

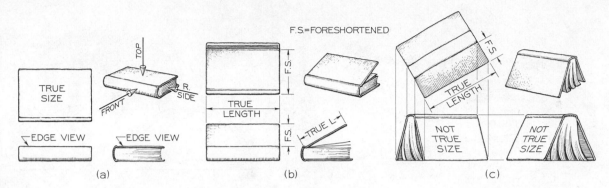

F.S.=FORESHORTENED

(a)　　　　(b)　　　　(c)

Fig. 7-16. Planes.

in true length. The front view shows one dimension true length and the other foreshortened, and the same is true in the top view.

Finally, set the book up on edge, as shown at (c), in an oblique position. In the top view, one dimension of the cover appears true length (because it is parallel to the top plane of projection), and one is foreshortened. Both dimensions are foreshortened in the front and side views.

7.14 Angles. You can use one of your triangles to demonstrate how angles appear when viewed from different directions. For example, Fig. 7–17 (a), if you look perpendicularly toward your 45° triangle, you will see it (front

view) in its true size and shape, all three angles being true size. The other two views show the angles as zero.

If you move the lower right corner away from you, (b), the front view remains true height, but the width is foreshortened. The 90° angle remains true size, but the other two angles appear larger or smaller. This process is carried further in (c) and (d), as shown. Notice what happens to the angles in each case.

7.15 Curved Surfaces. The various standard geometric shapes are illustrated in Fig. 5–1. The shapes used in engineering depend upon the requirements and upon shop processes.

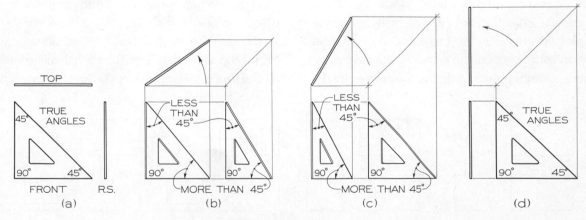

Fig. 7-17. Angles.

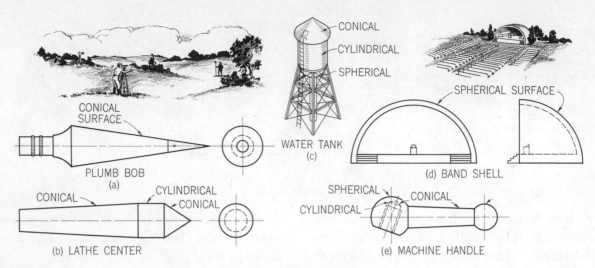

CONICAL SURFACE

PLUMB BOB
(a)

CONICAL   CYLINDRICAL
                 CONICAL

(b) LATHE CENTER

CONICAL
CYLINDRICAL
SPHERICAL

WATER TANK
(c)

SPHERICAL SURFACE

(d) BAND SHELL

SPHERICAL   CONICAL
CYLINDRICAL

(e) MACHINE HANDLE

Fig. 7–18. Curved Surfaces.

Rounded surfaces are quite common, and are easily produced on rotary-cutting machines, such as the *lathe* and the *drill press*, Secs. 10.11 and 10.12. The most common rounded shapes are the cylinder, cone, and sphere; some applications are shown in Fig. 7–18.

7.16 Plotting Points on Curve. When a plane intersects a cylindrical surface, a curved edge is produced. For example, consider three views of a piece of quarter-round molding, as shown in Fig. 7–19 (a). If the molding is cut at an angle by a plane, as shown at (b), a curved edge is formed that will appear as a plotted curve in the top view. As shown at (b), mark points on the curve in the side view, using enough points to define the curve clearly.

The points need not be spaced equally. Then, (c), project each point to the front and top views. Finally, (d), draw a smooth curve through the points in the top view, using the irregular curve, Sec. 3.30.

Another example of plotting a curve is shown in Fig. 7–20. The procedure is similar to that for Fig. 7–19.

7.17 Intersections and Tangencies. When a curved surface is *tangent* to a plane surface, no line should be shown where they join, Fig. 7–21 (a). When a curved surface *intersects* a plane surface, a definite edge is formed, (b). In the case shown at (c), no lines would appear in the top view, while at (d) a vertical surface in the front view produces a line in the top

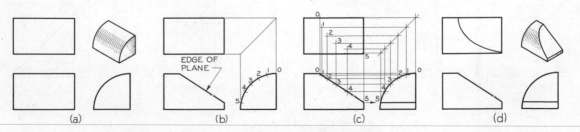

(a)   (b)   (c)   (d)

EDGE OF PLANE

Fig. 7–19. Plotting Points on Curve.

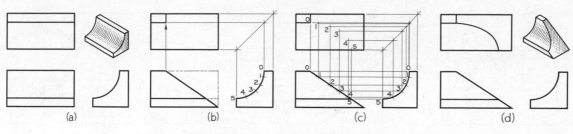

Fig. 7–20. Plotting Points on Curve.

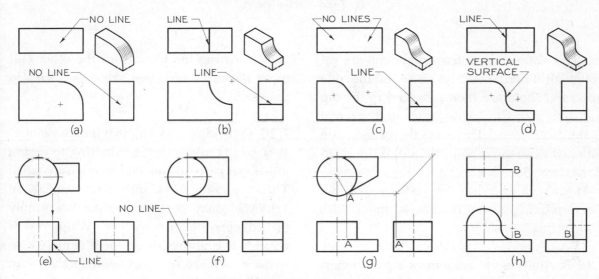

Fig. 7–21. Intersections and Tangencies.

view. Various applications of these principles are shown at (e) to (h). To locate the point of tangency A in (g), refer to Fig. 5–17 (c).

7.18 Intersections of Cylinders. In Fig. 7–22 (a) is shown the intersection of a small cylinder with a large cylinder. The intersection is so small that it is shown merely by a straight line. At (b) the intersection is larger, and is approximated by drawing an arc whose radius r is the same as the radius R of the large cylinder.

At (c) the intersection is still larger, and

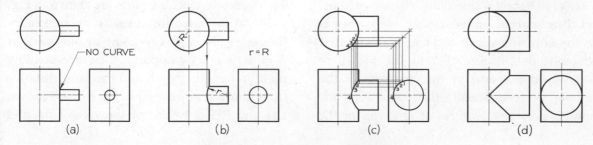

Fig. 7–22. Intersections of Cylinders.

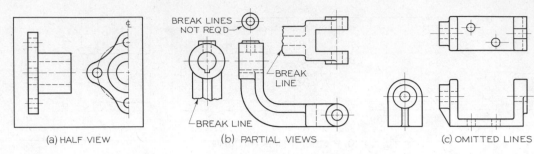

Fig. 7–23. Partial Views.

the true curve is constructed. Points are selected at random on the circle in the side view, and these are then projected to the top and front views to locate points on the curve in the front view. These are then connected with a smooth curve with the aid of the irregular curve, Sec. 3.30.

At (d) both cylinders are the same size, and the resulting true intersection is drawn with straight lines, as shown.

7.19 Partial Views. Sometimes a partial view (incomplete view) will serve the purpose as well as a full view, Fig. 7–23. At (a) the *half view* shown is understood to be symmetrical with the missing half. If a full view were drawn, the views would have to be drawn to a smaller scale to fit on the sheet.

At (b) are shown one complete view and several partial views. Break lines, Fig. 3–11, are used in two views, but are not needed in the circular view of the boss at the top.

At (c) are shown one complete view (front) and three partial or incomplete views. Study the top and side views and note the lines that are omitted. Observe that to include these lines would not add to the clearness of the drawing. Note especially that each side view includes only the features of one end of the

object, leaving the features of the other end to be shown by the other side view. See also Sec. 12.8.

7.20 Left-Hand and Right-Hand Drawings. It is common in industrial drafting to design individual parts to function in opposite pairs. These opposite parts, such as the pedals on a bicycle, may be exactly alike but simply placed opposite one another. When this is possible it is always done, for it is more economical to make two parts exactly alike than to make two different parts. But opposite parts often cannot be exactly alike, as, for example, the gloves in Fig. 7–24 (a), or a pair of shoes. Similarly, the right fender of an automobile cannot be the same shape as the left fender. Do not think for a moment that a left-hand part is simply a right-hand part turned around, for the two parts will be opposite and different.

A left-hand part is referred to as an LH part, and a right-hand part as an RH part. Two opposite machine parts are shown in Fig. 7–24 (a) on opposite sides of an imaginary reference plane. Every point in one part, as A or B, is exactly opposite the corresponding point in the other part, and the same distance from the reference plane. At (b) are shown LH and RH drawings of the same part, and

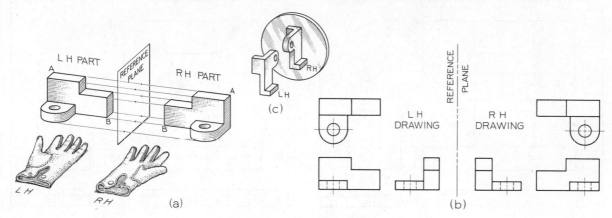

Fig. 7–24. Right-Hand and Left-Hand Parts.

it will be seen that the drawings are also symmetrical with respect to a reference plane line between them.

The draftsman can take any drawing or any object and see what the opposite part would look like by holding it to a mirror, (c). If you look at yourself in a mirror, you will see your hair parted on the opposite side from reality. A drawing can be held faced against a window pane; the reverse image can be traced on the back. This will be the drawing of the opposite part. A tracing can be run through a blueprint machine upside down, and although the lettering will be reversed and readable only in a mirror, the print will be that of the opposite part.

7.21 Problems. The following problems are primarily for mechanical drawing, but any of them would be equally suitable for freehand sketching. It is hoped that some of the earlier problems assigned will be drawn in pencil on tracing paper and that prints will be made to show the student whether his lines are dark enough. If desired, any of these problems,

after being drawn mechanically, may be traced in ink on vellum or tracing cloth. In that case the student should study the following chapter on inking before starting.

Any of the problems may be dimensioned, if desired by the instructor. In that case the student should study Sec. 4.11 or 4.13, on lettering, and Secs. 9.1 to 9.21, on dimensioning.

Figures 7–27, 7–28, and 7–29 consist of problems in which two views are given and a third view is to be added. You are to make sketches or mechanical drawings of problems assigned by your instructor. Use Layout C in either case, as shown in Figs. 7–25 and 7–26. See Layout C (Appendix, page 474). When sketches are required, isometric or oblique sketches may be included, if assigned, Fig. 7–25. For pictorial sketching, see Secs. 16.3, 16.4, 16.5, 16.18, and 16.19. In sketching, estimate all dimensions; use scale only for title strip and border.

If mechanical drawings are assigned, center the drawings in the working space, Secs. 7.9 and 7.10. Make lines clean and dark.

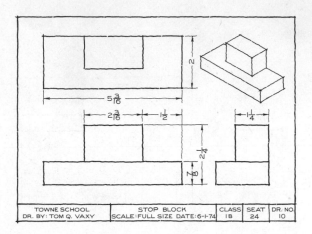

Fig. 7-25. Sketch.

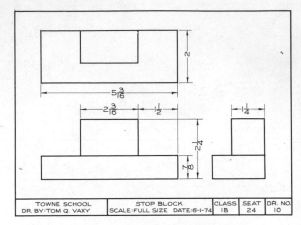

Fig. 7-26. Mechanical Drawing.

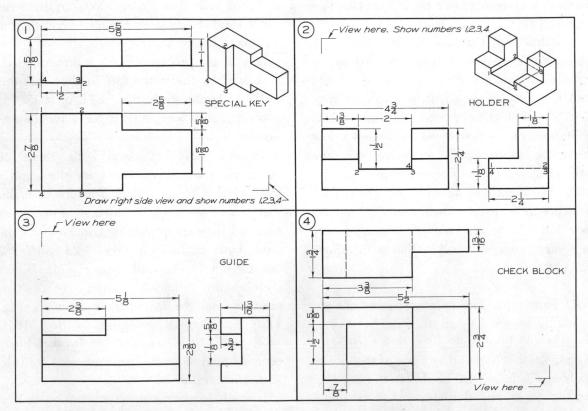

Fig. 7-27. Third-View Problems. Omit pictorial drawings and instructional notes. Move titles to title strip. Use Layout C (Appendix, page 474) for each problem.

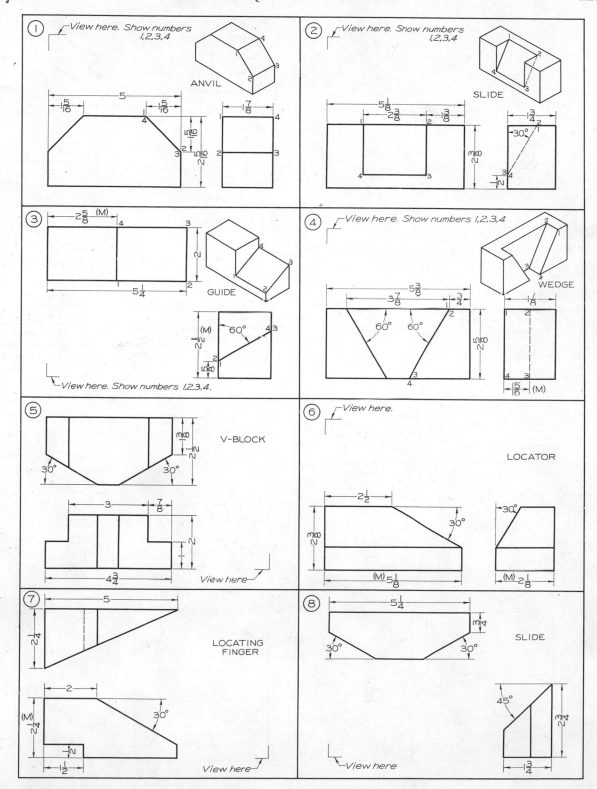

Fig. 7–28. Third-View Problems. See instructions on page 117. Omit pictorial drawings and instructional notes. Move dimensions marked (M) to the new views, and move titles to title strip. Use Layout C (Appendix, page 474) for each problem.

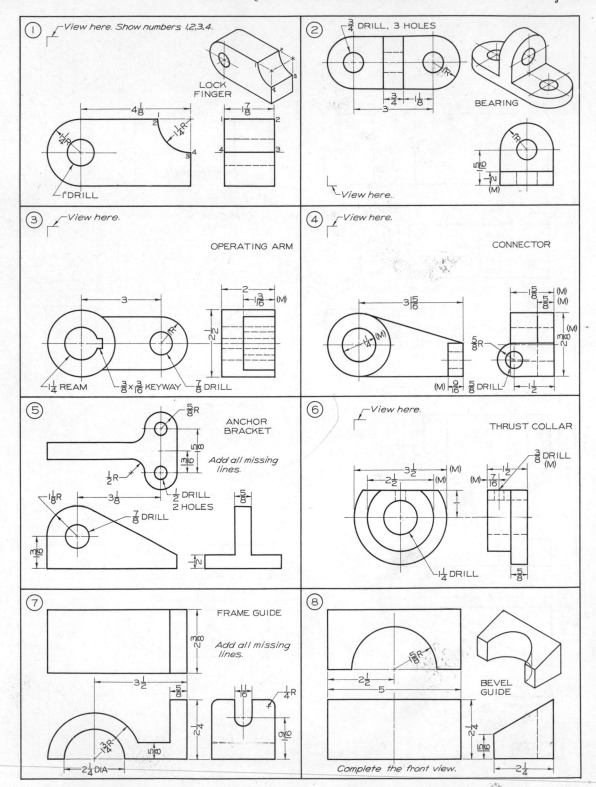

Fig. 7–29. Third-View Problems. See instructions on page 117. Omit pictorial drawings and instructional notes. Move dimensions marked (M) to the new views, and move titles to title strip. Use Layout C (Appendix, page 474) for each problem.

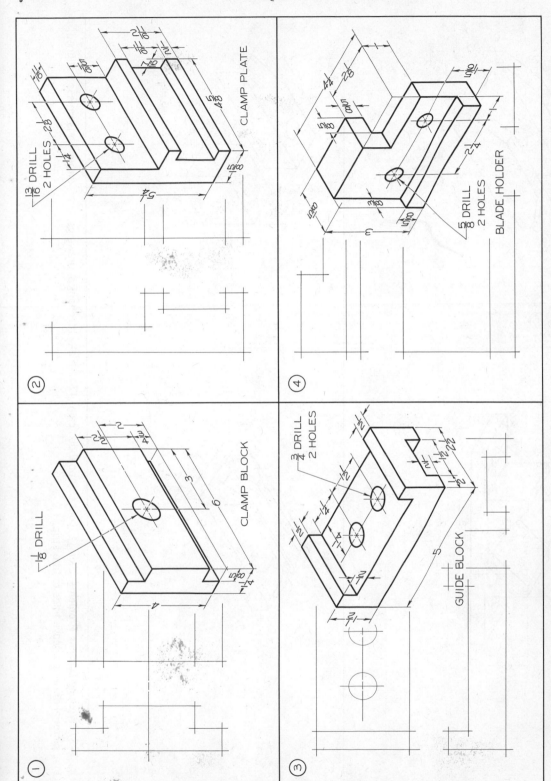

Fig. 7-30. Working Drawing Problems. Using Layout C (Appendix, page 474), make mechanical drawings. Omit dimensions unless assigned. Omit pictorial drawings. Move titles to title strip.

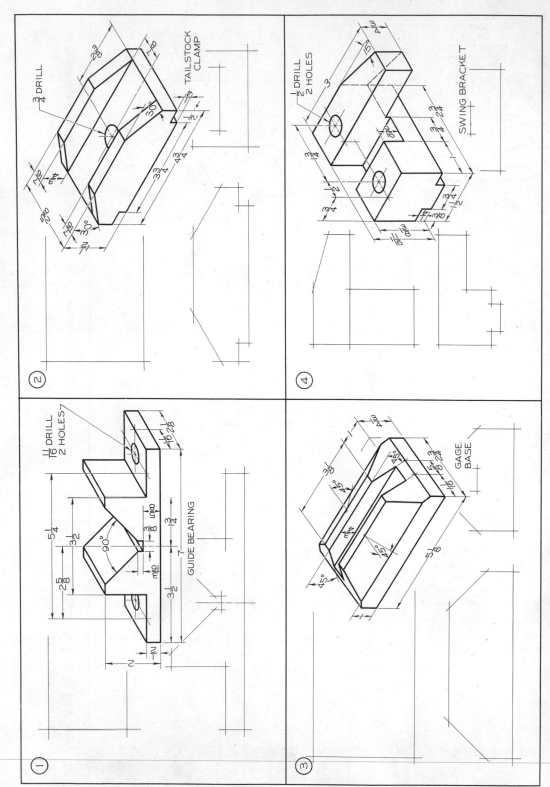

Fig. 7–31. Working Drawing Problems. Using Layout C (Appendix, page 474), make mechanical drawings. Omit dimensions unless assigned. Omit pictorial drawings. Move titles to title strip.

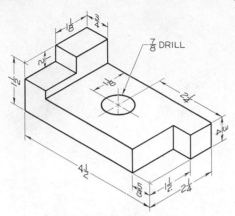

Fig. 7–32. Bearing Plate.*

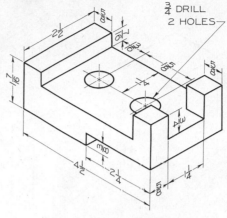

Fig. 7–33. Cross Base.*

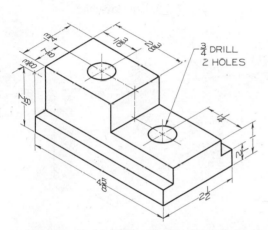

Fig. 7–34. Cutter Holder Shoe.*

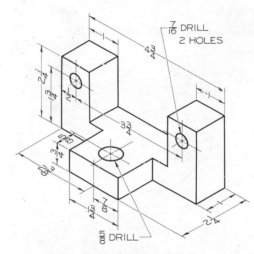

Fig. 7–35. Starting Catch.*

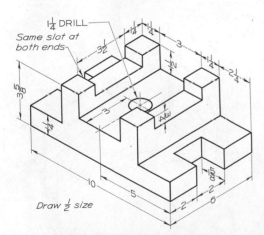

Fig. 7–36. Fixture Base.*

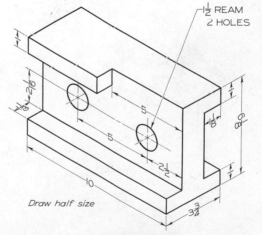

Fig. 7–37. Bed Plate.*

* Using Layout C (Appendix, page 474), draw necessary views with instruments. Omit dimensi
assigned.

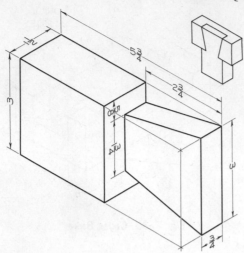

Fig. 7–38. Lap Dovetail.*

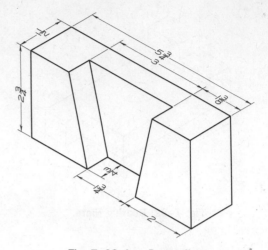

Fig. 7–39. Lap Dovetail.*

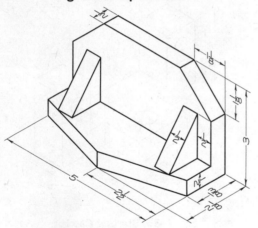

Fig. 7–40. Book End.*

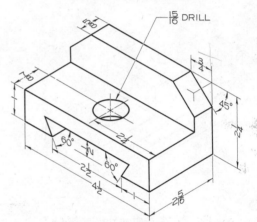

Fig. 7–41. Dovetail Slide.*

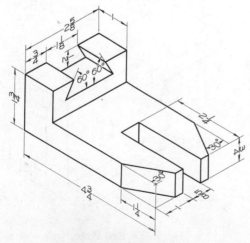

Fig. 7–42. Dovetail Finger.*

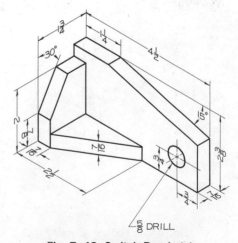

Fig. 7–43. Switch Bracket.*

* Using Layout C (Appendix, page 474), draw necessary views with instruments. Omit dimensions unless assigned.

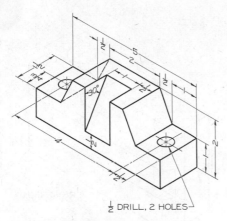

Fig. 7–44. Adjustor Block.*

½ DRILL, 2 HOLES

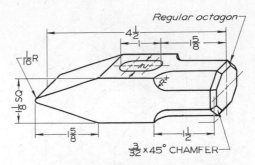

Regular octagon

Fig. 7–45. Riveting Hammer Head.*

³⁄₃₂ × 45° CHAMFER

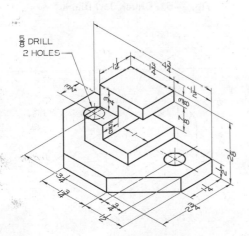

Fig. 7–46. Flipper Dog.*

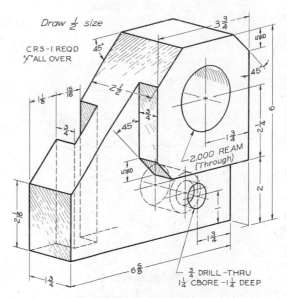

Draw ½ size

CRS–1 REQD
"f" ALL OVER

2.000 REAM
(Through)

¾ DRILL –THRU
1¼ CBORE –1¼ DEEP

Fig. 7–47. Guide Base.*

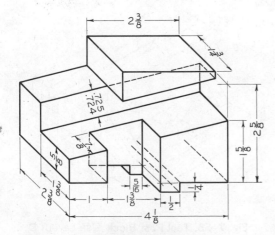

Fig. 7–48. Holder Base.*

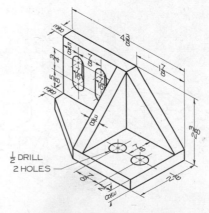

½ DRILL
2 HOLES

Fig. 7–49. Support Bracket.*

*Using Layout C (Appendix, page 474), draw necessary views with instruments. Omit dimensions unless assigned.

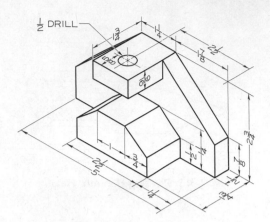

Fig. 7–50. Jig Block.*

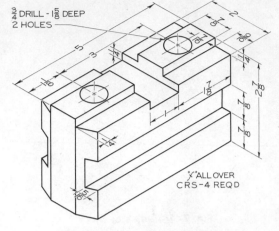

Fig. 7–51. Chuck Jaw Blank.*

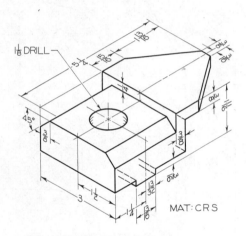

Fig. 7–52. Switch Dog.*

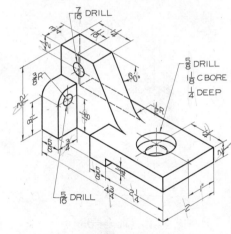

Fig. 7–53. Roller Rest Bracket.*

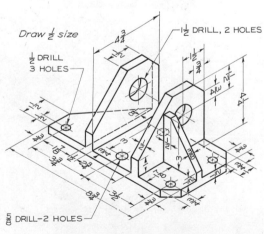

Fig. 7–54. Control Bracket.*

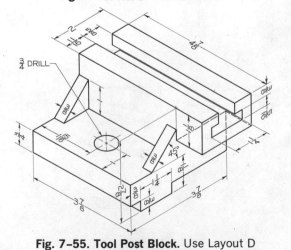

Fig. 7–55. Tool Post Block. Use Layout D (Appendix, page 475).

* Using Layout C (Appendix, page 474), draw necessary views with instruments. Omit dimensions unless assigned.

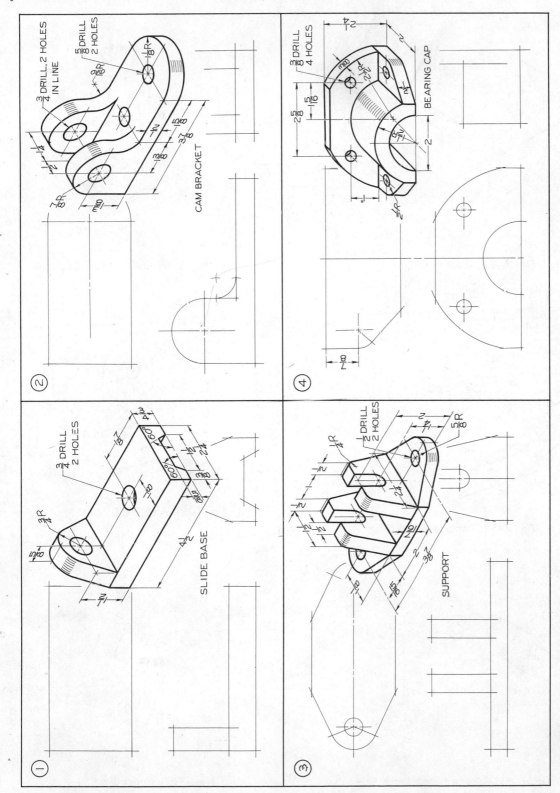

Fig. 7-56. Working Drawing Problems. Using Layout C (Appendix, page 474), make mechanical drawings. Move titles to title strip. Omit pictorial drawings. Omit dimensions unless assigned.

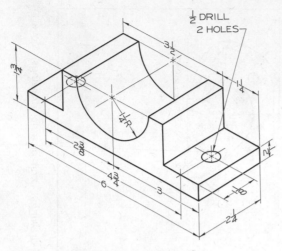

Fig. 7-57. Trunion Block.*

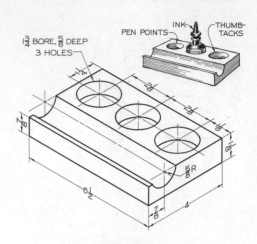

Fig. 7-58. Ink Stand.*

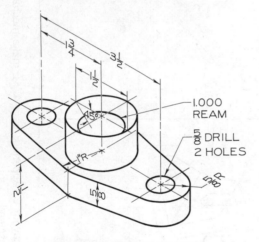

Fig. 7-59. Packing Gland.*

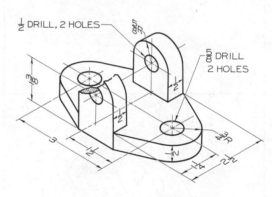

Fig. 7-60. Fastener Bracket.*

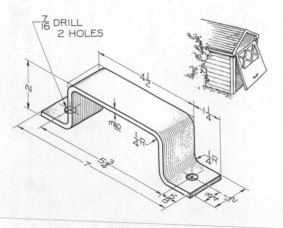

Fig. 7-61. Garage Door Handle.*

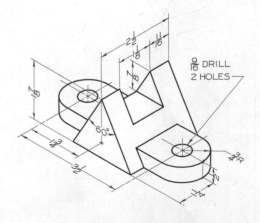

Fig. 7-62. Wedge Base.*

*Using Layout C (Appendix, page 474), draw necessary views with instruments. Omit dimensions unless assigned.

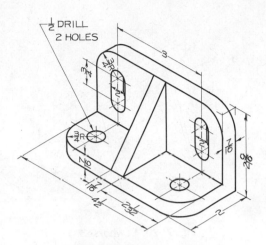

Fig. 7-63. Base Bracket.*

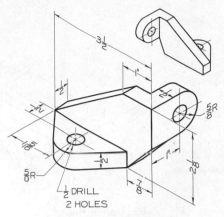

Fig. 7-64. Angle Bracket.*

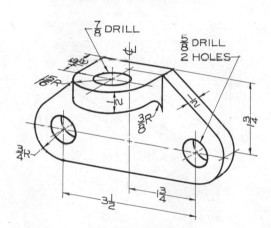

Fig. 7-65. Guide.*

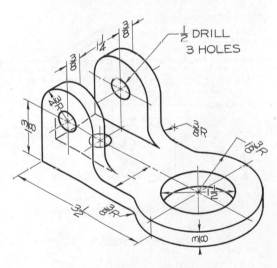

Fig. 7-66. Actuator Base.*

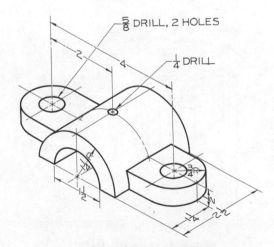

Fig. 7-67. Bearing Cap.*

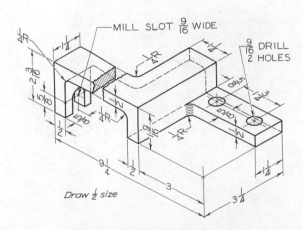

Fig. 7-68. LH Hook (for Buick).*

* Using Layout C (Appendix, page 474), draw necessary views with instruments. Omit dimensions unless assigned.

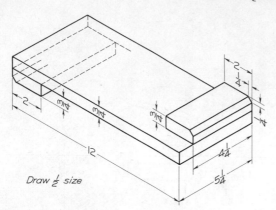

Draw ½ size

Fig. 7–69. Bench Hook.*

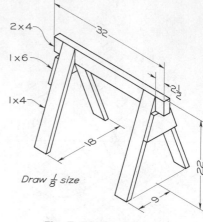

Draw ⅛ size

Fig. 7–70. Sawhorse.*

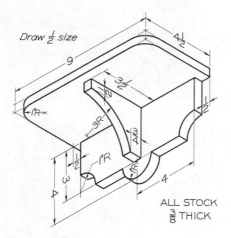

Draw ½ size

ALL STOCK ⅜ THICK

Fig. 7–71. Shelf.*

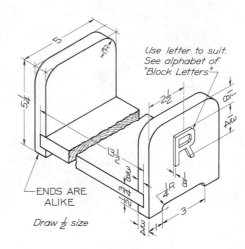

Use letter to suit. See alphabet of "Block Letters"

ENDS ARE ALIKE

Draw ½ size

Fig. 7–72. Book Rack.*

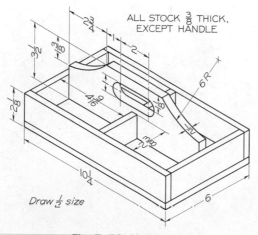

ALL STOCK ⅜ THICK, EXCEPT HANDLE

Draw ½ size

Fig. 7–73. Nail Box.*

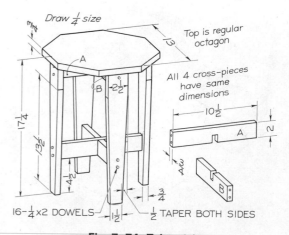

Draw ¼ size

Top is regular octagon

All 4 cross-pieces have same dimensions

16-¼x2 DOWELS   ½ TAPER BOTH SIDES

Fig. 7–74. Taboret.*

*Using Layout C (Appendix, page 474), draw necessary views with instruments. Omit dimensions unless assigned.

CHAPTER 8

INKING

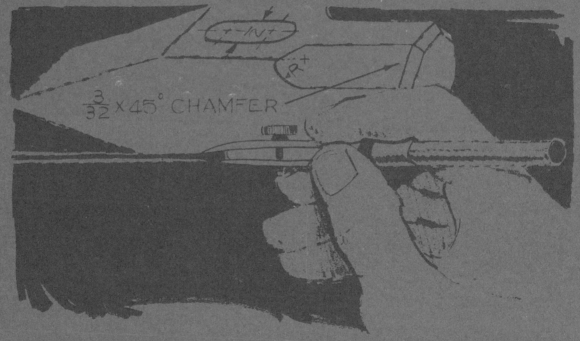

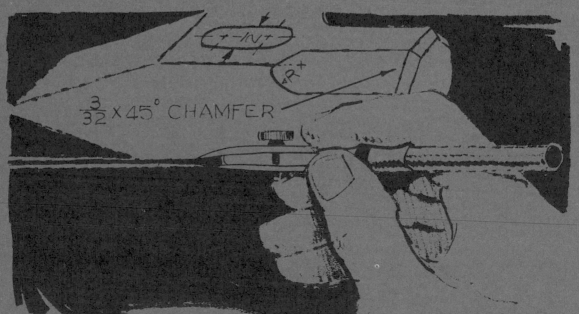

8.1 Drawing Ink. Black waterproof *drawing ink,* Fig. 8–1 (a), is used whenever it is necessary to give a fine appearance to a drawing or to make it more permanent. *Ink tracings* produce better prints than pencil tracings, but they take more time to make. For methods of reproduction, see Secs. 7.3 to 7.6.

The stopper should be *twisted* out—not pulled—to prevent breaking the cork. A small piece of cloth, or *penwiper,* is provided and should be kept for frequent use. Do not leave the ink bottle uncorked; the ink will evaporate and thicken.

Never allow drawing ink to remain and dry in any drawing instrument, as it will then be difficult to remove the dried ink and the acid in the ink will pit the metal and damage the instrument. Dried ink can be removed with a penknife; take care not to scratch, or score, the ruling pen. Never use sandpaper, as this will ruin the pen.

Be careful not to overturn the ink bottle. This extremely black and permanent ink can be removed from clothing or other materials only if they are washed with water while the ink is still wet. The ink bottle should be kept in the corner of a drawer, in a special holder, or fastened securely with a strip of paper as in Fig. 8–1 (a).

8.2 Ruling Pen. The *ruling pen,* Sec. 3.24, is used to rule ink lines. It should never be used freehand, as for lettering. Fill the pen to a height of about $\frac{1}{4}''$ with the *quill* of the ink stopper, Fig. 8–1 (a) and (b). Never dip the ruling pen into the ink. Use the pen at once; the ink dries rapidly and will soon clog the pen, especially when the pen is set for a fine line. The positions of the hands and instruments are the same for drawing ink lines as for pencil lines, Figs. 3–12 (a) and (b) and 3–13. If the ink does not flow readily from the pen,

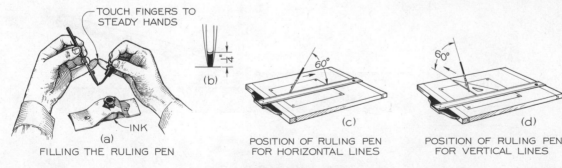

TOUCH FINGERS TO STEADY HANDS

(b)

(a)
FILLING THE RULING PEN

INK

(c)
POSITION OF RULING PEN FOR HORIZONTAL LINES

60°

(d)
POSITION OF RULING PEN FOR VERTICAL LINES

60°

Fig. 8–1. Use of Ruling Pen.

133

draw it across the back of your finger or a piece of drafting tape. The ink may still not flow; this means that it has dried too much, or that the nibs are completely closed. If the pen will not make a fine line, it probably needs sharpening. In this case get your instructor to help you. Amateur sharpening may well ruin your pen.

Horizontal lines are drawn from left to right along the T-square with the pen leaning at about 60° with the paper, Fig. 8–1 (c). The thumbscrew is on the side away from the draftsman. Adjust it with the thumb and forefinger of the hand holding the pen.

Vertical lines are drawn upward along the left-hand edge of the triangle, Fig. 8–1 (d).

The directions for drawing inclined lines are the same as for pencil lines, Fig. 3–18.

The correct method of drawing a horizontal line is shown in Fig. 8–2 (a). If the pen is leaned toward you as at (b), the line will be ragged because the outer nib does not touch the paper. If the pen is "toed in" as at (c), the ink is apt to run under the T-square. In such case, *blot the ink quickly* (use fingers if necessary), and let it dry. Erase with a Weldon Roberts Coral Pink (India) or an Eberhard Faber Ruby pencil eraser—not a gritty ink eraser. See Sec. 3.17 for erasing methods. If a portion of a heavy ink line has been erased, the ink line should be replaced by a series of fine lines drawn side by side until the desired thickness

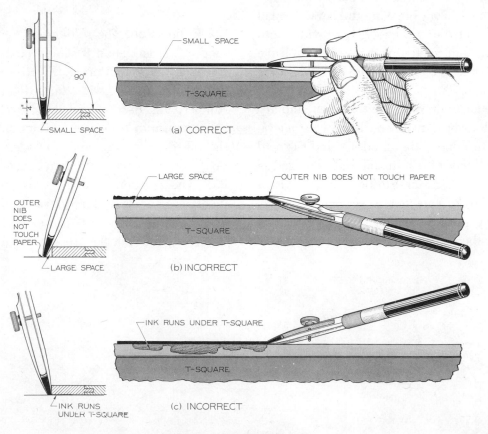

Fig. 8–2. Use of Ruling Pen.

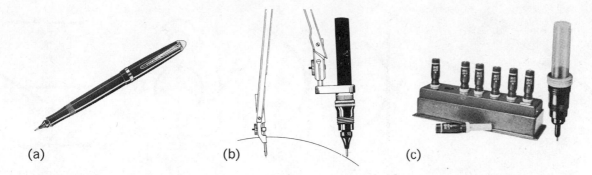

Fig. 8–3. Technical Fountain Pen and Accessories.

is obtained. If a heavy line is inked, it probably will not join smoothly with the line already drawn. Never use a blotter on an inked drawing except when an error is made and erasing is necessary.

A new instrument that has received widespread acceptance, and may be used instead of the ruling pen, is the technical fountain pen. See Sec. 4.18 on special pens. These pens are available as individual pens, Fig. 8–3 (a), or as sets consisting of one holder and several interchangeable point sections, Fig. 4–37. The various point sections can also be used to ink circles and arcs by simply attaching a special holder, Fig. 8–3 (b), to a giant bow compass.

Specially designed point sections, Fig. 8–3 (c), are also available for use with the Leroy, Wrico, and other similar lettering sets.

8.3 Use of Compass in Inking. The general method of using the compass is explained in Sec. 3.26. For inking, insert the *pen attachment* in the socket joint of your compass and tighten the screw, Fig. 8–4 (a). "Break" the leg so that the pen attachment will be approximately perpendicular to the paper, as shown. Adjust the needle point slightly longer than the pen, and fill the pen to about $\frac{1}{4}''$, as shown at (b). Then rotate the compass clockwise between your thumb and forefinger.

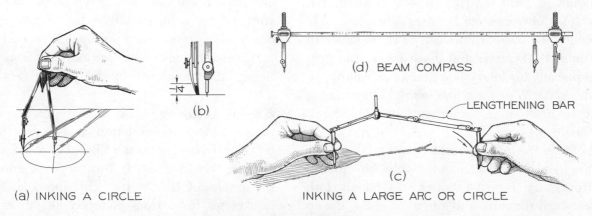

(a) INKING A CIRCLE

(b)

(d) BEAM COMPASS

LENGTHENING BAR

(c)

INKING A LARGE ARC OR CIRCLE

Fig. 8–4. Inking Arcs and Circles.

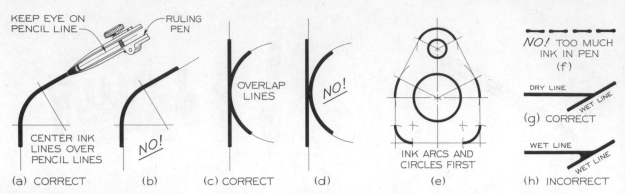

Fig. 8–5. Ink Lines over Pencil Lines.

For large arcs and circles, insert the *lengthening bar*, and move the compass as shown in Fig. 8–4 (c). A *beam compass*, made especially for large radii, is shown at (d).

For small circles and arcs (up to about 1″ radius), use the *bow pen*, Sec. 3.29. Never use the large compass for arcs and circles that can be drawn with the bow pen; the bow pen is more rigid.

8.4 Inking over Pencil Lines. Always center ink lines directly over the pencil lines, Fig. 8–5 (a), and not to one side, (b). If this is done properly in the case of a tangent arc, the lines will be on top of one another at the tangent point, (c), and not side by side as at (d). Always draw arcs and circles first, (e), after which the straight lines can be easily joined to them. If you have too much ink in the pen, especially for heavy lines, the ends will be too thick, (f). Wait for a line to dry before joining a wet line to it, (g); otherwise the wet ink tends to run together, (h).

Since the ink dries rapidly and tends to clog the pen, the instrument should be cleaned frequently. Insert a blotter or a folded cloth between the nibs, and also remove any ink on the outside. *Never lay a pen down with ink in it, or under any circumstances allow ink to dry in the pen.*

In most cases where the pen does not "feed" well, the pen needs cleaning and a supply of fresh ink. If this is not the cause, the nibs may be dull and need sharpening (see your instructor). Or you may have screwed the nibs entirely together; this cuts off the ink supply.

Regardless of the thumbscrew adjustment, an ink line will become thicker if (1) the pen contains too much ink or (2) the pen is leaned too far toward the paper in the direction in which the line is being drawn.

Before you draw an ink line on your drawing, draw a test line on a scrap of the *same kind of paper, using a straightedge.* In drawing heavy lines, do not run a wet line into another wet line, as the ink will merge and may "run." Be sure the other line is dry first.

8.5 Ink Lettering. Considerable practice is required for good ink lettering. The pens to be used depend upon the thickness of stroke desired; this thickness in turn depends upon the height of the letters. The approximate thicknesses for various common heights are

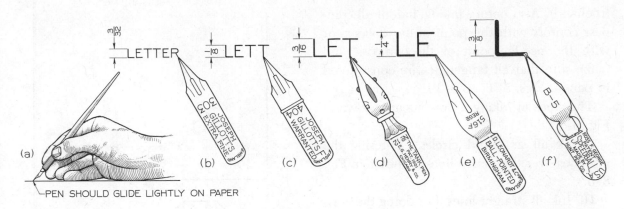

Fig. 8–6. Ink Lettering (Full Size).

shown in Fig. 8–6. The pens illustrated are widely used by draftsmen, but there are others on the market that will be found satisfactory. In general, ball-pointed pens will be suitable for lettering around $\frac{1}{4}''$ high, and above that the special pens, such as Speedball, Drawlet, Leroy, etc., are used, Secs. 4.16 and 4.18.

For the smaller letters, dip the pen into the ink about $\frac{1}{4}''$, then touch the side of the pen against the inside of the bottle to drain off the excess ink. Some draftsmen prefer to apply the ink to the under side of the pen with the quill of the stopper. Hold the pen naturally without cramping, as in Fig. 8–6 (a). Touch the pen *lightly* to the paper, and do not press down enough to spread the nibs of the pen apart. Move the pen with even pressure to produce strokes of uniform widths. Clean the pen frequently with a cloth, especially when you are finished.

New pen points are stiff and unsatisfactory, and must be "broken in" with use. Do not burn off the oil film on a new pen with a match, as this will damage the pen. Dip the pen in ink and wipe clean several times. Pro-

tect your lettering pen from damage by rolling it in a piece of paper to form a protecting tube. Fasten the paper with cellophane tape.

8.6 Inking a Drawing or Tracing. Display drawings and others that are not to be reproduced by blueprinting or similar methods may be inked directly on the pencil lines. The original pencil drawing should be drawn lightly. In such cases, the paper should be white and of a quality to take ink well. After all ink lines are dry, any uncovered pencil lines are erased with the Artgum, Fig. 3–24 (a).

Since most drawings are reproduced by blueprinting or by similar methods, the inking is done on tracing vellum or tracing cloth fastened tightly over the pencil drawing. Ink is applied to the dull side of tracing cloth because it takes ink better. Most tracing cloths are easily damaged by water or perspiration, and care should be taken to avoid this.

Thicknesses of ink lines are shown at the right of Fig. 3–11. In general, the heavier lines are much heavier in ink than in pencil. Before inking a line, draw a test line on a scrap of the same kind of paper, using a straightedge (not freehand). Also before inking, indent all compass centers with the point of the dividers or with the needle point of the compass, and make sure that all tangencies are constructed in pencil, Figs. 5–17 to 5–19.

The steps in inking a drawing are shown in Fig. 8–7.

I. Ink all arcs and circles, centering the ink lines over the pencil lines as shown in Fig. 8–5.

II. Ink all straight lines, first doing the horizontal, then the vertical, and finally the inclined lines. It is much easier to join straight lines to arcs than the reverse.

III. Ink all center lines, dimension lines, and extension lines.

IV. Ink in all arrowheads and lettering. It is best to draw guide lines for lettering directly on the tracing paper or cloth.

Some draftsmen prefer to ink center lines before indenting compass centers. This keeps ink from running through the holes in the tracing paper or cloth.

Be patient and allow ink to dry before inking adjacent lines. You can raise the triangle above wet ink by placing another triangle under it—but back a short distance from the working edge. For the use of irregular curves in inking, see Sec. 3.30.

8.7 Tracing Problems. The problems in the previous chapter will be suitable for inking practice. Do not ink in any drawings directly on detail drawing paper; make tracings instead. It is recommended that vellum be used, at least in the beginning, because tracing cloth is expensive.

The first tracing problems you select should be composed of straight lines and full circles, without tangencies. Later tracing problems should include tangencies.

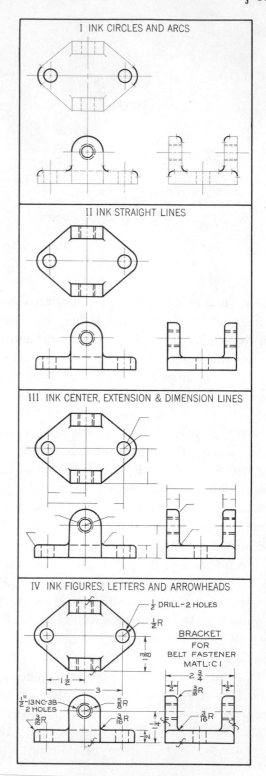

Fig. 8–7. Order of Inking.

CHAPTER 9

DIMENSIONING

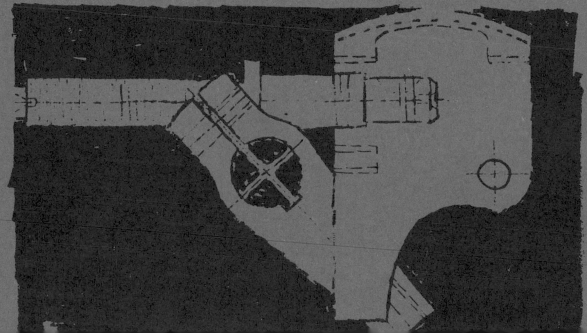

9.1 Complete Description of Objects. If a drawing is to be complete so that the object represented can be made from it exactly as intended by the draftsman or designer, it must tell two complete stories. It tells these by means of (1) *views,* which describe the shape of the object, and (2) *dimensions and notes,* giving sizes and other shop information.

The drawing shows the object in its completed state, and, whether the views are drawn full size or to scale, Sec. 3.19, the dimensions must be the actual dimensions of the completed object. The job of the shop is to produce the object exactly as shown on the drawing. If the drawing is wrong, the object will be made wrong. Such errors may be highly expensive (suppose 10,397 pieces are made before the error is caught) and may cost the draftsman his job.

Although the professional draftsman must understand the fundamental shop processes (Chapter 10) so he can issue correct instructions to the workmen through his dimensions and notes, most of the problems of elementary dimensioning are covered by a few simple rules. You will have little difficulty if you carefully follow these rules, apply them with common sense, and always keep in mind that someone must actually make the object from your drawing.

Remember, the dimensions are at least as important as the views of the object, and correctness is absolutely necessary. Do not make the mistake of simply giving the dimensions you use to make the drawing. Give the dimensions you want the workmen to use in making the part. A large number of examples of good dimensioning are shown in Figs. 16–48 to 16–53.

9.2 Learning to Dimension. First, you must learn the correct *conventions,* or standard practices of dimensioning: the types of lines, the spacing of dimensions, correct arrowheads, and so forth, Fig. 9–1. Note the strong *contrast* between the thick visible lines of the drawing and the thin lines used in dimensioning. The views should stand out clearly from the dimensions.

The four types of lines used (see the Alphabet of Lines, Fig. 3–11) are the *extension line, dimension line, center line,* and *leader.* These are all drawn thin but dark, with a medium-hard pencil, such as a 2H.

The extension line, Fig. 9–1, "extends" from the object, with a gap of about $\frac{1}{16}$" next to the object, and continues to about $\frac{1}{8}$" beyond the outermost arrowhead. Never leave a gap where extension lines cross any other line.

A dimension line, Fig. 9–1, has an arrowhead at each end indicating the extent of the dimension. A gap is left (except in architectural and structural drawing) near the middle for the dimension figure. On small drawings, dimension lines are spaced at least $\frac{3}{8}$" from the

141

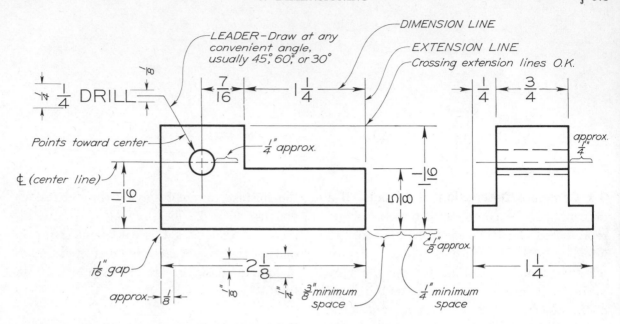

Fig. 9-1. Dimensioning Conventions.

object and at least $\frac{1}{4}''$ apart. The spacing must be uniform throughout the drawing.

Center lines, Fig. 9–1, are used to indicate axes of symmetry and in place of extension lines for locating holes and other features. "Wild" center lines spoil a drawing. Make center lines end about $\frac{1}{4}''$ outside the hole or feature. See also Sec. 6.14.

A leader, Figs. 9–1 and 9–2, is a thin solid line that "leads" from a note or dimension and is terminated by an arrowhead touching the part to which attention is directed. Leaders are straight *inclined* lines (never vertical or horizontal), which are usually drawn at 45°, 60°, or 30° with horizontal, but may be drawn at any convenient angle, Fig. 9–2. A short horizontal "shoulder" should extend out from the mid-height of the lettering. The inclined line

should be drawn so that if extended it will pass through the center of the circle. Leaders may extend from either the beginning or end of a note, Fig. 9–2.

9.3 Arrowheads. *Arrowheads,* Fig. 9–3, are drawn with two sharp strokes toward or away from the point. The length should be about $\frac{1}{8}''$ and the width about one-third the length. For better appearance, they may be filled in, as shown at (b). Avoid sloppy, careless arrowheads such as those marked NO at the bottom of the figure.

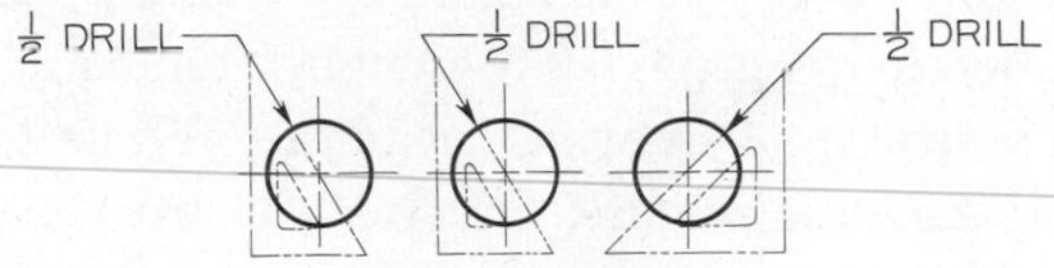

Fig. 9-2. Leaders.

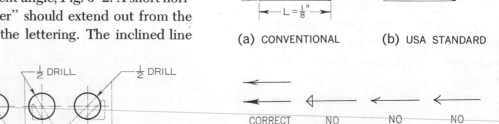

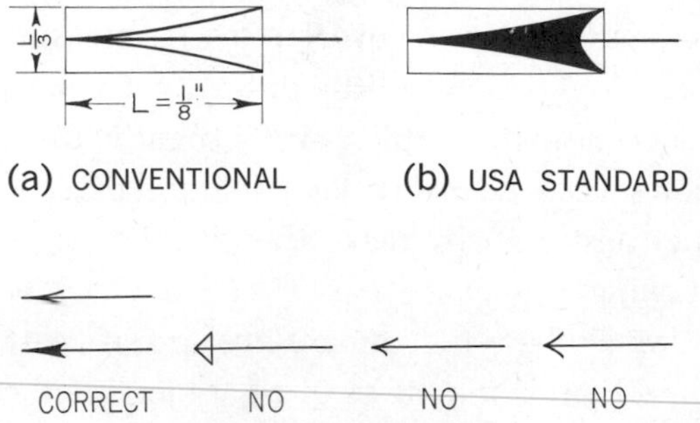

Fig. 9-3. Arrowheads.

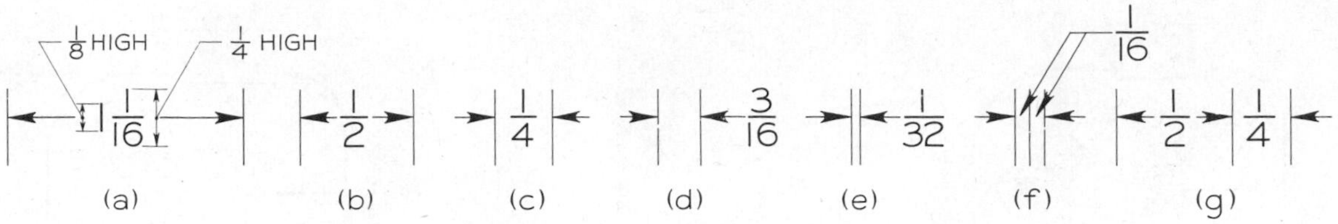

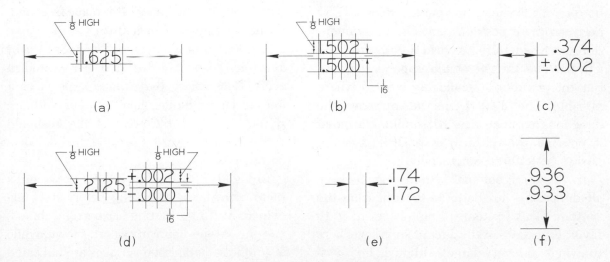

Fig. 9–4. Common Fraction Dimension Figures (Full Size).

9.4 Dimension Figures. The correct lettering of dimension figures is explained in Secs. 4.11 and 4.13, and should be studied thoroughly. Since incorrect or unclear dimension figures can lead to costly mistakes in the shop, great care should be exercised to letter them properly. The standard height for whole numbers is $\frac{1}{8}''$, and for fractions double this, or $\frac{1}{4}''$, as shown in Fig. 9–4 (a). Clear spaces should be left on each side of the fraction bar, Fig. 4–19 (a). A typical dimension is shown in Fig. 9–4 (a), and methods of giving dimensions to avoid crowding in small places are illustrated at (b) to (g). *Never letter a dimension figure over any line of the drawing.*

The correct methods of lettering decimal dimension figures are shown in Fig. 9–5. The numerals in all cases are made $\frac{1}{8}''$ high, re-gardless of whether they appear on one line or two lines. The space between lines of numerals is $\frac{1}{16}''$, or $\frac{1}{32}''$ above and below the dimension line, as shown at (b) and (d).

When the maximum and minimum limits are specified, one directly above the other, the maximum limit is always placed above the minimum limit; and that portion of the dimension line between the lines of numerals is omitted, as shown at (b) to (f). When both limits are specified on one horizontal line, as with a leader or note, the minimum limit is always placed first and a short dash placed between the limits.

Guide lines should be used for all dimension figures, and all numerals and decimal points should be carefully lettered and properly spaced.

Fig. 9–5. Decimal Dimension Figures.

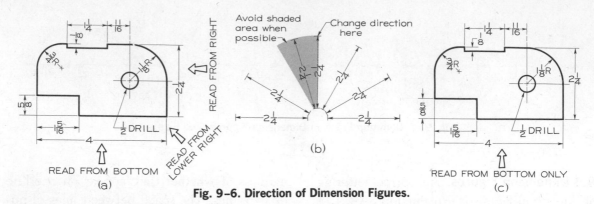

Fig. 9–6. Direction of Dimension Figures.

9.5 Inch Marks. As explained in Sec. 3.19, dimensions of machine parts are expressed in inches *understood*. It is common practice to omit all inch marks on a drawing when all dimensions are expressed in inches, except in situations where this may lead to misinterpretation of dimensions or notes. A one-inch dimension would be shown as 1″, and a one-inch drill note would be shown as 1″ DRILL, etc.

9.6 Direction of Dimension Figures. The two systems of reading direction for dimensions on drawings are the *aligned* and *unidirectional* systems.

In the aligned system, Fig. 9–6 (a), dimensions are lettered so as to be read from the bottom of the sheet, or from the right side, or between these positions. Sometimes a dimension must be lettered to read slightly from the left, such as the $\frac{3}{4}$R at the upper left at (a), but this condition should be avoided where possible. The shaded zone at (b) shows the directions of dimensions that should be avoided if possible. Notes, such as $\frac{1}{2}$ DRILL, should always be lettered horizontally.

In the unidirectional system, Fig. 9–6 (c), all dimensions are lettered to read from the bottom. This system is rapidly gaining in favor, especially in the aircraft industry where drawings are very large. Although both systems are in use in industry, the USASI indi-

cates that the unidirectional system is preferred.

9.7 Finish Marks. A *finish mark* is a symbol to indicate that a surface is to be *finished*, or machined. A finish mark on a drawing tells the patternmaker, Sec. 10.2, that he must allow extra material on the pattern which will provide extra metal on the casting to be removed in machining. The same finish mark tells the machinist to machine the surface.

Two types of finish marks are approved, Fig. 9–7 (a) and (b), the V-type being the newer form coming into increasing use. The V is made $\frac{1}{8}$″ high, with the aid of a triangle. The old form of finish mark, which is still widely used, is the italic *f*. It is made freehand, about $\frac{3}{16}$″ high, as shown at (b).

A finish mark is shown on the edge view of a surface to be finished, and is repeated in every view where the surface appears as a line, including hidden lines and curved lines. If the V is used, the point of the V should point inward toward the solid metal like a cutting tool.

In Fig. 9–7 (c) are shown three views and a pictorial of a rough casting before it is machined. At (d) the casting has been machined, and the V-type finish marks are shown, while at (e) the same finished surfaces are indicated by the *f*-type finish marks. The material re-

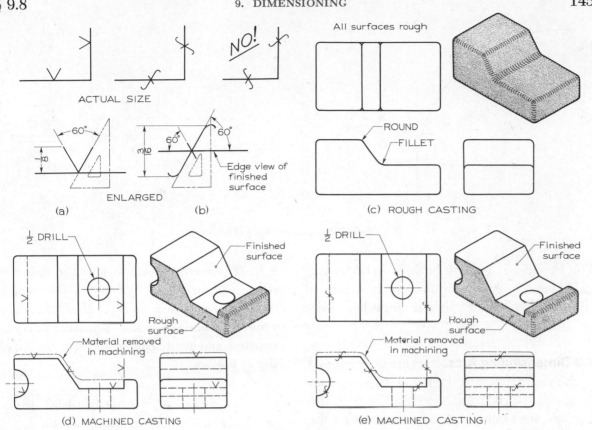

Fig. 9–7. Finish Marks.

moved in machining at (d) and (e) is shown for illustration purposes. As explained in Sec. 9.1, the drawing shows the part in its completed state.

Finish marks are not needed for drilled holes or for any other holes where the machining operations are clearly shown in a note, Sec. 9.24. If a part is to be finished all over, omit all finish marks and letter a general note on the drawing, such as FINISH ALL OVER or FAO. No finish marks or notes are needed if a part is machined from cold finished stock.

9.8 Dimensioning Angles. Angles are drawn with the triangles or with the aid of the protractor, Fig. 3–19. They are indicated by degrees, Fig. 9–8 (a) to (f), or by two dimensions as shown at (g). When degrees are given, the circular dimension lines are drawn with the compass center at the vertex of the angle, (f).

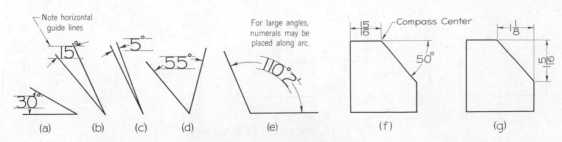

Fig. 9–8. Dimensioning Angles.

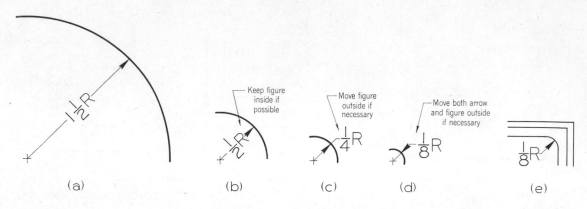

Fig. 9–9. Dimensioning Arcs (Full Size).

The lettering is preferably done on horizontal guide lines, but an exception may be made for larger angles, where circular guide lines may be used, (e).

9.9 Dimensioning Arcs. Arcs are dimensioned in the views in which their true shapes appear by giving the radius, as shown in Fig. 9–9. The letter R, for radius, is always lettered after the figures, as shown.

The dimension figure and the arrowhead should be inside the arc, as shown at (a) and (b), where there is sufficient space. If the space is too crowded, the figure is moved outside the arc, (c), and if necessary the arrowhead is also moved outside, (d). In some cases, where lines of the drawing interfere, the method at (e) is used.

9.10 Fillets and Rounds. Fillets and rounds, Sec. 10.3, are dimensioned as shown in Fig.

9–9. It is not necessary to give the radius of every fillet or round, but only a few typical radii, as shown in Fig. 10–2. If all fillets and rounds are uniform in size, dimensions may be omitted, and a note can be added to the drawing as follows:

ALL FILLETS $\frac{1}{4}$R AND ROUNDS $\frac{1}{8}$R

or   ALL FILLETS & ROUNDS $\frac{1}{8}$R

9.11 Placement of Dimensions. Correct and incorrect methods of placing dimensions are shown in Figs. 9–10 and 9–11. As shown in Fig. 9–10 (a), the smaller dimensions should be placed nearest to the object and lined up in chain fashion. Note that a complete chain of these is unnecessary. The overall dimensions are always farthest from the view. Avoid placing the dimensions, as at (b), where dimension lines must cross extension lines.

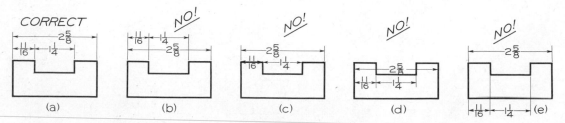

Fig. 9–10. Placement of Dimensions.

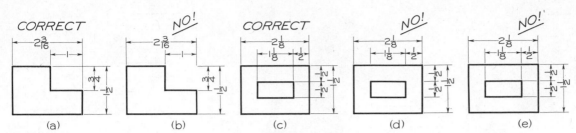

Fig. 9–11. Crossing Lines.

Never place a dimension to coincide with a line of the drawing or join end to end with a line of the drawing, (c). Avoid placing dimensions on a view, (d), unless some advantage is gained thereby. In certain cases, such as the radii in Fig. 9–6 (a), and on complicated drawings where it is difficult to find places for dimensions, it is necessary to place some dimensions within the view, Fig. 15–14. As a general rule, however, *never place a dimension on a view unless something in clearness or directness of application is gained thereby.* Avoid placing dimensions so that extension lines cross lines of the drawing unnecessarily, (e).

Extension lines may cross each other freely, as in Fig. 9–11 (a). They should never be shortened as at (b). In many cases, extension lines must cross the lines of the drawing as at (c). The extension lines should never be shortened as at (d) or have gaps at crossing points as at (e).

9.12 Steps in Applying Dimensions. The steps in placing dimensions on a view are shown in Fig. 9–12.

I. Draw extension lines dark and sharp, using a medium-hard pencil (as 2H) with a sharp conical point. Extend the center lines of the hole, to be used in the same manner as extension lines.

II. Use the scale to space the dimensions at least $\frac{3}{8}''$ from the object and $\frac{1}{4}''$ apart ($\frac{3}{8}''$ if space is available). Draw the dimension lines dark and sharp, leaving gaps for the dimension figures.

III. Draw all arrowheads about $\frac{1}{8}''$ long and very narrow.

IV. Add all dimension figures and lettering. Draw guide lines as shown in Sec. 4.11.

Note the contrast at IV between the heavy visible lines of the view and the dimensions. The view should stand out clearly from the dimensions.

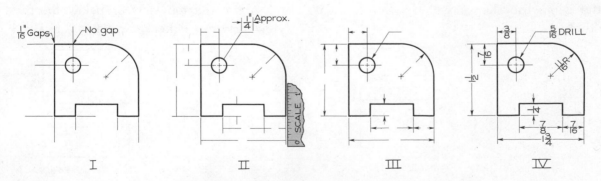

Fig. 9–12. Steps in Applying Dimensions.

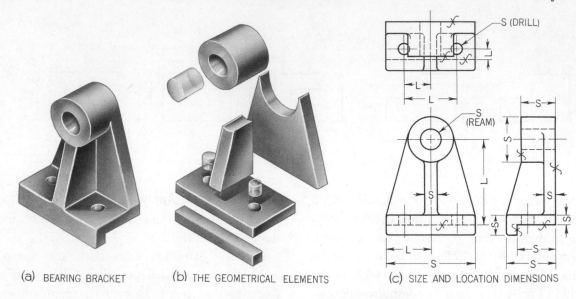

(a) BEARING BRACKET   (b) THE GEOMETRICAL ELEMENTS   (c) SIZE AND LOCATION DIMENSIONS

Fig. 9–13. Size (S) and Location (L) Dimensions.

9.13 Geometric Breakdown. Any machine or structure, when broken down into its basic shapes, is found to be made up largely of simple geometric shapes, such as the prism, cylinder, cone, and sphere. For example, the Bearing Bracket in Fig. 9–13 (a) is made up of simple basic shapes, as shown at (b). Even the holes are regarded as cylinders. Three views of the Bearing Bracket are shown at (c), which illustrates the two main types of dimensions indicated by the letters S and L:

S = Size Dimensions, which give the sizes of the geometric shapes.

L = Location Dimensions, which locate the geometric shapes with respect to each other.

In actual practice, of course, the dimension figures are given instead of the letters S and L.

9.14 Size Dimensions of Prisms. The simplest geometric shape is the prism, Fig. 9–14 (a), or modifications of it. Three dimensions are required, the *width, height,* and *depth.* Only two views are required to give these dimensions, as shown at (b) and (c). In either case, the width ($4\frac{1}{2}''$) and the height ($3''$) are given on the front view, and the depth ($1\frac{3}{4}''$) is given on the top view or side view.

In the example at (b), the $4\frac{1}{2}''$ dimension should be between the views, as shown, and not above the top view or below the front view, or across either view, Sec. 9.11. The $3''$

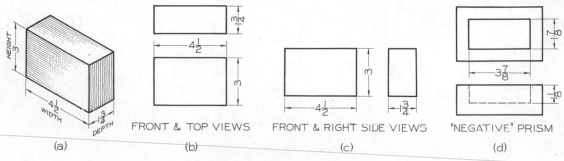

FRONT & TOP VIEWS   FRONT & RIGHT SIDE VIEWS   "NEGATIVE" PRISM

(a)   (b)   (c)   (d)

Fig. 9–14. Size Dimensions of Prisms.

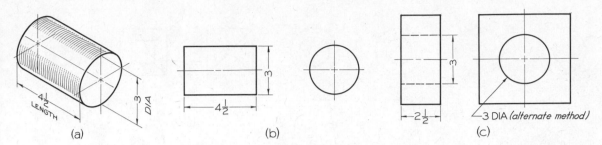

Fig. 9-15. Size Dimensions of Cylinders.

and $1\frac{3}{4}''$ dimensions should be lined up together, as shown. Similarly, at (c) the 3″ dimension should be between views, and the $4\frac{1}{2}''$ and $1\frac{3}{4}''$ dimensions should be lined up together. However, these may be placed above the views, if desired.

9.15 Size Dimensions of Cylinders. The cylinder, Fig. 9–15 (a), is the next most common shape, and is usually seen as a shaft or a hole. Dimension a cylinder by giving both its diameter and length in the *rectangular* view, as shown at (b) and (c). Never give the radius of a cylinder, Sec. 10.10, or give the diameter in the circular view, either diagonally across the circle or between extension lines from the

circle. An exception is made when dimensioning a hole where the shop operation is given in a note, as in Fig. 9–16 (c) and (e), or in dimensioning a large hole or a bolt circle, where the diagonal diameter is permitted, Fig. 9–19. See also Sec. 9.24.

Several applications of size dimensions of cylinders are shown in Fig. 9–16. At (a) the pulley is made up of a number of concentric cylinders. All diameters, except one, are given in the rectangular view, some at the right and some at the left so as to place the diameters as close to the shapes as possible. Note how the dimension figures are "staggered" to provide ample space for each figure.

At (b) is shown a case where DIA is given

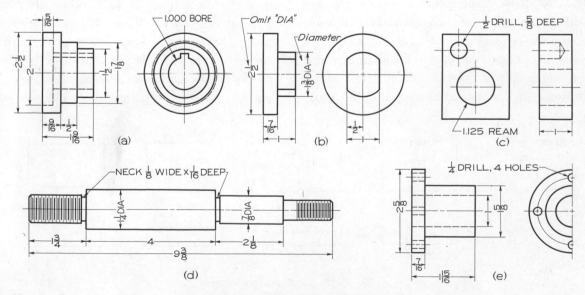

Fig. 9-16. Size Dimensions of Cylinders.

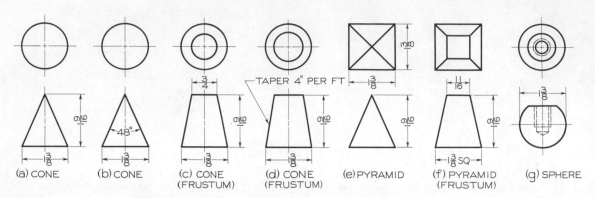

Fig. 9–17. Size Dimensions of Cones, Pyramids, and Spheres.

to show clearly that the small piece is cylindrical in spite of the two flat surfaces. At (c) is shown the use of notes where it is desired to specify the operations. In all such cases, the *diameter*, and not the radius, is given. At (d) is shown a cylindrical shaft in which the abbreviation DIA on each diameter is given to make an end view unnecessary. At (e) is shown a bearing in which a half-view is used to save space. This can be done only when both halves are the same.

9.16 Size Dimensions of Cones, Pyramids, and Spheres. Dimension a cone by giving both the diameter of the base and the altitude in the triangular view, Fig. 9–17 (a). Sometimes it is desirable to give the diameter and the angle, (b), or give two diameters and the altitude, (c), or one diameter and the *taper*

by note, (d). *Taper per foot* means the difference in diameter in one foot of length.

Dimension a pyramid by giving the dimensions of the base in the square or rectangular view, and the altitude in the other view, (e). If the base is square, only one dimension for the base is necessary if marked SQ, as shown at (f).

Dimension a sphere by giving its diameter, as shown at (g).

9.17 Location Dimensions. Location dimensions, Sec. 9.13, are used to locate geometric shapes with respect to each other after the size dimensions for these shapes have been given. Rectangular shapes are located from surface to surface, Fig. 9–18 (a) and (b). Cylinders are located from their center lines, (c) and (d). Location dimensions for holes should

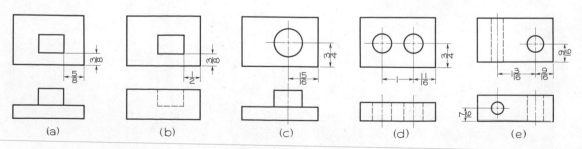

Fig. 9–18. Location Dimensions.

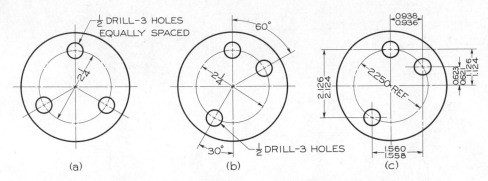

Fig. 9–19. Locating Holes about Common Center.

be given in the circular view of the holes, if possible, as at (d); otherwise, as at (e).

If holes are equally spaced about a common center, Fig. 9–19 (a), the diagonal diameter of the *circle of centers*, or *bolt circle*, is given and the equal spacing is indicated in the note. If the holes are unequally spaced, it is necessary to give angles to one of the two main center lines, (b). Where great accuracy is required, the holes should be located as shown at (c), where the bolt circle diameter is given only for reference.

Symmetry of an object about one or two center lines is very common, in which case dimensions are given from those main center lines, Fig. 9–20. At (a) one small hole is located

from the main center line A, and the other small hole is located from it. The two small holes must be tied together, and no dimension at B should be given.

At (b), in addition to the location dimensions from the two main center lines, the diagonal diameter is given for reference in laying out the holes in the shop. At (c) the vertical center line is the main axis of symmetry. Note the omission of the unnecessary dimension at C.

A typical example of location dimensions on a three-view drawing of a bracket is shown in Fig. 9–21. Location dimensions should lead to finished surfaces or to important center lines, as shown.

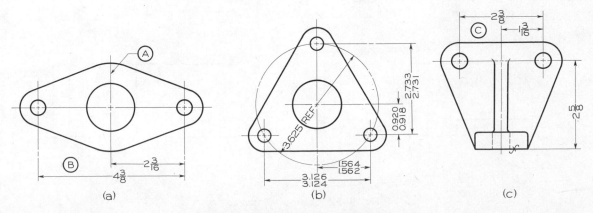

Fig. 9–20. Locating Holes on Center.

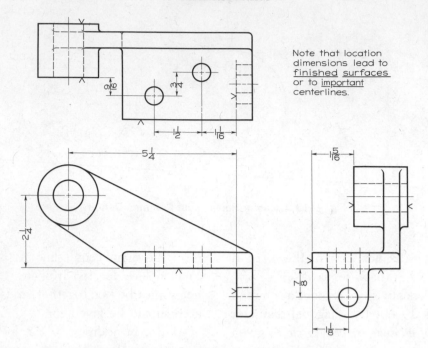

Note that location
dimensions lead to
finished surfaces
or to important
centerlines.

Fig. 9–21. Location Dimensions to Finished Surfaces.

9.18 Contour Dimensioning. *Dimensions should be given in the views where the shapes are shown,* Fig. 9–22 (a). When the shop man reads the drawing, he will naturally look for the size description close to the shape description. The meaning of a dimension is always clearest in the view where the corresponding shape is given. Violations are shown at (b). In general, avoid dimensioning to hidden lines where possible.

9.19 Rounded Ends. Rounded-end shapes are dimensioned according to the shop methods used, Fig. 9–23. At (a) and (b) the shapes, to

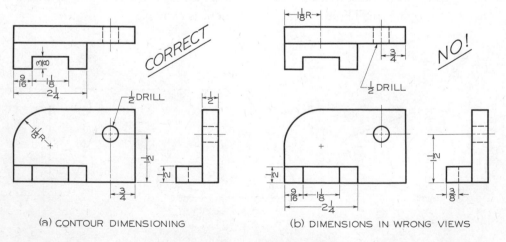

(a) CONTOUR DIMENSIONING   (b) DIMENSIONS IN WRONG VIEWS

Fig. 9–22. Contour Dimensioning.

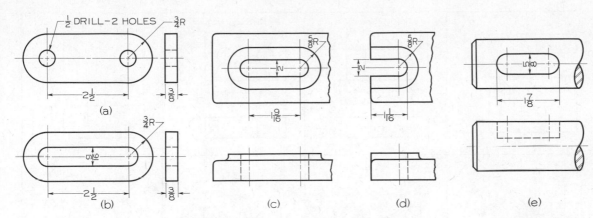

Fig. 9–23. Dimensioning Rounded-End Shapes.

be cast or to be cut from sheet or plate, are dimensioned as they would be laid out in the shop, by giving the center-to-center distance and the radii of the ends. At (c) the pad on a casting with a milled slot is dimensioned from center to center conforming to the total travel of the milling cutter. At (d) the full length of the milled slot represents the total travel of the milling cutter. At (e) the method of dimensioning a Pratt and Whitney keyseat is shown. Here the width and the total length are given in conformity with the standard key sizes. In general, where great accuracy is required, the overall length of rounded-end shapes should be given. As a general rule, give diameters of slots rather than radii of the ends, (b) to (e).

9.20 Superfluous Dimensions. Before giving any dimension, ask yourself the question, "What is this dimension for?" Is it really needed? Is it the best way to tell exactly how the part is to be made? In Fig. 9–24 are shown several cases of superfluous dimensioning. In practically every case additional and completely unnecessary dimensions are given because the draftsman was not thinking of shop requirements. Give dimensions in the most

direct and simple way, and never give a dimension more than once in the same view or in different views. Avoid duplication. *If you cannot give a definite reason for a dimension, omit it.*

9.21 Dimensioning Curves. If a continuous curve is made up of a series of circular arcs, the various radii are given, Fig. 9–25 (a). Note that for extremely large radii, such as the radius $12\frac{1}{4}$, the center may be drawn closer to the arc as shown. In such cases, the main part of the dimension line is drawn toward the actual center.

If a curve is not made up of circular arcs or if the use of radii is not desired, the curve is dimensioned by a series of dimensions at regular intervals, (b). Note accumulation of values in the upper line of dimensions.

9.22 Mating Dimensions. When two or more parts fit together, they are called *mating parts*. On two mating parts certain dimensions must correspond to make the parts fit together. These are called *mating dimensions*. For example, Fig. 9–26 (a), a bracket fits into a base and two bolts are used to fasten the parts

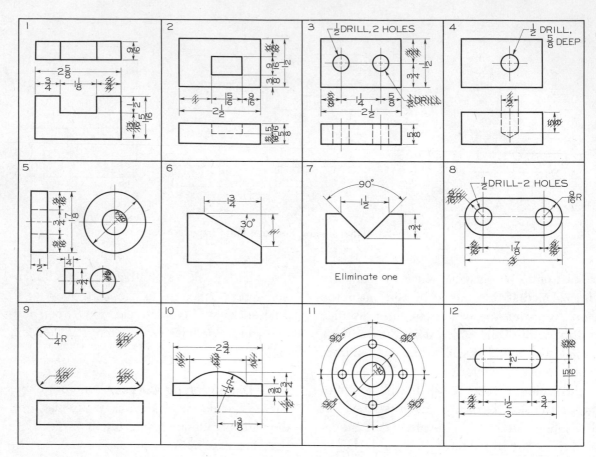

Fig. 9–24. Superfluous Dimensions.

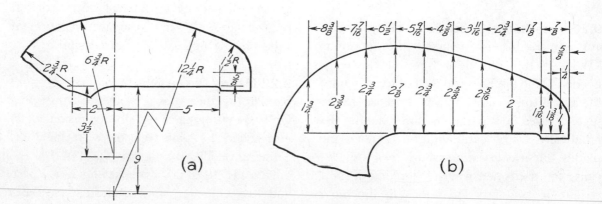

Fig. 9–25. Dimensioning Curves.

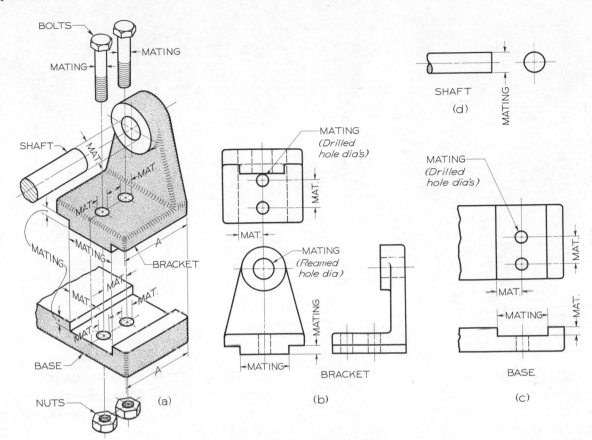

Fig. 9–26. Mating Dimensions.

together. The dimensions of the projection under the bracket must correspond to those on the slot in the base. Also, the diameters of the bolts must correspond to the sizes of the drilled holes in the bracket and in the base, the spacing of the drilled holes must correspond in the two mating parts, and the diameter of the shaft must correspond to the reamed hole in the bracket.

The drawings of the principal parts are shown at (b) to (d). Mating dimensions are those that correspond in the separate drawings of the mating parts and are needed for the accurate fitting of the parts. Dimensions that agree on two mating parts, such as dimension A in Fig. 9–26 (a), but which are not essential for accurate fitting, are not mating dimensions.

Also, keep in mind that the actual *values* of two *corresponding* mating dimensions, while close, may not be exactly the same. For example, the width of the slot in the base may be $\frac{1}{32}''$ or $\frac{1}{64}''$ or several thousandths larger than the width of the projection of the base that fits in the slot, but these are mating dimensions figured from a single basic width.

9.23 Pattern and Machine Dimensions. The bracket in Fig. 9–26 was machined from a rough casting in a manner similar to that for the casting in Fig. 9–7 (c) to (e). Bear in mind that the patternmaker and the machinist both follow the same working drawing, which shows the completed object, some dimensions being used by the patternmaker, some only

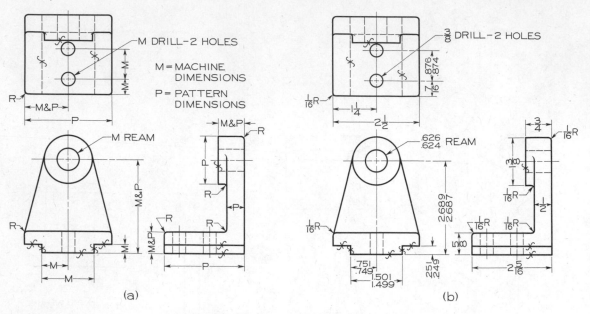

(a)   (b)

Fig. 9–27. Pattern and Machine Dimensions.

by the machinist, and some by both, Fig. 9–27 (a).

The patternmaker uses only those dimensions needed to make the wood pattern from which the rough workpiece is molded in sand, Sec. 10.2. Castings are not exactly uniform or accurate, so pattern dimensions are rough, to about the nearest $\frac{1}{16}''$, and given in whole numbers and fractions.

The machinist is interested only in the rough workpiece handed to him and in the dimensions he needs to machine the various holes and surfaces. If accuracy no greater than the nearest $\frac{1}{64}''$ is required, which corresponds to the $\frac{1}{64}''$ divisions on the machinists scale, the dimensions are given in whole numbers and fractions. If greater accuracy is required, as for the size of the hole for the shaft, the dimensions are given in decimals, Sec. 9.25. The completely dimensioned working drawing as used by both the patternmaker and the machinist is shown in Fig. 9–27 (b).

9.24 Notes. To supplement the ordinary dimensions, *notes* are often necessary to supply additional information, usually in connection with required shop processes. Notes should be brief and carefully worded so as to permit only one interpretation. The wording and form of such notes are pretty well standardized, and these standards should be carefully followed. *Notes should always be lettered horizontally on the sheet.* There are two kinds of notes, as follows:

General Notes—These give general information about the drawing as a whole, as:

FINISH ALL OVER

FILLETS & ROUNDS $\frac{1}{8}$R

BREAK SHARP EDGES

General notes like these should be lettered where space is available in the lower portion of the sheet. In machine drawings the title strip will carry many general notes, such as material, number of pieces required, general tolerances, etc.

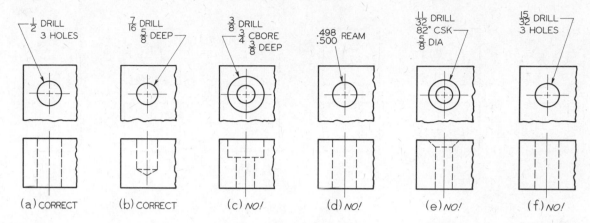

Fig. 9–28. Leaders to Notes.

Local Notes—These apply to specific operations to be performed, and are connected by a leader, Fig. 9–2, to the appropriate point on the drawing. Typical examples are:

$\frac{1}{4}$ DRILL—2 HOLES
$\frac{1}{8}$ × 45° CHAMFER
MEDIUM KNURL

Always attach leaders at the front of the first word of a note, or after the last word as shown in Fig. 9–28 (a) and (b), and never to any other part of a note as shown from (c) to (f).

Methods of forming the most common types of holes are discussed in Secs. 10.11 and 10.12. Of these, the most common is the *drilled hole*. A hole cut entirely through a piece is called a *through hole*, Fig. 9–29 (a), and one not through is called a *blind hole*, (b). The bottom of a drilled hole is conical, and is drawn with the 30° × 60° triangle, as shown. Drill depth, usually given in the drill note, *does not include the conical point.*

"Fractional-size" drills are available in drill sizes of $\frac{1}{16}$" diameter to $3\frac{1}{2}$" diameter. Diameter increments are $\frac{1}{64}$" for drills of $\frac{1}{64}$" to $1\frac{3}{4}$" diameter, $\frac{1}{32}$" for drills of $1\frac{3}{4}$" to $2\frac{1}{4}$" diameter,

and $\frac{1}{16}$" for drills $2\frac{1}{4}$" to $3\frac{1}{2}$" diameter. In addition, there are a great many "numbered" and "lettered" drills in the small-diameter range. See Table 2 (Appendix, page 456).

It is common practice to give fractional-size drills in fractions, as in Fig. 9–29 (a) and (b), and to give the decimal sizes in parentheses for lettered and numbered drills, (c) and (d). Practice is tending toward giving the decimal sizes for all diameters under $\frac{1}{2}$" as at (e). This is recommended by the United States of America Standards Institute.

Typical notes for reamed, bored, counterbored, spotfaced, and countersunk holes are shown from (f) to (k). Note that countersunk holes, (k), are drawn conventionally at 90° to approximate the angle of 82° (actual). The method of specifying several operations on a large hole is shown at (m).

The USA Standards* on dimensioning practices recommend that, when practicable, a finished part should be defined without indicating any specific manufacturing methods. For example, only the diameter of a hole would

* USAS Y14.5—1966.

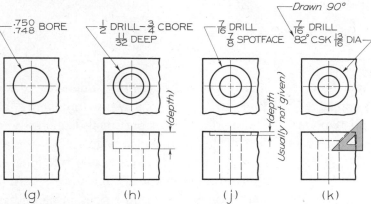

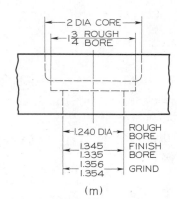

Fig. 9–29. Notes for Holes.

be given, without indicating whether it should be drilled, reamed, etc. However, where such information is essential to the description of engineering or manufacturing requirements, it should be clearly indicated on the drawing or other accompanying documents.

Methods of producing holes in the shop are discussed in Secs. 10.11 and 10.12. Examples of many notes commonly used on drawings are shown in Fig. 9–30. See Technical Terms (Appendix, pages 445–450) for definitions of various shop operations.

9.25 Decimal Dimensions. Some parts, even though machined, do not need to be more accurate than the nearest $\frac{1}{64}''$, such as certain locomotive parts, railroad car parts, or agri-

cultural machinery parts. These drawings are dimensioned entirely with whole numbers and common fractions. Other parts, such as certain parts of a milling machine or a sewing machine, must be so accurate that every dimension must be given in decimals. Most parts, however, require only certain key dimensions to be highly accurate while the others can be "rough." The average drawing, then, contains some rough dimensions in common fractions and some accurate dimensions in decimals, Fig. 9–27 (b). In architectural drawing or drawings of woodwork, all dimensions are given in whole numbers and fractions.

A system that is increasingly used, especially in the automotive and aircraft industries, is the complete decimal system—that is, all di-

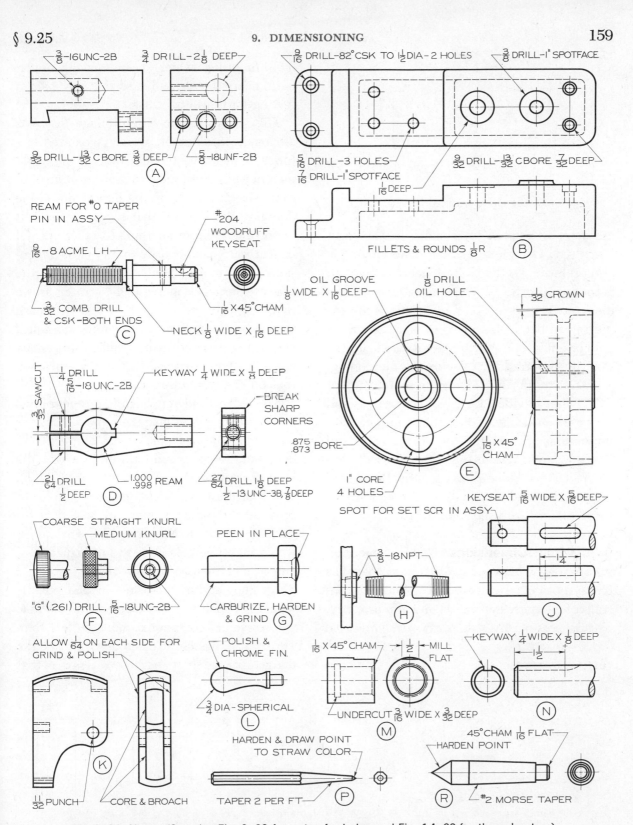

Fig. 9–30. Notes. (See also Fig. 9–29 for notes for holes and Fig. 14–23 for thread notes.)

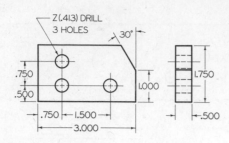

Fig. 9–31. Complete Decimal Dimensioning.

mensions are stated in decimals as shown in Fig. 9–31. Complete decimal dimensioning is also preferred by the United States of America Standards Institute.* Thus, measurements formerly expressed in common fractions are expressed in decimals to two or more places, as 2.33 or 5.625, etc. A two-place decimal dimension is preferred for decimal dimensioning, as 2.12 or 7.62, but always in increments of .02"; when divided by two, the result is still a two-place decimal. For still more accurate dimensions, in which the micrometer and other accurate measuring tools are used, two-place decimal dimensions in other than .02" increments, and decimals to three, four, or more places, are used, Sec. 10.10.

9.26 Limit Dimensions.

When something goes wrong with your car and some part has to be replaced, you can easily obtain and install a new part that will fit and run properly. This is possible by means of *interchangeable*

* USAS Y14.5—1966.

manufacturing, which means making similar parts nearly enough alike so that any pair of mating parts will fit satisfactorily.

This sounds simple. Even though the parts are made in different, widely separated factories, it might be thought that all the workmen have to do is make each measurement to exact size as shown by the dimension on the drawing. But it is not that easy, because it is impossible to make anything to *exact size.* Of course, parts can be made to very close dimensions, even to a millionth of an inch, but the expense becomes increasingly greater as we require greater accuracy. It would be foolish to require airplane-engine accuracy for a part of a hand-operated meat grinder, simply because the grinder would then cost so much that no one would buy it.

The old method of giving dimensions of two mating parts was to give the same dimensions on both parts in simple whole numbers and common fractions, and indicate in a note the kind of fit desired, as shown in Fig. 9–32 (a). The machinist was then depended upon to produce the parts so that they would fit together properly. In the example, the machinist would make the hole close to $1\frac{1}{2}$" in diameter, and would then make the shaft, say .003" (three thousandths inch) less in diameter. It would not matter if the diameter of the hole were several thousandths more or less than $1\frac{1}{2}$", for the machinist could always make the diameter of the shaft about .003" less than that

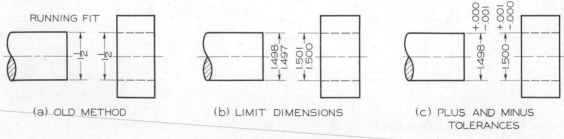

(a) OLD METHOD   (b) LIMIT DIMENSIONS   (c) PLUS AND MINUS TOLERANCES

Fig. 9–32. Tolerance Dimensions.

of the hole and thus obtain the desired "running fit." But in quantity production this would not work, since the sizes would vary considerably, and the parts would not be interchangeable; that is to say, not every shaft would fit in every hole.

In order to make sure that every part will fit properly in assembly, it is necessary to indicate a permissible amount of "oversize" or "undersize" for each dimension. Let us suppose that the hole in Fig. 9–32 (b) may be 1.500″ or 1.501″ or anywhere in between these. These figures are called *limits*. The difference between them is called *tolerance,* in this case .001″. Similarly, the shaft is assigned limits of 1.498″ and 1.497″, and the largest shaft (1.498″) will fit in the smallest hole (1.500″) with a minimum air space of .002″. This minimum space is called *allowance*. Notice at (b) that for both the shaft and hole the larger figure is on top.

Another way to indicate the same tolerances is to give a decimal figure followed by plus and minus values, as in Fig. 9–32 (c). The hole is given the basic size of 1.500″, with the provision that it may be as much as .001″ larger. The shaft is given a basic size of 1.498″ (which leaves an allowance of .002″ between the parts) with the provision that it may be as much as .001″ smaller. Thus, the loosest fit

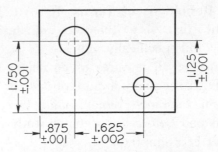

Fig. 9–33. Plus or Minus Tolerances.

would be 1.501″ minus 1.497″, or .004″, which is thought to be close enough.

Tolerances are often given in both directions (plus or minus) from a basic size when a variation in one direction is no more critical than in another, such as in locating the holes in Fig. 9–33.

A typical drawing containing limit dimensions is given in Fig. 9–34. Note that the dimensions at the bottom are all given from one common surface. This is called *baseline dimensioning;* it is often used to avoid difficulties resulting from the accumulation of tolerances in a chain of dimensions. Never give a complete chain of tolerance dimensions and also an overall tolerance dimension. One dimension in the chain, or the overall dimension, should be omitted or changed to common fractions.

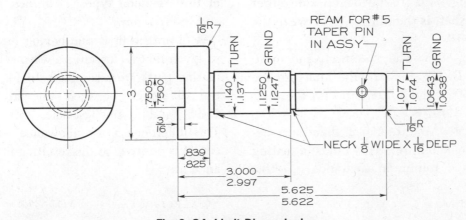

Fig. 9–34. Limit Dimensioning.

9.27 Definitions of Terms. It is vitally important that any tolerance dimensioning clearly and specifically indicate the sizes and locations of all features of an object or assembly. It is equally important that the terms used in tolerance dimensioning be clearly understood before any detailed study of this subject is attempted.

The following definitions have been adopted from the USA Standards:*

Nominal Size—The designation which is used for the purpose of general identification. In Fig. 9–32 (b) the nominal size of both hole and shaft is $1\frac{1}{2}''$.

Basic Size—That size from which the limits of size are derived by the application of allowances and tolerances. In Fig. 9–32 (b) the basic size is the decimal equivalent of the nominal size $1\frac{1}{2}''$, or 1.500".

Actual Size—An actual size is a measured size.

Design Size—That size from which the limits of size are derived by the application of tolerances. When there is no allowance, the design is the same as the basic size. In Fig. 9–32 (b) the design size of the hole is 1.500" (diameter of smallest hole), and that of the shaft is 1.498" (diameter of largest shaft).

Tolerance—The total permissible variation of a size, or the difference between the limits of size. In Fig. 9–32 (b) the tolerance on either the hole or shaft is the difference between the limits, or .001".

Limits of Size—The maximum and minimum sizes. In Fig. 9–32 (b) the limits for the hole are 1.500" and 1.501", and for the shaft 1.498" and 1.497".

Allowance—An intentional difference between the maximum material limits of mating parts. It is a minimum clearance (positive allowance) or maximum interference (negative allowance) between mating parts. In Fig. 9–32 (b) the allowance is the difference between the smallest hole, 1.500", and the largest shaft, 1.498", or +.002", and thus a *clearance fit*. If the allowance is negative, the result will be an *interference fit*. See Sec. 9.28.

Unilateral Tolerance—A tolerance in which variation is permitted only in one direction from the design size. In Fig. 9–32 (c) the design size of the hole is 1.500". The tolerance .001" is all in one direction—toward a larger size. The design size of the shaft is 1.498". The tolerance .001" is again all in one direction, but in this case toward a smaller size. In a unilateral tolerance either the plus or the minus value must be zero.

Bilateral Tolerance—A tolerance in which variation is permitted in both directions from the design size. In Fig. 9–33 the horizontal distance between the holes is 1.625 ± .002. The design size is 1.625" and the tolerance is .004", or a variation of .002" in each direction from the design size.

9.28 General Types of Fits.* A *fit* is the general term used to signify the range of tightness that may result from the application of a specific combination of allowances and tolerances in the design of mating parts. Fits are of three general types: *clearance, transition,* and *interference.*

A clearance fit is one having limits of size so given that a clearance always results when mating parts are assembled. In Fig. 9–32 (b) the difference between the largest shaft, 1.498", and the smallest hole, 1.500", is .002". This difference is the allowance, and in this case it is positive, so the resulting fit is a clearance fit.

---

* USAS B4.1—1955 and USAS Y14.5—1966.

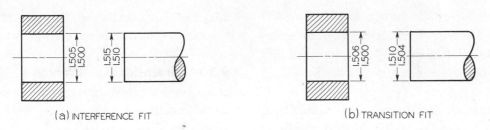

Fig. 9-35. Fits Between Cylindrical Parts.

An interference fit is one having limits of size so given that an interference always results when mating parts are assembled. In Fig. 9–35 (a) the smallest shaft is 1.510″, and the largest hole is 1.505″, resulting in an interference of metal (negative allowance) of a minimum of .005″.

A transition fit is one having limits of size so given that either a clearance or an interference may result when mating parts are assembled. In Fig. 9–35 (b) the smallest shaft, 1.504″, will fit in the largest hole, 1.506″, with a clearance of .002″. However, the largest shaft is 1.510″, and the smallest hole is 1.500″, resulting in an interference of metal (negative allowance) of .010″.

9.29 Basic Hole and Basic Shaft Systems.* To specify the dimensions and tolerances of

* USAS B4.1—1955 and USAS Y14.5—1966.

an internal and an external cylindrical surface so that they will fit together as desired, it is necessary to begin calculations by assuming either the minimum hole size or the maximum shaft size.

A *basic hole system* is a system of fits in which the design size of the hole is regarded as the basic size, and the allowance is applied to the shaft.

In Fig. 9–36 (a) the minimum hole size, 1.500″, is considered as the basic size. An allowance of .003″ is assumed, and *subtracted* from the basic hole size to obtain the diameter of the maximum shaft, 1.497″. A tolerance of .002″ is decided upon and applied to both the hole and the shaft to obtain the maximum hole of 1.502″ and the minimum shaft of 1.495″. The minimum clearance is the difference between the smallest hole, 1.500″, and the largest shaft, 1.497″, or .003″. The maximum clearance is the difference between the largest hole,

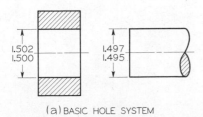

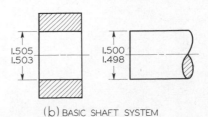

Fig. 9-36. Basic Hole and Basic Shaft Systems.

1.502″, and the smallest shaft, 1.495″, or .007″. To obtain the maximum shaft size of an interference fit, the desired allowance, or maximum interference, would be *added* to the basic hole size.

A *basic shaft system* is a system of fits in which the design size of the shaft is the basic size, and the allowance is applied to the hole.

In Fig. 9–36 (b) the maximum shaft size, 1.500″, is considered as the basic size. An allowance of .003″ is decided upon and *added* to the basic shaft size to obtain the diameter of the minimum hole, 1.503″. A tolerance of .002″ is decided upon and applied to the hole and shaft to obtain the maximum hole, 1.505″, and the minimum shaft, 1.498″. The minimum clearance is the difference between the smallest hole, 1.503″, and the largest shaft, 1.500″, or .003″. The maximum clearance is the difference between the largest hole, 1.505″, and the smallest shaft, 1.498″, or .007″. To obtain the minimum hole size of an interference fit, the desired allowance would be *subtracted* from the basic shaft size.

To convert a basic hole size to a basic shaft size, simply subtract the allowance for a clearance fit, or add it for an interference fit. To convert a basic shaft size to a basic hole size, add the allowance for a clearance fit, or subtract it for an interference fit.

9.30 Dimensioning Problems. Problems given on the following pages are presented to afford practice in elementary dimensioning. They are to be drawn, and then the full-size dimensions are to be added, as shown in Figs. 9–37 and 9–38. Various lines have been omitted, and the student is to add these to make a complete drawing.

For more practice in dimensioning, use the problems assigned from any previous chapter, especially Chapter 7.

Figures 9–39, 9–40, and 9–41, which follow, consist of drawings in which certain lines are missing and in which dimensions are omitted. They are to be drawn mechanically, as assigned by your instructor. Add all missing lines and dimension fully, as shown in Fig. 9–38, using Layout C (Appendix, page 474). These problems are printed one-fourth size. Make your drawings full size by stepping off each measurement four times with the dividers. Add dimensions and notes to agree with your full-size drawings. Use heavier visible lines in the finished drawing, Fig. 9–38. Move all titles to title strips.

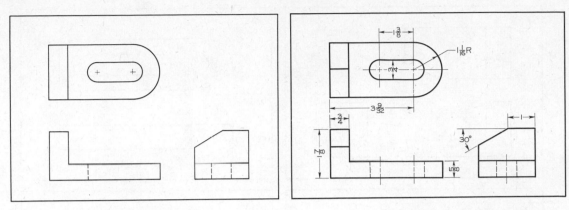

Fig. 9-37. Given Views.   **Fig. 9-38. Completed Drawing.**

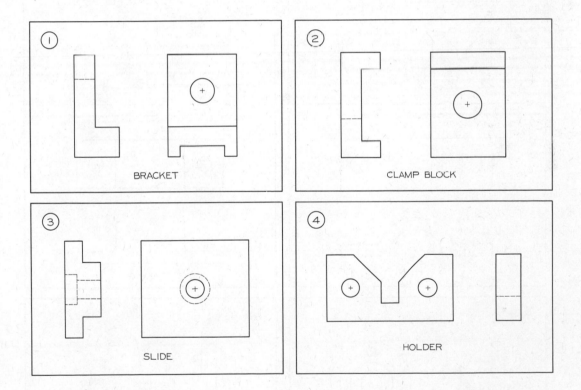

Fig. 9-39. Dimensioning Problems. Redraw with instruments. Add missing lines and dimension fully.

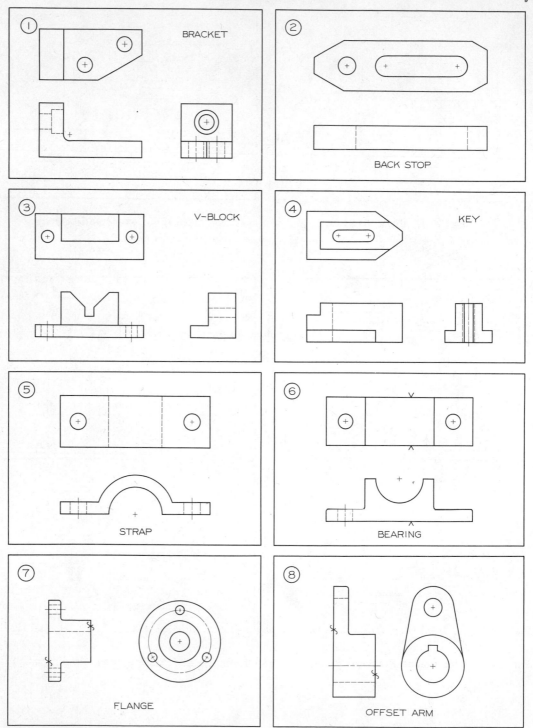

Fig. 9–40. Dimensioning Problems. Redraw with instruments. Add missing lines and dimension fully. See instructions on page 164.

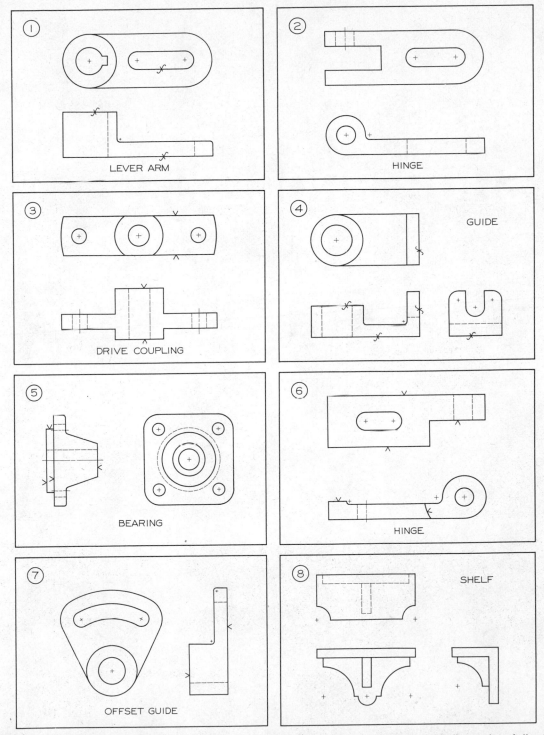

Fig. 9–41. Dimensioning Problems. Redraw with instruments. Add missing lines and dimension fully. See instructions on page 164.

CHAPTER 10

SHOP PROCESSES

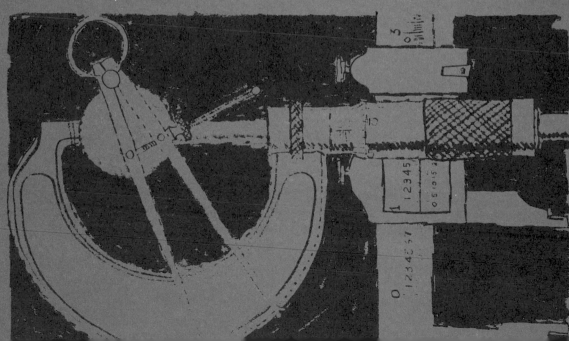

10.1 The Shops. The working drawing is received in the shops in the form of a print, as shown in Fig. 10–1. Inasmuch as the drawing gives complete instructions to the shop concerning how an object is to be made, it follows that the draftsman or designer must have at least a working knowledge of shop processes. The best training is, of course, extensive and varied experience in the shop. However, a great deal can be learned through observation and through study of periodicals and books. The purpose of this chapter is to provide the basic knowledge of shop processes needed by the young draftsman or machine designer.

In general, the production of a machine part consists of three main stages: *rough forming*, *finishing*, and *assembling*. The three most important types of rough forming are *casting* (in a *foundry*), *forging* (in a *forge shop*), and *weld-*

Fig. 10–1. Machinist Reading Blueprint.

ing (in a *welding shop*). Finishing is largely done in the *machine shop*. In the *assembly shop*, all the various parts are put together to form the completed machine, involving in many cases further machining operations that are best done in assembly.

In general, the drawing shows the part in its completed state, and each of the shops "picks off" from the drawing the information it needs. However, in the case of complicated castings, a special *pattern drawing* may be made that gives only the information needed in the pattern shop. Since forging dies are expensive, it is common to make special forging drawings for the forge shop, Fig. 10–10.

10.2 Sand Casting. A typical blueprint as received in the shops is shown in Fig. 10–2. In order to produce the part shown, a rough *sand casting* must be made, as shown in Fig.

10–3 (a), and then *machined*, as shown at (b). The casting is made by pouring molten metal (iron, steel, aluminum, brass, or some other metal) into a cavity in damp sand. The metal is allowed to cool until it hardens, whereupon it is removed from the sand. The cavity is made by placing a model of the object, called a *pattern*, in the sand and then withdrawing it, leaving an imprint of the model in the sand. You do somewhat the same thing when you press your hand into the sand at the beach and leave an imprint of it there.

Patterns are usually made of white pine or other woods. In some cases, duplicate patterns are made of metal. A wood pattern of the bearing in Fig. 10–2 is shown in Fig. 10–3 (c).

Since shrinkage occurs when metals cool, patterns are made slightly oversize. The patternmaker uses a *shrink rule* whose units are slightly oversize. Since various metals shrink

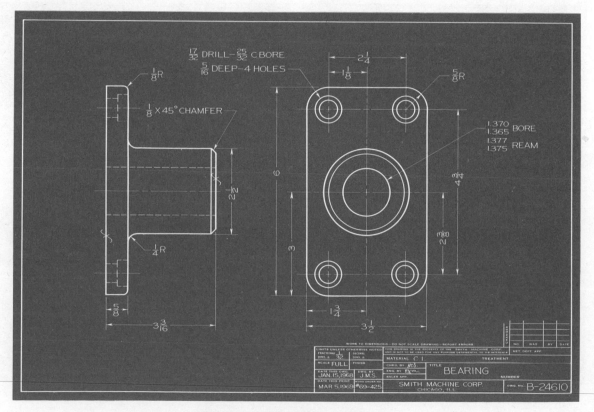

Fig. 10–2. A Detail Working Drawing.

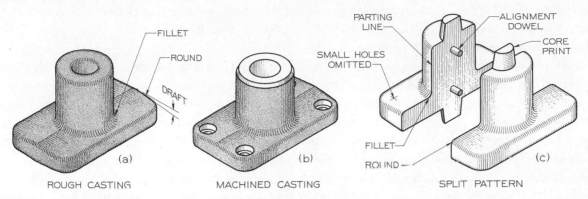

Fig. 10–3. Casting and Pattern.

different amounts, a different rule is used for each metal. For example, cast iron shrinks $\frac{1}{8}''$ per foot. The detail drawing, Fig. 10–2, shows the object in its final state and does not allow for shrinkage. This is taken care of entirely in the pattern shop by making the pattern oversize.

Draft is the taper given to a pattern to permit it to be easily withdrawn from the sand without damaging the shape of the mold. Draft is taken care of by the patternmaker, and is not shown on the drawing unless it is also a feature of the design. Only a slight draft on each side of the flat base is needed for this object, Fig. 10–3 (a).

The patternmaker must also make the pattern oversize in certain places to provide additional metal for each surface that is to be machined. For example, the bottom surface of the base of the casting of Fig. 10–2 is to be *finished,* or machined, and the patternmaker must provide extra thickness to the bottom of the base so that there will be from $\frac{1}{16}''$ to $\frac{1}{8}''$ of metal to be removed. See Sec. 9.7.

The sand is contained in a two-part box called a *flask,* Fig. 10–4 (e), the upper part being called the *cope* and the lower part the *drag.* For more complicated work, one or more intermediate boxes, called *cheeks,* are used between the cope and drag. Patterns are often

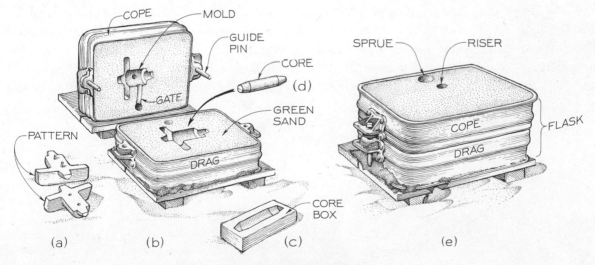

Fig. 10–4. Sand Molding.

split, Fig. 10–3 (c), so that one half can be placed in the cope and the other in the drag. The molds in the cope and drag are shown in Fig. 10–4 (b), these having been formed by ramming sand around each half of the pattern. A *sprue stick*, or round peg, is placed in position during the ramming process and then removed to leave a hole through which the molten metal is poured. A trench, called the *gate*, is formed to lead from the hole to the casting. The sand packed around the pattern is called *green sand*.

Green sand is often not strong enough to form some shapes, in which cases *dry sand cores* are used. Dry sand cores are made by ramming a prepared mixture of sand and a binding substance into a *core box*, Fig. 10–4 (c). The core is then removed and baked in a core oven to make it hard. The most common use of a core is to extend it through a casting to form a *cored hole*. The *core prints* of the pattern, Fig. 10–3 (c), form openings in the sand to support the ends of the core in Fig. 10–4 (b). When the metal is poured into the mold, it flows around the core, leaving a hole in the casting. When the casting has cooled, it is removed from the sand, and the dry sand core is broken out. The casting is then cleaned and any projecting fins are ground off on a grinder. The rough casting is now ready for the machine shop, where the cored hole will be enlarged by *boring* and *reaming*, the top and bottom surfaces will be machined, and

the small holes will be drilled and counterbored, Fig. 10–3 (b).

10.3 Fillets and Rounds. As shown in Figs. 10–3 (a) and 10–5, a rounded inside corner on a casting is called a *fillet*, and a rounded outside corner a *round*. Rounds are formed on a pattern by simply rounding the edges with a plane or sandpaper. Fillets are formed by gluing on leather or wooden strips, or by forming wax into the corners with a round-nosed hot tool similar to a soldering iron. Sharp corners should be avoided on a casting, and inside corners in particular should have as large fillets as possible. Not only is it difficult to obtain sharp corners in a sand mold, but such corners are weak and are apt to result in failure in the casting.

All fillets and rounds are shown on the drawing, Fig. 10–2, drawn to scale. If they are all $\frac{1}{8}''$ radius or larger, they should be drawn with the compass, and if under $\frac{1}{8}''$ radius they should be carefully drawn freehand. For methods of dimensioning, see Sec. 9.10.

10.4 Runouts. In Sec. 7.17, we have seen the various ways of showing intersecting and tangent surfaces. If fillets and rounds are present, an exact representation is not only difficult but usually unnecessary. Conventional methods should be used in such cases, as shown in Fig. 10–6. In the cases shown from (a) to (c), the points of tangency must be found in the top

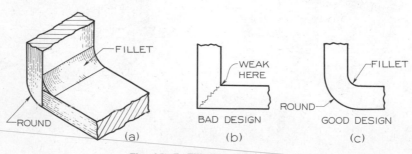

Fig. 10–5. Fillets and Rounds.

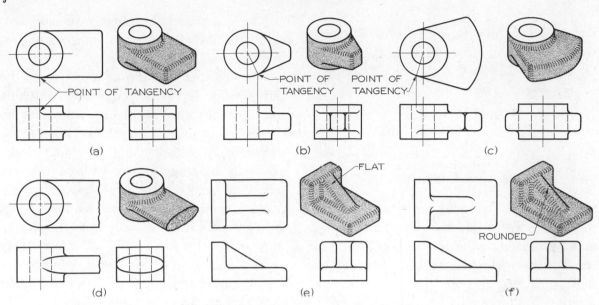

Fig. 10–6. Runouts.

view and projected to the other views where freehand *runouts* are carefully drawn. At (d) the intersecting member is elliptical, and the runouts are turned inward. At (e) and (f) the runouts depend upon the shape of the top of the triangular rib. If runouts are large, as at (d), an irregular curve should be used; otherwise they should be drawn freehand.

10.5 Conventional Edges. It is often necessary to draw lines representing rounded edges when, strictly speaking, no edges exist. For example, the upper top views in Fig. 10–7

would in each case be completely blank as shown, if lines were not used to represent rounded edges, and the drawings would be misleading. Therefore, if an edge has only a small radius, a line should be shown. If the radius is large, as X in (b), no line should be shown. It is best to follow this rule: *Draw lines for rounded edges whenever such lines make the drawing clearer.*

10.6 Forging. A *forging* is produced by hammering heated bars or billets of metal between *dies.* Forging presses, or hammers, are used

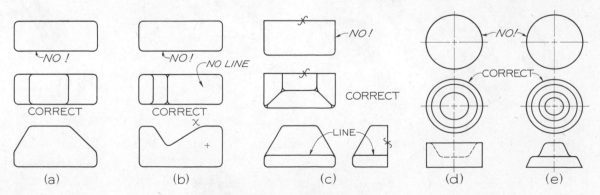

Fig. 10–7. Conventional Edges.

Courtesy Interstate Drop Forge Co.

Fig. 10–8. Forging Dies.

to exert the tremendous pressure that forces the plastic metal into the dies. The pressure may be exerted by impact of heavy blows; the result is then *drop forging*. If the pressure is a slow squeezing action, the result is *press forging*. The forging dies have machined cavities and projections of the shapes desired.

The dies for forging an automobile connecting rod are shown in Fig. 10–8. The several stages of forging are shown in Fig. 10–9. The rough stock, shown at (a), is gradually formed through the shapes shown to (e), where large fins (called *flash*) are formed. These must be trimmed off as shown at (f). The final forging, shown at (g), is ready for machining.

The advantages of forgings over sand castings are that forgings are much stronger, tougher, and less brittle.

The working drawing as it comes to the forge shop may show the finished machine part, in which case the diemaker provides the necessary forging and machining allowances. Separate forging drawings are often made, showing the rough forging with only the dimensions needed in the forge shop. As in sand casting, a draft must be provided for forgings so they will withdraw easily from the dies. In most cases, this is from 5° to 7°, but it may be less under special conditions. Separate machining and forging drawings of a Pan Cleaner Sprocket are shown in Fig. 10–10.

10.7 Die Casting. *Die casting* is the fastest of the casting processes. It is often used where rapid production and economy are essential, as in automobile carburetors, door handles, and radio and television chassis. Die castings are formed by forcing molten metal, usually a zinc alloy, into cavities between metal dies. Die castings are much more accurate than

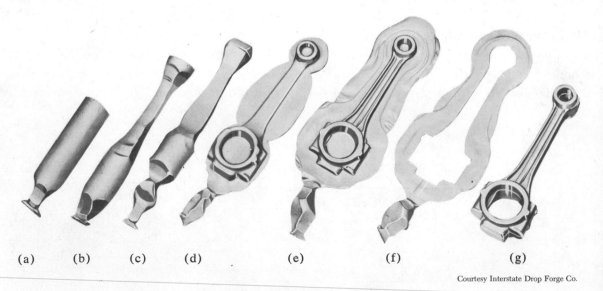

(a)   (b)   (c)   (d)   (e)   (f)   (g)

Courtesy Interstate Drop Forge Co.

Fig. 10–9. Forging of Connecting Rod.

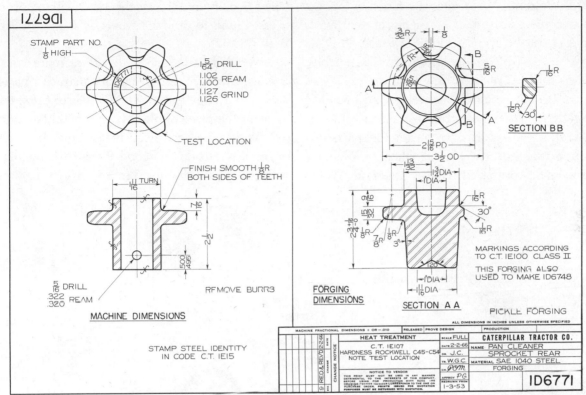

Courtesy Caterpillar Tractor Co.

Fig. 10–10. Machining Drawing and Forging Drawing.

sand castings and usually require little or no machining.

10.8 Stock Forms. Many forms of metal, plastic, wood, or other materials are so often used that their composition and sizes have been standardized and may be obtained readily from the manufacturer in stock sizes. Among these are bars of various shapes, flat stock, and rolled structural shapes, Fig. 10–11. Many of the smaller machine parts, especially screw-machine work (which require the toughness and strength of rolled metal), are machined altogether from these stock forms. Some examples are shown in Fig. 15–11. On drawings of such parts, finish marks (or "finish all over" notes) are usually omitted, as the parts are understood to be finished.

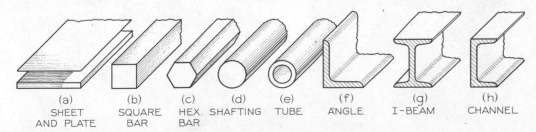

| (a) | (b) | (c) | (d) | (e) | (f) | (g) | (h) |
| SHEET AND PLATE | SQUARE BAR | HEX. BAR | SHAFTING | TUBE | ANGLE | I-BEAM | CHANNEL |

Fig. 10–11. Common Stock Forms.

10.9 Welding. *Welding* is the fusion or joining of two pieces of metal by means of heat, with or without the application of pressure. *Welded* structures are usually built up from stock forms, such as plate, tubing, and angles. These parts are cut to shape, assembled, and welded together. Often heat-treating and machining operations are performed after the parts are joined together. A "Master Chart of Welding Processes," prepared by the American Weld-ing Society and listing the various basic processes used to produce welds, is shown in Fig. 10–12.

Welded structures are often more inexpensive and more satisfactory than cast parts, and to a large extent welded construction has taken the place of older types of construction.

Welding is an extensive science in itself, and it is outside the scope of this text. Welding drawings are discussed in Sec. 15.16.

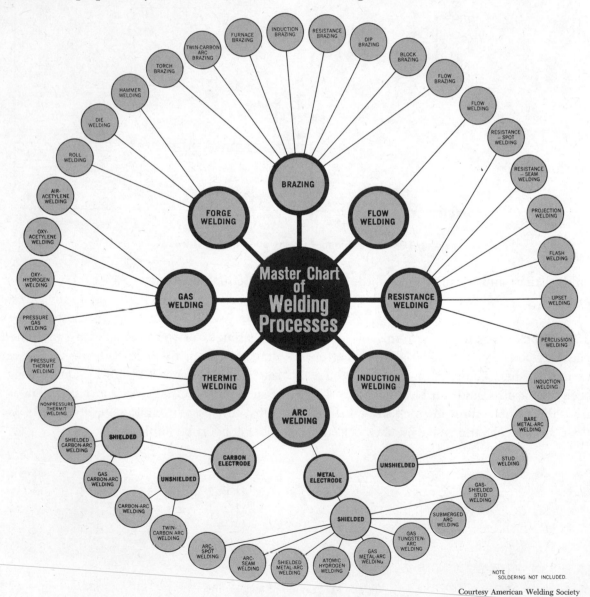

NOTE
SOLDERING NOT INCLUDED.

Courtesy American Welding Society

Fig. 10–12. Master Chart of Welding Processes.

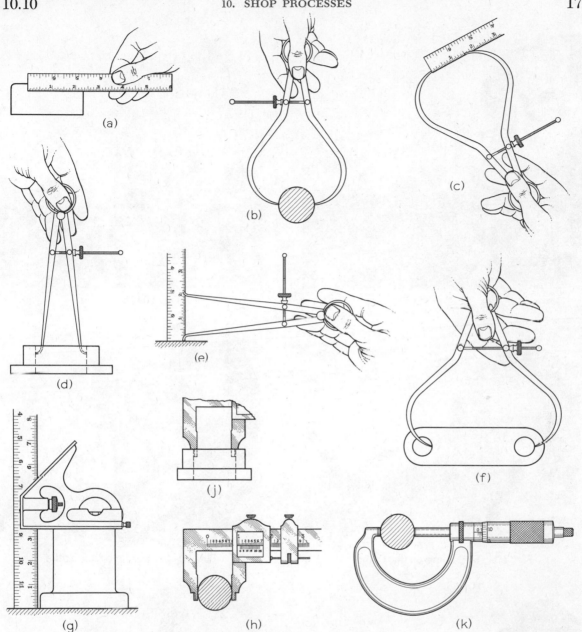

Fig. 10–13. Measurements in the Shop.

10.10 Measurements. The *machinists steel rule,* or *scale,* is the most commonly used measuring tool in the shop, Fig. 10–13 (a). The smallest division is $\frac{1}{64}''$; therefore this scale is used when measurements require no greater accuracy. The *outside spring calipers* are used to check outside diameters, (b), and the measurement is then read off on the steel scale, (c). The *inside spring calipers* are used to measure inside diameters, (d) and (e). The outside calipers can also be used to measure the distance between holes (center to center), as shown at (f). The *combination square* may be used in a variety of ways. One way is shown at (g). The *vernier caliper* can be used to measure outside diameters, (h), or inside diameters,

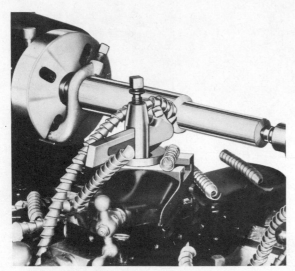

Fig. 10–14. Turning.*

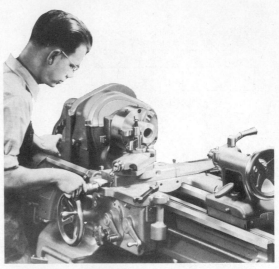

Fig. 10–15. Facing.*

Fig. 10–16. Drilling on the Lathe.*

Fig. 10–17. Boring on the Lathe.*

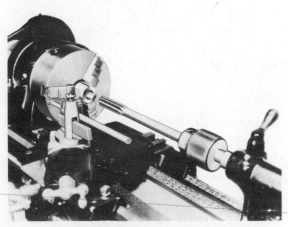

Fig. 10–18. Reaming on the Lathe.*

Figs. 10–19. Knurling on the Lathe.*

* Courtesy South Bend Lathe Works.

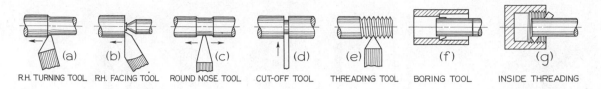

Fig. 10–20. Lathe Operations.

(j). The *micrometer caliper*, (k), is one of the most commonly used precision measuring devices in the shop. By means of *vernier* graduations, measurements may be made to four decimal places, or $\frac{1}{10,000}''$.

As shown in Fig. 10–13 (b), (d), (h), and (k), diameters are easily measured in the shop. Therefore, *diameters* of cylinders, and not radii, should be given on drawings. See Sec. 9.15.

10.11 The Lathe. Cylinders, cones, and other rounded shapes are often machined on a *lathe* (pronounced lāyth). A long piece of stock is usually held between centers, the left end being fastened securely to a *face plate* by means of a "dog," and the right end turning freely, Fig. 10–14. A short piece of stock is usually held in a *chuck* (essentially a revolving vise), Fig. 10–15. The cutting of an external cylindrical surface is called *turning*, Fig. 10–14, and the cutting of a flat surface is called *facing*, Fig. 10–15. Other common operations performed on the lathe include *drilling*, Fig. 10–16, *boring*, Fig. 10–17, and *reaming*, Fig. 10–18. In addition, threads are often cut on the lathe, Fig. 14–3. In all of these, the stock rotates and the cutting tool is fed into or along the stock as required.

Knurling, Fig. 10–19, is also done on the lathe by using a knurling tool in the tool holder and forcing it against the revolving work. The result is a roughened surface composed of crossing diagonal grooves or of parallel grooves lengthwise of the piece. These are common on thumb screws and handles of various kinds

to provide a better grip. Knurls may be fine, medium, or coarse, depending upon the roughness required.

Some of the most common lathe operations performed by a tool bit held in a tool holder or boring bar are shown in Fig. 10–20.

10.12 How Finished Holes Are Made. The *drilling* of a common drilled hole on a *drilling machine*, or *drill press*, is shown in Fig. 10–21. The revolving drill, Fig. 10–22 (a), is fed into the work, which remains stationary. The drill is also used in the lathe, Fig. 10–16, the milling machine, and other machines. The drill does not produce a very accurate hole either in roundness or in straightness. For greater accuracy, the drill is followed by *boring*, Figs.

Courtesy South Bend Lathe Works

Fig. 10–21. Drilling.

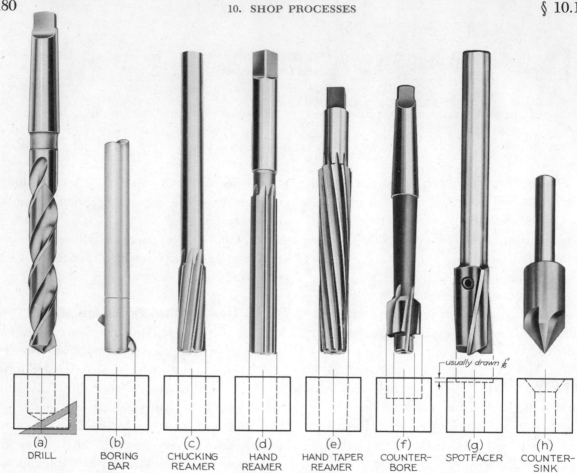

usually drawn $\frac{1}{16}''$

(a)	(b)	(c)	(d)	(e)	(f)	(g)	(h)
DRILL	BORING BAR	CHUCKING REAMER	HAND REAMER	HAND TAPER REAMER	COUNTER-BORE	SPOTFACER	COUNTER-SINK

Fig. 10–22. Types of Holes.*

10–17 and 10–22 (b), or by *reaming*, Figs. 10–18 and 10–22 (c), (d), and (e). Boring enlarges the drilled hole slightly and makes it rounder and straighter. Reaming enlarges the drilled or bored hole slightly and improves the surface quality. Good practice is to drill, bore, and then ream to produce an accurate hole.

Counterboring is the cutting of an enlarged cylindrical portion at the top of a hole, Fig. 10–22 (f), usually to receive the head of a fillister-head or a socket-head screw, Figs. 14–31 to 14–33.

Spotfacing is simply the cutting of a shallow counterbore, usually about $\frac{1}{16}''$ deep or deep enough to get under the "scale" on the rough surface, Fig. 10–22 (g), or to finish off the top of a boss to form a bearing surface. The depth of a spotface is usually not indicated in the note on the drawing, but is left to the shop. It is commonly drawn $\frac{1}{16}''$ deep. A spotface provides an accurate bearing surface for the under side of a bolt or screw head.

Countersinking is the cutting of an enlarged conical portion at one end of a hole, Fig. 10–22 (h), usually to receive the head of a flat-head screw, Figs. 14–31 and 14–33. To simplify drafting, the countersink is usually drawn with a 90° included angle even though the note calls for, say, 80° or 82°.

* (a), (c), and (d) courtesy National Twist Drill & Tool Co.; (e), (f), (g), and (h) courtesy Morse Twist Drill & Machine Co.

Fig. 10-23. The Shaper.

Fig. 10-24. The Milling Machine.

Tapping is the threading of a small hole with one or more *taps*, Fig. 14–18.

10.13 The Shaper. In the *shaper,* Fig. 10–23, the stock is held in a vise while a single-pointed cutting tool, similar to that used in the lathe, cuts as it is forced forward in a straight line past the stationary work. In the figure, the tool is moving forward on a cutting stroke. The work remains stationary as the tool cuts. After this, the tool returns to the starting position to take another cut. Then the work is "fed" or moved slightly to the side so that a fresh portion of the work will be in the path of the tool for the next cut.

10.14 The Milling Machine. In the *milling machine* the work is fastened to the table by means of clamps or a vise, and then fed into a rotating milling cutter, Fig. 10–24. By using cutters of many different shapes, a wide variety of cuts is possible on the milling machine. Some of the most common operations are shown in Fig. 10–25.

10.15 The Grinding Machine. The *grinding machine* is used to remove a small amount of metal to bring the work to a very fine and accurate finish. In *surface grinding,* Fig. 10–26, the work is moved past the revolving grinding wheel. In *cylindrical grinding,* the work revolves slowly on centers and is moved past the grinding wheel. On a drawing, a surface to be ground is indicated with a letter G on the edge view of the surface, or by a note GRIND, Fig. 15–10.

10.16 Broaching. In the *broaching machine,* a long cutting tool called a *broach,* which has a series of teeth that gradually increase in size, is forced through a hole or over a surface to produce a desired shape. Thus, a drilled hole may be changed to triangular, square, hexagonal, or some other shape, Fig. 10–27; or a surface may be machined with a flat *surface broach.* As the broach is forced through or across the work, each succeeding tooth bites deeper and deeper until the final teeth form the required hole or surface.

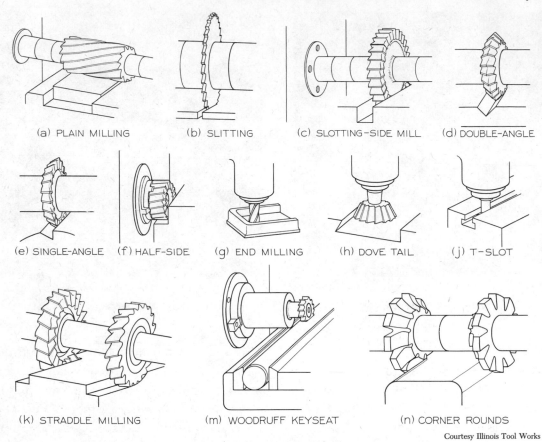

(a) PLAIN MILLING   (b) SLITTING   (c) SLOTTING–SIDE MILL   (d) DOUBLE–ANGLE

(e) SINGLE–ANGLE   (f) HALF–SIDE   (g) END MILLING   (h) DOVE TAIL   (j) T–SLOT

(k) STRADDLE MILLING   (m) WOODRUFF KEYSEAT   (n) CORNER ROUNDS

Courtesy Illinois Tool Works

Fig. 10–25. Milling Machine Operations.

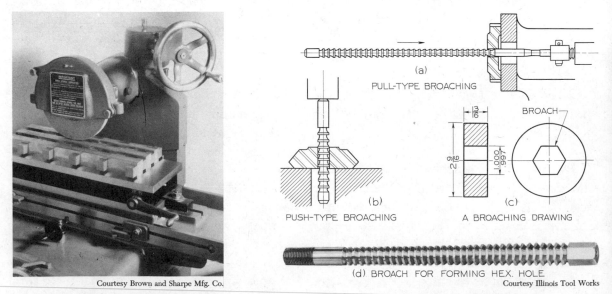

(a) PULL-TYPE BROACHING

(b) PUSH-TYPE BROACHING

(c) A BROACHING DRAWING

(d) BROACH FOR FORMING HEX. HOLE

Courtesy Illinois Tool Works

Courtesy Brown and Sharpe Mfg. Co.

Fig. 10–26. The Grinder.   **Fig. 10–27. Broaching.**

10.17 Heat-Treating. The process of changing the properties of metals by heating and cooling is called *heat-treating. Annealing* and *normalizing* are generally used to soften metal and involve heating followed by slow cooling. *Hardening* requires heating and then rapid cooling (quenching) in oil, water, or other substances. There are many other kinds of heat treatment, such as *tempering, casehardening, carburizing,* etc., but space does not permit their explanation here.

10.18 Shop Notes. The various operations described in this chapter are specified on the drawing in the form of notes. For a complete discussion of notes, see Sec. 9.24. A large variety of notes are illustrated in Figs. 9–29 and 9–30.

10.19 Automatically Programmed Tools. Since the beginning of recorded history, tools have served as an extension of man's capabilities, and as a result, tools of all types have suffered design limitations. As tools were developed and refined, new designs had to be based on the physical limitations of the man who was to use them.

When power was applied to tools, some but not all of the limitations were removed. As tools became more and more complicated, more automatic features were incorporated, each designed to overcome one or more of the limitations imposed by the need for a human operator. But until the advent of *numerical control,* the operator was still a large factor in machine tool design.

Numerical control in a sense moves the operator from his post in front of the machine to a desk from which he issues commands to

* From *Automatically Programmed Tools* by Dr. Shizuo Hori, Scientific Advisor, IIT Research Institute, with permission of the author.

the tool. In so doing, numerical control has achieved a major ambition of tool designers, which has been to permit the machine to do the work and the man to do the thinking. In numerical control, the tool is completely automatic. Each function of the tool is separately driven, and all functions are controlled through the use of a punched tape (somewhat like a miniature player piano roll) in which numerical signals have been cut to correspond to the draftsman's or designer's instructions. The tool does all of its work in response to these signals. A typical numerically controlled machining department is shown in Fig. 10–28.

In manual control, the machine operator usually receives his information for controlling the machine from a blueprint or a template. When machining a part, particularly one with a complex contour, the operator first sets up the raw stock in a jig and then proceeds with a long sequence of simple cuts. Customarily, he makes a rough cut first, and then makes several finer cuts until the desired tolerance is achieved. As each simple cut is finished, the work is repositioned in the jig.

Under numerical control, most of the setups of the work-piece required in hand control are eliminated. The "commands" for the tool are punched into the tape, which is fed into the tool's controller. Two or more axes are synchronized so that the cutter, with the work held in one position, will go through a complex motion. Under numerical control, many cuts can be made to the desired accuracy on the first pass; others require a rough cut and then a finish cut. In either event, the total number of passes is much less than that needed by hand control.

Numerical control has proved useful with machine tools of a basic nature such as drill presses and boring mills, but its greatest use has been with continuous-path tools, where the tool is continuously in contact with the work and requires continuous control. However, in the case of complex parts to be cut on a continuous-path tool, the preparation of the punched tape is a very tedious and expensive process. Since the tool requires continuous control, a steady stream of signals must be fed to it. A programmer, in preparing a tape for such a part, has to punch each of these individual signals into the tape—and on some pieces, hundreds of thousands of signals might be required. As a result, the saving of time and money resulting from the use of numerical control is in some instances dissipated in the cost of preparing the tape.

To solve the problem of tape preparation, research workers went to digital computers. They developed methods by which computers instead of programmers could put together the detailed sequence of signals needed to actuate a machine tool. A number of different systems to do this have been developed, the most advanced of which is *APT*.

The APT (Automatically Programmed Tools) system originated at the Massachusetts Institute of Technology in 1952. Since then, the system has been modified and improved several times, and recently the IIT Research Institute was selected to assume responsibility for the existing APT system, Fig. 10–29, and

Courtesy Sundstrand Machine Tool Division, Belvidere, Ill.

Fig. 10–28. Numerically Controlled Machining Department.

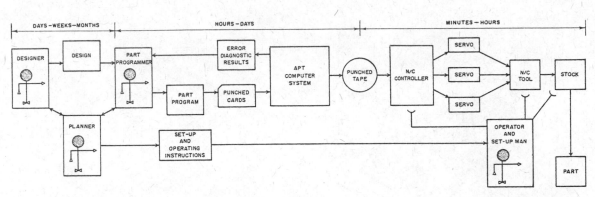

Fig. 10–29. The APT System.

to direct the future course of the long-range developmental program.

The significant step provided by APT in the evolution of machine tool control is that it supplants numerical control with *symbolic control*. The APT language contains over 250 word-symbols through which instructions are given to a computer. Since the APT system uses a language rather than signals, the programmer can communicate with his machine tool in much the same way he communicates with his fellow workers. The computer serves as a translator, changing the word-symbols given to it by the programmer into numerical signal commands. Each word-symbol causes the computer to punch into the tape as many numerical signal commands as are required

to carry out the action described by the programmer. Then the punched tape is fed into the controller of the tool. Research is currently being carried on to adapt the APT system to various other manufacturing and engineering methods.

10.20 Problems. The following problems are given to provide practice in drawing objects that have rough surfaces and machined surfaces, and to give experience with dimensioning and shop notes. Add finish marks on all machined surfaces.

Most problems can be drawn full size on Layout C (Appendix, page 474). If a reduced scale or a larger sheet, or both, are needed, it is so noted on each problem.

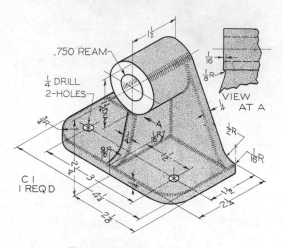

Fig. 10–30. Shaft Bracket.*

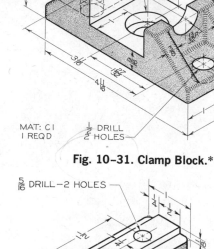

Fig. 10–31. Clamp Block.*

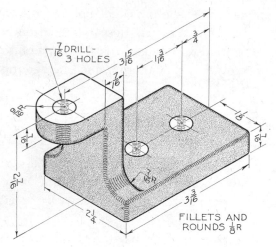

Fig. 10–32. Swivel.*

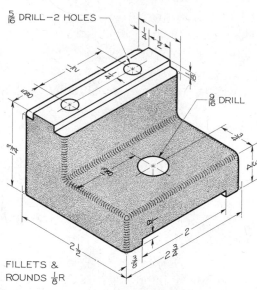

Fig. 10–33. Control Base.*

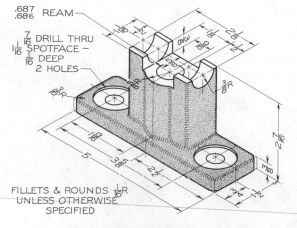

Fig. 10–34. Plunger Bracket.*

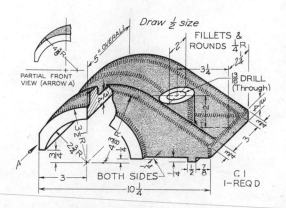

Fig. 10–35. Tailstock Cap.*

* Use Layout C (Appendix, page 474).

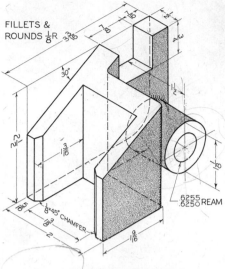

Fig. 10-36. Double Shifter Yoke.*

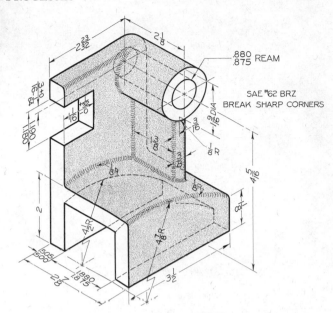

Fig. 10-37. Double Shifter.*

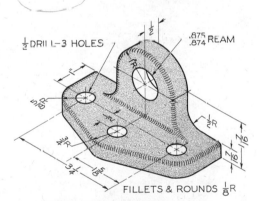

Fig. 10-38. RH Pipe Bracket.*

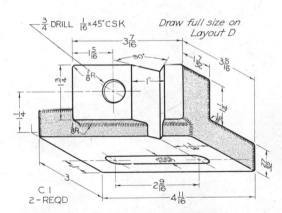

Fig. 10-39. Control Bracket.

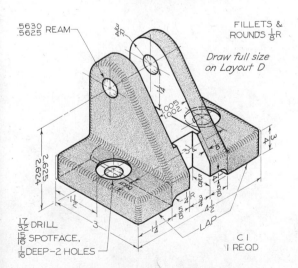

Fig. 10-40. Shifter Link Bracket.

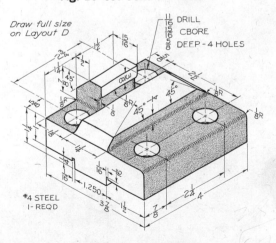

Fig. 10-41. Rest Block.

* Use Layout C (Appendix, page 474).

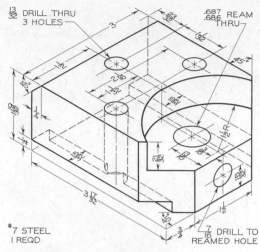

Fig. 10-42. Lock Bolt Block.*

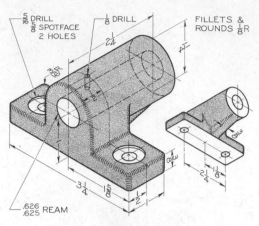

Fig. 10-43. Gear Bracket, RH.*

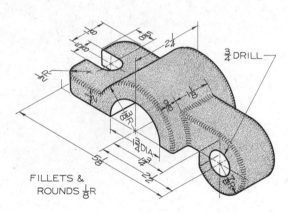

Fig. 10-44. Gum Roll Cap.*

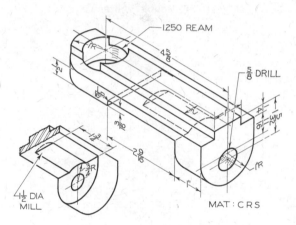

Fig. 10-45. Tie Bracket.*

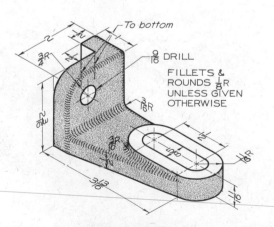

Fig. 10-46. Adjustable Arm.*

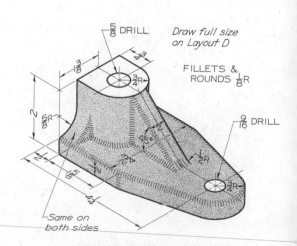

Fig. 10-47. Support.

* Use Layout C (Appendix, page 474).

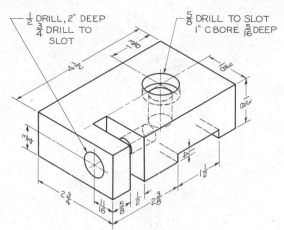

Fig. 10–48. Drive Holder.*

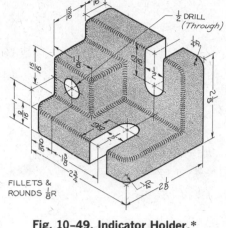

Fig. 10–49. Indicator Holder.*

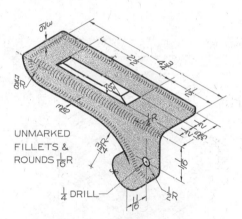

Fig. 10–50. Grinder Guide.*

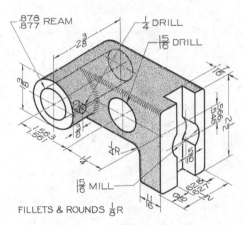

Fig. 10–51. Shifter Block.*

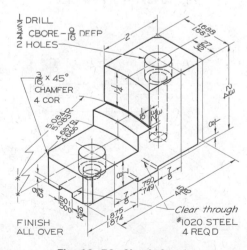

Fig. 10–52. Chuck Jaw.*

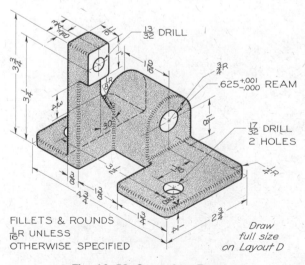

Fig. 10–53. Secondary Base.

* Use Layout C (Appendix, page 474).

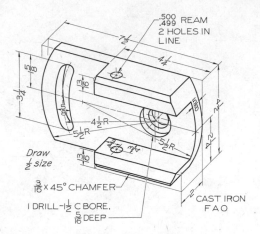

Fig. 10–54. Clapper Box.*

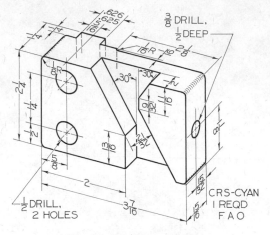

Fig. 10–55. Control Dog.*

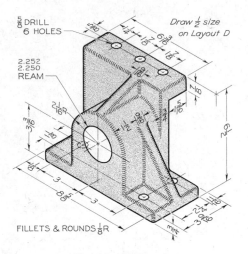

Fig. 10–56. Bracket.

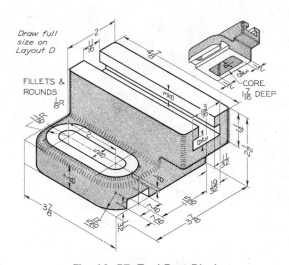

Fig. 10–57. Tool Post Block.

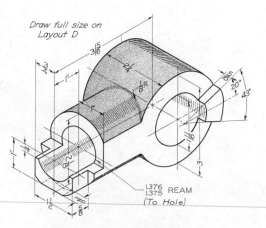

Fig. 10–58. Adjustable Head.

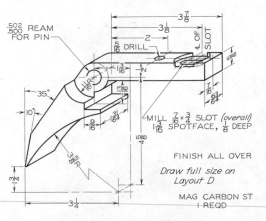

Fig. 10–59. Chip Breaker Shoe.

* Use Layout C (Appendix, page 474).

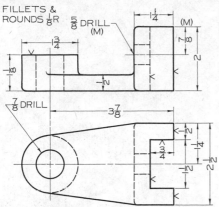

Fig. 10–60. Upper Guide.*

Given: Front and bottom views.
Req'd: Front, top, and right-side views.

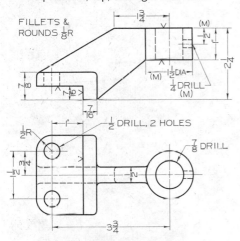

Fig. 10–62. Guide Bearing.*

Given: Front and bottom views.
Req'd: Front, top, and right-side views.

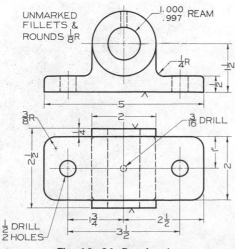

Fig. 10–64. Bearing.*

Given: Front and bottom views.
Req'd: Front, top, and right-side views.

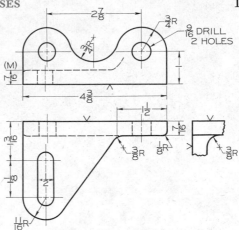

Fig. 10–61. Bracket.*

Given: Front and top views.
Req'd: Front to be top view; new front
and right-side views.

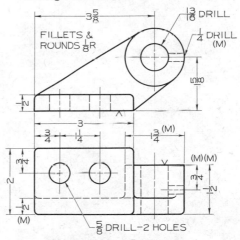

Fig. 10–63. LH Bearing.*

Given: Front and bottom views.
Req'd: Front, top, and right-side views.

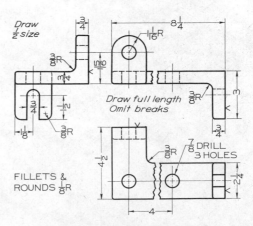

Fig. 10–65. Holder Strap.*

Given: Front, bottom, and left-side views.
Req'd: Front, top, and right-side views.

* Use Layout C (Appendix, page 474). Move dimensions (M) to better locations.

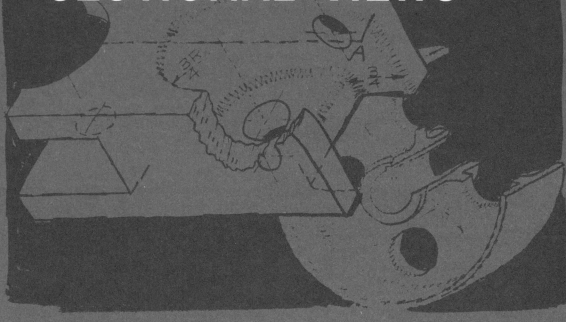

CHAPTER 11

SECTIONAL VIEWS

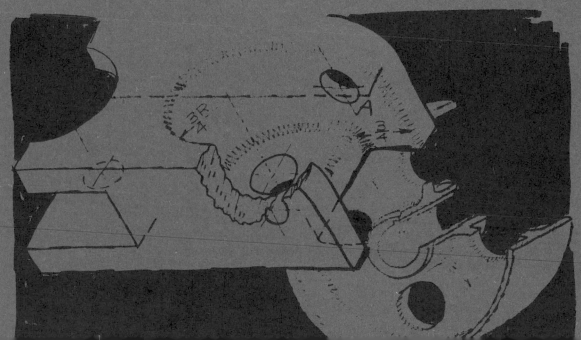

11.1 Full Sections. The basic method of representing an object by its views has been described in Chapter 6. Whenever necessary, hidden parts of the object have been shown by means of hidden lines, Secs. 6.4 and 6.15. However, if an object has internal shapes that would require hidden lines, particularly if it has a complicated interior, as an automobile engine block does, we can obtain a clearer picture of the internal shapes by drawing the object cut apart.

For example, imagine an object to be cut in half, as shown in Fig. 11–1 (a). The half nearest the observer is then pulled away, as shown at (b), exposing the back half. Here we see an imaginary *cutting plane* that might have been passed through the object instead of the hack saw blade. At (c) the back half of the object is shown in front-view and top-view positions. The corresponding two-view drawing with the front view in section is shown at (d). The top view shows the entire object with the cutting plane shown edgewise. The front view shows the back half of the object only, with the front half removed. The sectional view is called a *full section,* because the cutting plane has passed fully through the object. The edge view of the cutting plane is represented in the top view by a *cutting-plane line,* Fig. 3–11; the arrowheads at the ends show the direction of sight for the section. The parallel section lines indicate the surfaces cut by the cutting plane.

The cutting-plane line is shown in Fig. 11–1

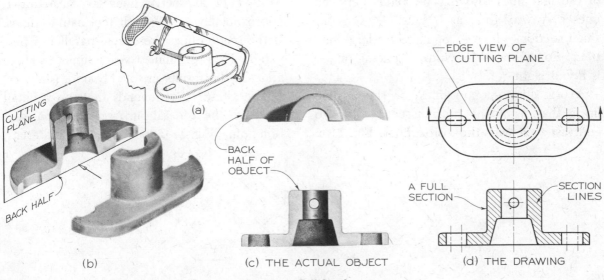

Fig. 11–1. Full Section.

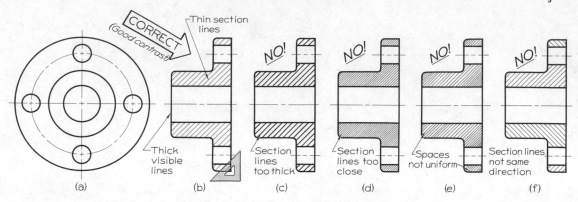

Fig. 11–2. Section-Lining Technique.

(d) for illustration. In practice, it is omitted in cases such as this where the location of the cutting plane is obvious.

11.2 Section Lining. Section lines should be drawn at an angle of 45°, as shown in Fig. 11–2 (b), unless there is an advantage in using a different angle. Use a medium-grade pencil, such as a 2H, with a *sharp* conical point, Fig. 3–8 (c) and (d), and make section lines dark, and *very thin,* to contrast well with the thick visible lines. *Space the section lines by eye,* keeping the spacing uniform by continually looking back and comparing with the spacing of the first lines drawn. The spaces between lines may vary from as little as $\frac{1}{32}''$ for very small sections up to $\frac{1}{8}''$ or more for large sections. For average drawings, a spacing of $\frac{3}{32}''$ is about right.

Avoid thick section lines (or thin visible lines) that would not permit section lines to contrast well with the visible lines, Fig. 11–2

(c). In general, space the lines well apart as at (b), and not crowded closely as at (d). Keep the spacing uniform, and especially avoid the tendency to crowd the lines closer together as the section lining approaches a corner or a small area, (e). All sectioned areas in a view of a single piece must be sectioned in the same direction, as at (b), and not as at (f). Section lining in different directions is understood to indicate entirely different parts, as in assembly drawing, Sec. 15.13.

The direction of visible outlines around a sectioned area sometimes makes it necessary to change the direction of section lines, Fig. 11–3. At (a) the section lines have been drawn approximately at 45° with the main outlines. If the section lines are drawn parallel or perpendicular to prominent visible lines, as at (b) and (c), the results are highly unsatisfactory.

The USASI recommends that, on a detail drawing, the general-purpose (cast-iron) section lining, Fig. 11–2, be used for all materials,

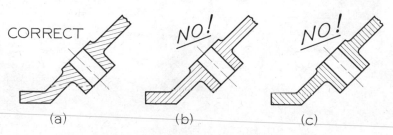

Fig. 11–3. Direction of Section Lining.

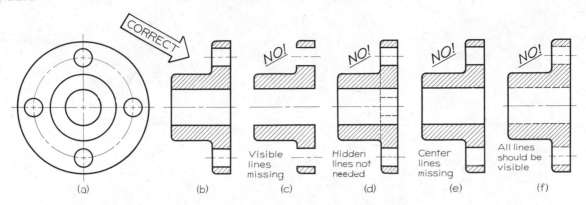

Fig. 11–4. Visible, Hidden, and Center Lines.

since the exact material specifications will generally be more detailed than by section-lining symbol alone, and thus will be given in a note or title strip. Symbolic section lining, Fig. 15–18, may be used on assembly drawings in situations where it may be desirable to show a distinction between different materials; otherwise, the cast-iron symbol would be used for all materials. For assembly sections, see Sec. 15.13.

11.3 Visible, Hidden, and Center Lines. As a rule, visible edges and contours behind the cutting plane should be shown, Fig. 11–4 (b). If all visible lines are not shown, a disconnected and confusing drawing results, as shown at (c). Since sections are drawn to replace hidden-line representation, hidden lines

are unnecessary and tend to confuse the drawing, (d). Hidden lines are always omitted except in special cases that require them for clearness. Center lines should not be omitted as at (e). No sectioned area can ever be bounded by hidden lines as at (f), as the edges of the cut surfaces are always visible.

11.4 Half Sections. If the cutting plane is passed only halfway through an object, Fig. 11–5 (a), and then the quarter of the object in front of the cutting plane is removed, (b), the resulting section is a half section, (c). Thus, a half section has the advantage of showing in a single view both inside and outside shapes; hence, its usefulness is limited largely to symmetrical objects. Hidden lines are usually omitted from the unsectioned half, as shown

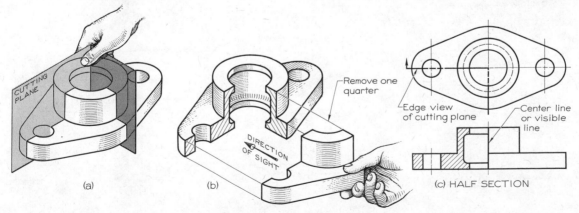

Fig. 11–5. Half Section.

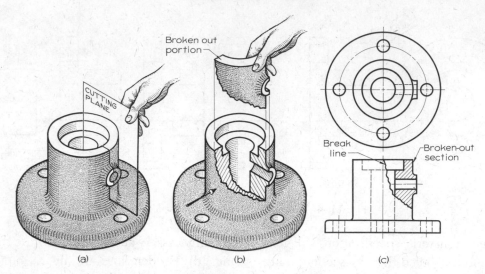

Fig. 11–6. Broken-Out Section.

at (c), unless they are needed for clearness or for dimensioning of the inside shapes. The greatest usefulness of the half section is in assembly drawing, Fig. 15–17 (b) and (c), where it is often desirable to show both inside and outside construction of a symmetrical assembly in a single view.

Notice, in Fig. 11–5 (c), how the cutting-plane line is shown, with one arrow to indicate the direction of sight. Usually, the location of the cutting plane is obvious, and it is not necessary to show the cutting-plane line. It is shown here only for illustration. The line separating the sectioned half from the unsec-

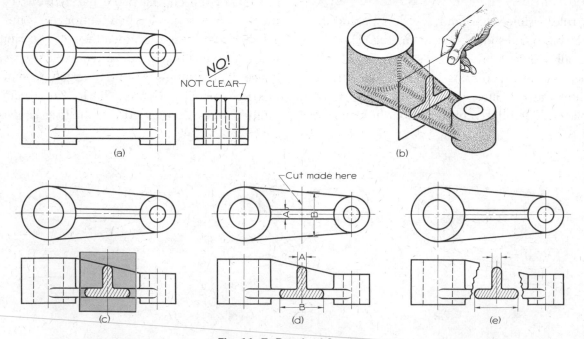

Fig. 11–7. Revolved Section.

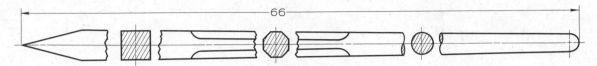

Fig. 11–8. Revolved Sections.

tioned half may be either a center line or a visible line.

11.5 Broken-Out Sections. Often only a small part of a view needs to be sectioned in order to show some detail of inside construction. For example, Fig. 11–6 (a), an imaginary cutting plane is passed through the part to be sectioned. At (b) the plane and the broken-out portion of the object in front of it are removed. At (c) is shown the broken-out section, limited by a freehand break line, Fig. 3–11. The cutting-plane line need not be shown in the top view, as the location of the cut is obvious.

11.6 Revolved Sections. In Fig. 11–7 (a) are shown three views of a Lever Arm in which the right-side view shows the T-shape of the central portion, but the view is very confusing because of the number of lines. If a cutting plane is passed at right angles through the T-shaped central portion, as shown at (b), and then revolved into position as shown at (c), the result is a *revolved section*, (d). Note that distances A and B on the section are taken from A and B in the top view. The revolved section shows the T-shape clearly.

If desired, the view may be broken away on each side of a revolved section, as shown at (e). This often makes the section stand out better and provides clearer dimensioning, as shown.

Revolved sections applied to a Lining Bar are shown in Fig. 11–8. In this case, breaks were used not only to make the sections stand out, but also to shorten the object so that the drawing would fit on the paper. Note that the actual full length is indicated by the 66″ dimension. For a further discussion of breaks, see Sec. 11.12.

In Fig. 11–9, a correct revolved section for an I-shaped arm is shown at (a), and some common errors are shown at (b), (c), and (d).

11.7 Removed Sections. A *removed section* is a section that is moved from its normal position to some more convenient position on the sheet, Fig. 11–10. In such cases, the removed section may be drawn to a larger scale if desired. In this way some small detail can be magnified and dimensioned more clearly than at the scale of the main drawing. We need not show all visible lines behind each cutting plane, but only those near the section and

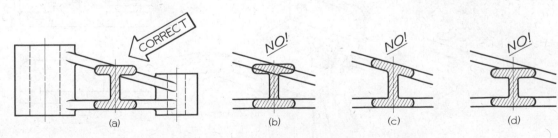

Fig. 11–9. Correct and Incorrect Revolved Sections.

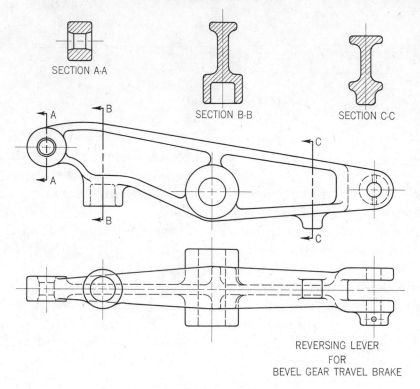

Fig. 11–10. Removed Sections.

needed for clearness. For example, in Section B–B, the back rim of the hole is shown, but the rest of the object to the left of the cutting plane is not shown. Similarly, in Section C–C, all visible lines beyond the cutting plane are omitted.

Since a removed section may be drawn in any convenient place on the sheet, it is necessary to label each cutting plane by means of capital letters at the ends, as A–A, B–B, etc., and to place a note under each section, as SECTION A–A, SECTION B–B, and so on.

A removed section should not be rotated on the paper; that is, it should always be drawn with its lines parallel to what they would be in the normal position.

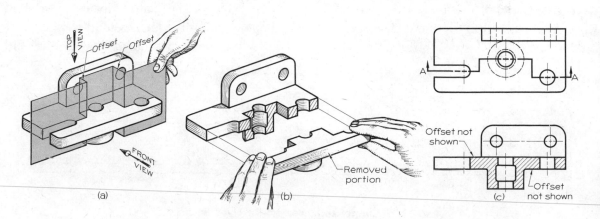

Fig. 11–11. Offset Section.

11.8 Offset Sections. In order to include in a single section several features of an object that are not in a straight line, the cutting plane may be bent or "offset" so as to pass through these features. Such a section is called an *offset section.* For example, in Fig. 11–11 (a), it is desired to include in the section the slot at the left end, the central hole, and the hole at the right end; the cutting plane is shown with offsets to include these. The front portion of the object is then removed, as shown at (b), and the offset section is shown at (c). The edge view of the bent cutting plane is shown by the cutting-plane line A–A in the top view. Note that the *bends, or offsets, are not shown in the sectional view.*

11.9 Ribs and Spokes in Section. A *rib* or *web* is a thin flat part of an object, used for bracing or adding strength, as shown in Fig. 11–12 (a). If a cutting plane, A–A, is passed crosswise through a rib, (b), the rib will be section-lined as shown in the top view at (c). If a cutting plane B–B is passed "flatwise" through the rib, (b), the rib should not be section-lined, but left entirely blank as at (d). The reason for this is that if the rib is section-lined, as at (e), there is a false impression of thickness or solidity.

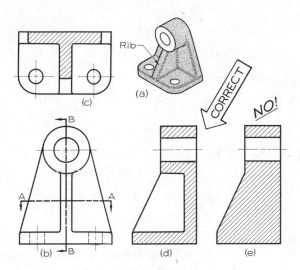

Fig. 11–12. Ribs in Section.

In Fig. 11–13 (a) and (b) are shown a front view and a section of a wheel that has a solid web between the hub and rim. The web is sectioned because the cutting plane cuts crosswise through it, and the section gives a correct appearance of solidity. A similar wheel is shown at (c), but it has spokes instead of a solid web. In the correct section, (d), the *spokes are not section-lined.* The reason for this is that if the spokes are section-lined, as at (e), the section gives a false appearance of solidity and not the open effect of spokes. Note that (e) and (b) are exactly alike.

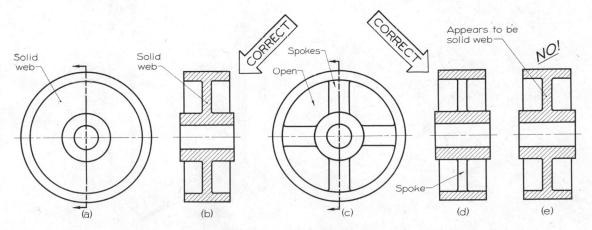

Fig. 11–13. Solid Web and Spokes in Section.

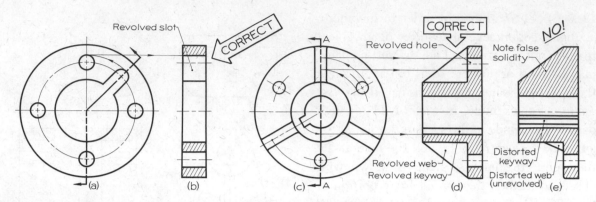

Fig. 11–14. Revolved Features.

11.10 Revolved Features. Clearness can often be improved, or a section may include more information, if certain features are sectioned in a revolved position. For example, in Fig. 11–14 (a), the cutting plane can be bent so as to pass through the slot; the slot is then revolved to the upright position and projected across to the section, as shown at (b).

At (c) is shown a flange with three ribs, three holes, and a keyway. The straight cutting plane A–A includes only one rib and one hole and not the keyway. However, we can revolve a rib, a hole, and the keyway into the cutting plane as shown, and then project across to the section, as at (d). The incorrect section is shown at (e), where the upper rib is section-lined, and the various features are not revolved. The view is misleading, does not give enough information, and the lower rib and the keyway are shown in unsatisfactory and distorted positions. Note that features may be revolved into the straight cutting plane, or the plane may be bent to pass through these features and then revolved.

When a wheel has an odd number of spokes, such as three or five, Fig. 11–15 (a), a straight

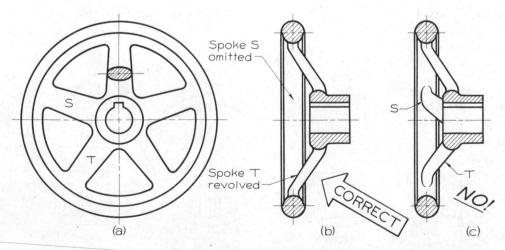

Fig. 11–15. Spokes in Section.

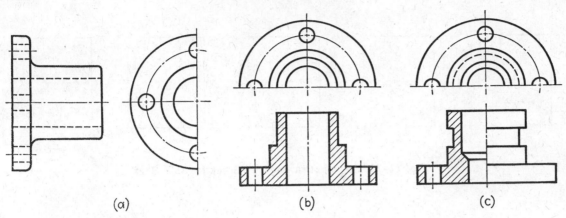

Fig. 11–16. Half Views.

cutting plane will produce a confusing section, as shown at (c), in which spokes S and T are distorted. The correct section, (b), is made by revolving spoke T into the imaginary cutting plane, and eliminating spoke S from the section.

11.11 Half Views. If space is limited, a half view may be drawn of a symmetrical object, Fig. 11–16. Note that the "near half" of the circular view is drawn at (a), while at (b) and (c) the "far half" is shown in each case. See also Sec. 7.19.

11.12 Conventional Breaks. Long objects, such as a garden rake, Fig. 11–17 (a), may appear too small on the drawing or require a very large sheet of paper. A larger scale can easily be used if the long portion is "broken," and a considerable length removed, (b). The broken ends are drawn in a standard way and are called *conventional breaks*. Parts thus broken must have the same section throughout or be tapered uniformly. Note that the full-length dimension at (b) shows the actual length. The steps in drawing the "S" break used in Fig. 11–17 are shown in Fig. 11–18.

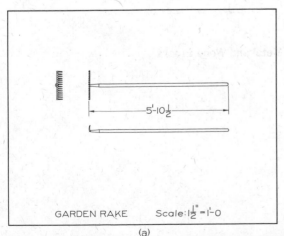

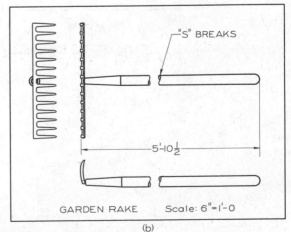

Fig. 11–17. Use of Conventional Breaks.

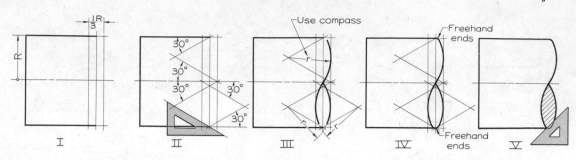

Fig. 11–18. Steps in Drawing "S" Break for Solid Shaft.

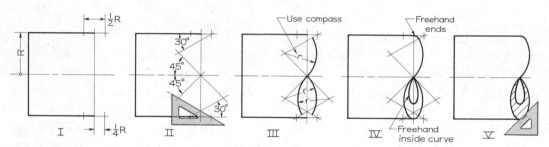

Fig. 11–19. Steps in Drawing "S" Break for Tubing.

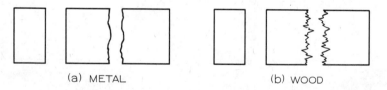

(a) METAL   (b) WOOD

Fig. 11–20. Rectangular Metal and Wood Breaks.

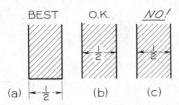

Fig. 11–21. Dimensions and Section Lines.

The steps in drawing the "S" break for tubing are shown in Fig. 11–19. The methods shown in Figs. 11–18 and 11–19 are to be used for diameters of approximately 1″ or over where the compass can easily be used. For smaller diameters the "S" breaks should be carefully drawn freehand.

Ordinary breaks for metal and wood are shown in Fig. 11–20.

11.13 Dimensions across a Section. Dimensions for a section should be placed outside the section if possible, Fig. 11–21 (a). Where it is necessary to have a dimension on a sectioned area, an opening in the section lining should be left for the dimension figure, as at (b). Dimensions should never be lettered over section lining, as shown at (c).

11.14 Sectioning Problems. The following problems are intended to be drawn with instruments or freehand, as required by the instructor. All are to be drawn on Layout C (Appendix, page 474), except Figs. 11–35, 11–38, and 11–39, which require larger paper. All problems are to be dimensioned unless otherwise assigned by the instructor.

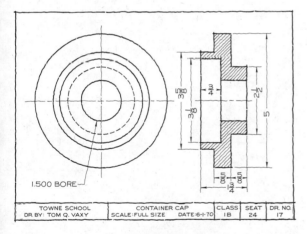

Fig. 11–22. Sketch.

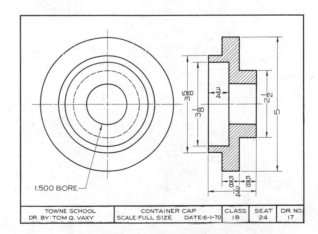

Fig. 11–23. Mechanical Drawing.

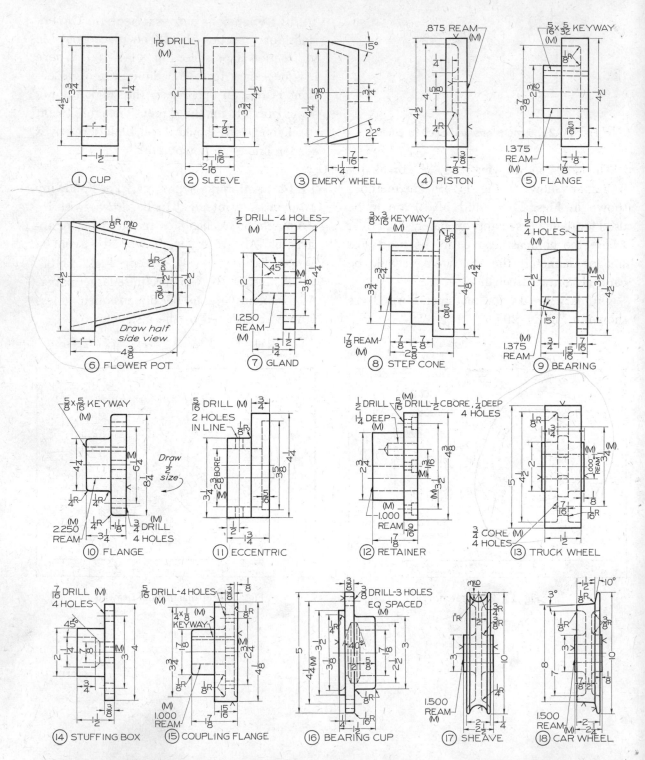

Fig. 11–24. Sectioning Problems. Using Layout C (Appendix, page 474), make sketch as shown in Fig. 11–22 or mechanical drawing as shown in Fig. 11–23 of problem assigned from above. In each case draw a circular view and a full or half section as assigned. Vertical dimensions are diameters. Dimensions marked (M) should be moved to circular view. Holes are equally spaced. Give part names in title strips.

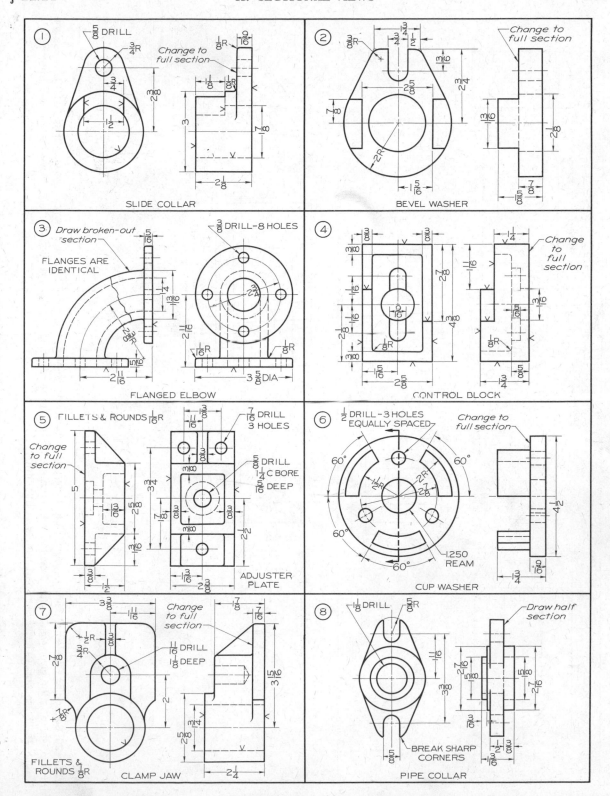

Fig. 11–25. Sectioning Problems. Using Layout C (Appendix, page 474), make sketch, Fig. 11–22, or mechanical drawing, Fig. 11–23. Section as indicated. Omit instructional notes. Give part names in title strips.

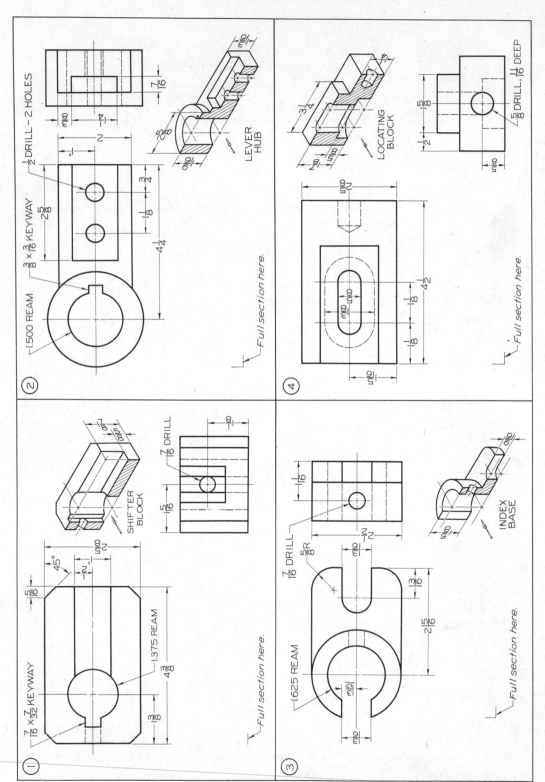

Fig. 11-26. Sectioning Problems. Using Layout C (Appendix, page 474), make sketch, Fig. 11-23, of assigned problem, showing the given views plus a section as indicated. Omit pictorial drawings and instructional notes. Move dimensions from pictorial drawings to the sectional views. Give part names in title strips.

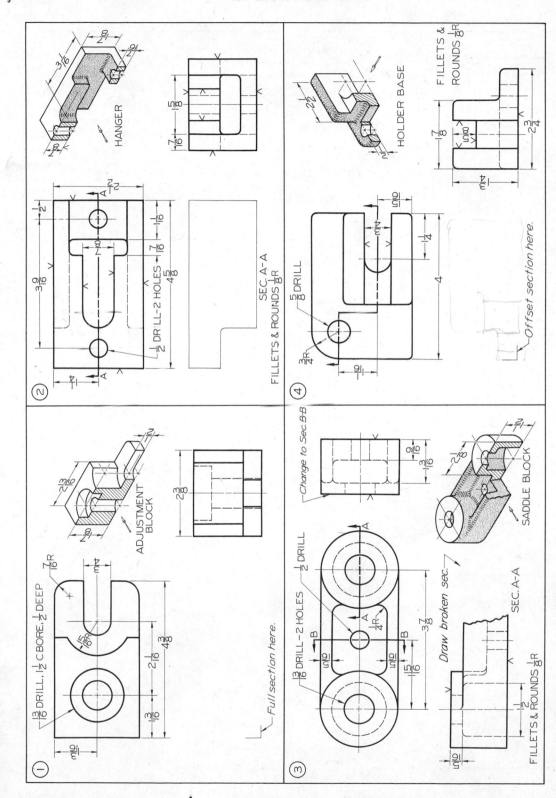

Fig. 11-27. Sectioning Problems. Using Layout C (Appendix, page 474), make sketch, Fig. 11-22, or mechanical drawing, Fig. 11-23, of assigned problem, showing the given views plus a section as indicated. Omit pictorial drawings and instructional notes. Move dimensions from pictorial drawings to the sectional views. Give part names in title strips.

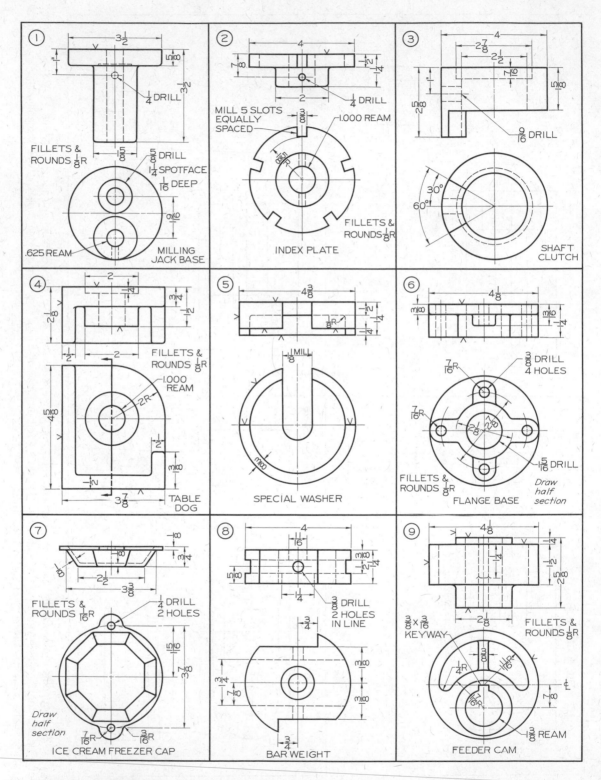

Fig. 11-28. Sectioning Problems. Using Layout C (Appendix, page 474), make a sketch, Fig. 11-22, or mechanical drawing, Fig. 11-23. Omit given top view, draw front view as shown, and draw side view in full section, except in Problems 6 and 7, which require half sections. Omit instructional notes. Give part names in title strips. Move dimensions to new locations as necessary.

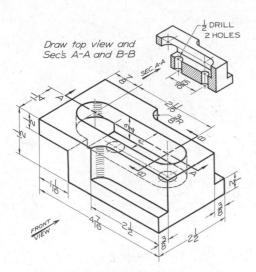

Draw top view and
Sec's. A-A and B-B

Fig. 11-29. Guard Block (Layout C).

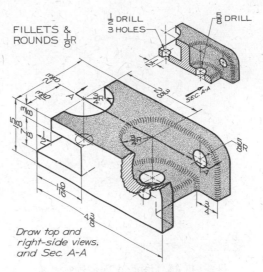

FILLETS &
ROUNDS $\frac{1}{8}$R

Draw top and
right-side views,
and Sec. A-A

Fig. 11-30. Pivot Base (Layout C).

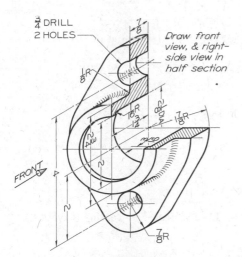

$\frac{3}{4}$ DRILL
2 HOLES

Draw front
view, & right-
side view in
half section

Fig. 11-31. Packing Gland (Layout C).

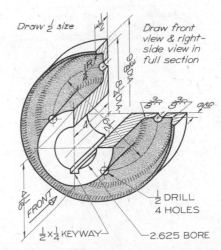

Draw $\frac{1}{2}$ size

Draw front
view & right-
side view in
full section

$\frac{1}{2}$ DRILL
4 HOLES

$\frac{1}{2} \times \frac{1}{4}$ KEYWAY

2.625 BORE

Fig. 11-32. Flange (Layout C).

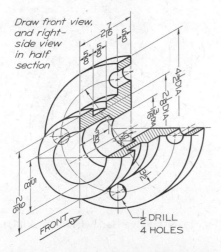

Draw front view,
and right-
side view
in half
section

$\frac{1}{2}$ DRILL
4 HOLES

Fig. 11-33. Stuffing Box (Layout C).

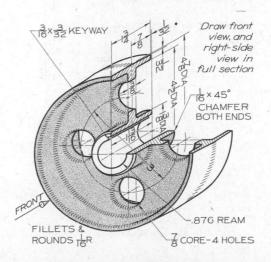

$\frac{3}{16} \times \frac{3}{32}$ KEYWAY

Draw front
view, and
right-side
view in
full section

$\frac{1}{16} \times 45°$
CHAMFER
BOTH ENDS

.876 REAM

$\frac{7}{8}$ CORE-4 HOLES

FILLETS &
ROUNDS $\frac{1}{16}$R

Fig. 11-34. Pulley (Layout C).

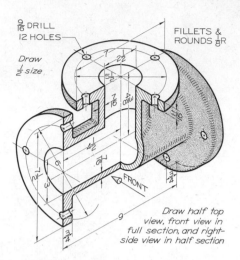

Fig. 11–35. Flanged Tee (Layout D).

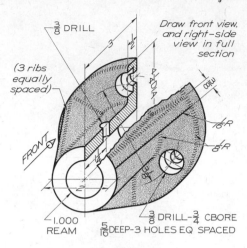

Fig. 11–36. Bearing (Layout C).

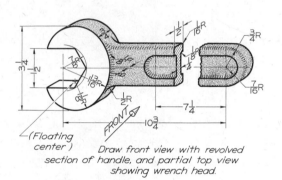

Fig. 11–37. Wrench (Layout C).

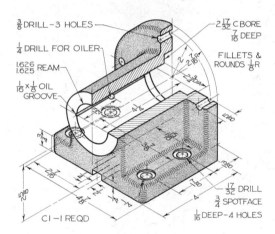

Fig. 11–38. Head-End Bearing (Layout E).

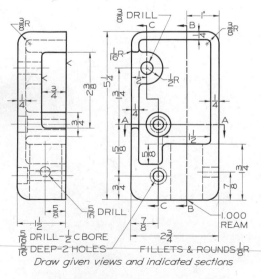

Fig. 11–39. Clamp Guard (Layout D).

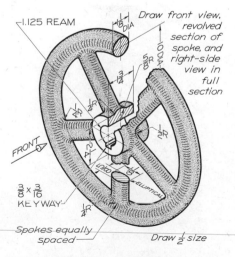

Fig. 11–40. Valve Handwheel (Layout C).

CHAPTER 12

AUXILIARY VIEWS

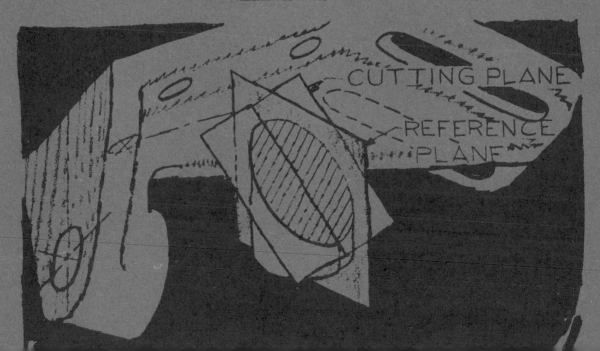

12.1 Inclined Surfaces. Up to this point we have considered the various *regular views* of objects. These views, projected upon the

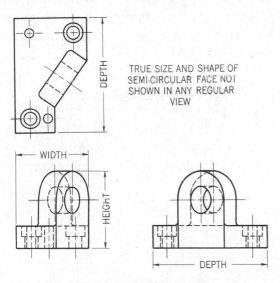

TRUE SIZE AND SHAPE OF SEMI-CIRCULAR FACE NOT SHOWN IN ANY REGULAR VIEW

Fig. 12–1. Regular Views.

planes of the "glass box," Fig. 6–5, were enough to describe clearly the shape of the object.

However, if the object has an inclined surface that needs to be shown true size, the regular views are not enough. For example, Fig. 12–1, the true size and shape of the semicircular face of the angle bracket is not shown in any of the views. Since the inclined surface is not parallel to any plane of the glass box, the circles project as ellipses and the result is not only confusing but difficult to draw.

12.2 The Auxiliary View. In order to obtain a true-size view of the inclined face, we must view the object at right angles to that face, Fig. 12–2 (a), through a special inclined *auxiliary plane* parallel to it. A view obtained in this manner is an *auxiliary view*. In this case, the

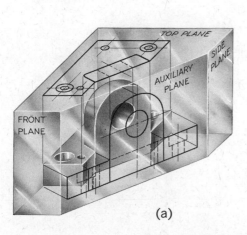

(a)

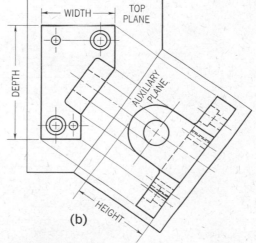

(b)

Fig. 12–2. The Auxiliary Plane.

213

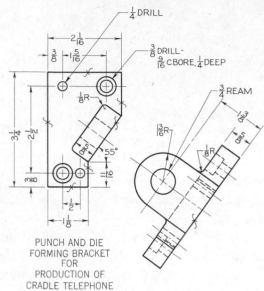

PUNCH AND DIE
FORMING BRACKET
FOR
PRODUCTION OF
CRADLE TELEPHONE

Fig. 12–3. Minimum Required Views.

auxiliary plane is perpendicular to the top plane of the glass box and hinged to it. When this auxiliary plane is unfolded, Fig. 12–2 (b),

the auxiliary view is shown in its correct position on the drawing. Note that the true size and shape of the inclined face are shown in the auxiliary view. Actually, only these two views are needed to describe the shape of this object. The complete auxiliary-view drawing is shown in Fig. 12–3. We know it is complete because all dimensions can be shown on these views, and the object can be built in the shop.

12.3 The Three Auxiliary Views. The three types of ordinary auxiliary views are the *depth auxiliary*, *height auxiliary*, and *width auxiliary*, Fig. 12–4. They are named according to the principal dimensions of the object shown in the auxiliary view, and are also frequently referred to as *primary auxiliary views*.

Thus, the auxiliary view in Fig. 12–3 is a height auxiliary view because the auxiliary view shows the principal dimension, *height*. The auxiliary plane is always hinged to the

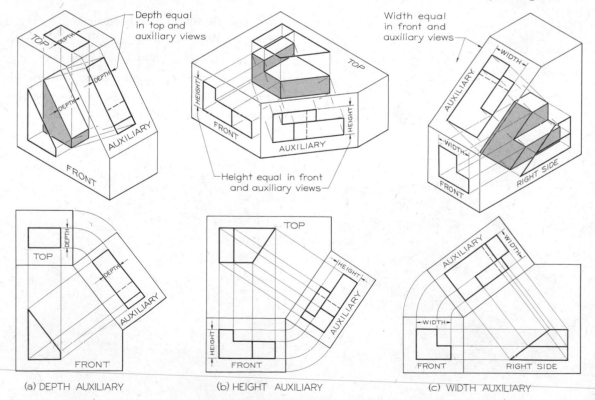

(a) DEPTH AUXILIARY   (b) HEIGHT AUXILIARY   (c) WIDTH AUXILIARY

Fig. 12–4. The Three Auxiliary Views.

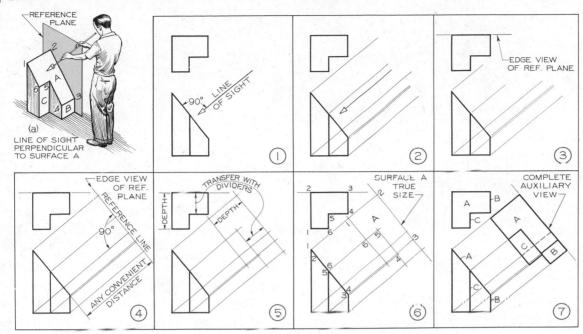

Fig. 12-5. Steps in Drawing a Depth Auxiliary View.

regular plane to which it is perpendicular. Note that in each auxiliary view, one main dimension is projected directly from a regular view, and one is transferred with dividers (or scale) from another regular view.

12.4 To Draw a Depth Auxiliary View. The
steps in drawing a simple front auxiliary view are shown in Fig. 12-5. As shown at (a), the line of sight is assumed perpendicular to surface A, in order to show the true size and shape of that surface. The steps are:

1. Draw arrow perpendicular to inclined surface A in the front view.

2. Draw light projection lines from all points of the object parallel to the arrow. Slide the triangle on the T-square, as shown in Fig. 12-6 (a).

3. Assume a *reference plane* to contain the back face of the object. Draw *reference line* in top view. This is the edge view of the reference plane.

4. Draw reference line (edge view of reference plane) perpendicular to projection lines.

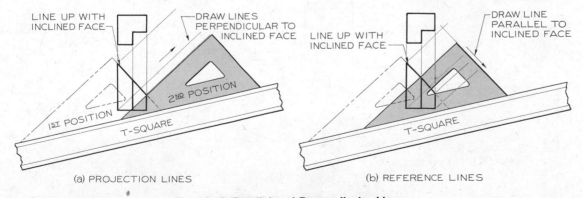

Fig. 12-6. Parallel and Perpendicular Lines.

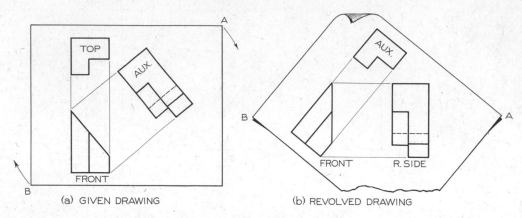

Fig. 12–7. Revolving a Drawing.

Slide triangle on T-square as shown in Fig. 12–6 (b).

5. Transfer, with dividers or scale, the depth dimensions from the top view to the auxiliary view. Measurements are made in both cases perpendicular to the reference line.

6. Project points 1, 2, 3, 4, 5, and 6, and join them to produce the auxiliary view of the inclined face A.

7. Complete the auxiliary view by projecting the remaining points to obtain surfaces B and C.

12.5 Revolving the Drawing. Fig. 12–7 (a) is an auxiliary-view drawing. At (b) the same

drawing is revolved until the auxiliary view is directly to the right of the front view, becoming a right-side view, and the top view then becomes an auxiliary view. The views at (b) are exactly the same as at (a); only the names of the views are changed. To understand an auxiliary view, or actually to draw one, it may be helpful to revolve the drawing as shown. In this way the auxiliary view becomes a regular view.

12.6 Reference Planes. Locate the reference plane in the position where it can be most conveniently used. For the object shown in Fig. 12–8 (a), the reference plane may coincide

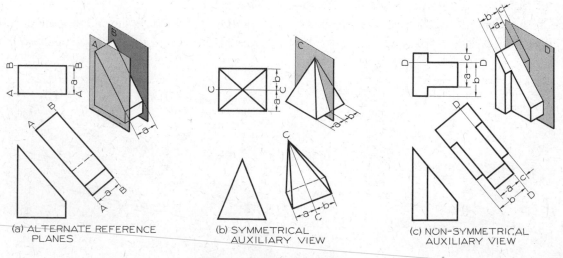

(a) ALTERNATE REFERENCE PLANES

(b) SYMMETRICAL AUXILIARY VIEW

(c) NON-SYMMETRICAL AUXILIARY VIEW

Fig. 12–8. Reference Planes.

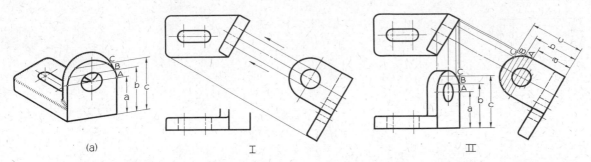

Fig. 12–9. Plotted Curves.

with either the front or the back surface, as shown. Usually, where the object is symmetrical, the reference plane is assumed to run through the center of the object, (b). In other cases it may be convenient to assume the reference plane to be coinciding with some inside surface, (c). Remember that the reference line represents the edge view of the reference plane, and in the auxiliary view it must always be perpendicular to the projection lines.

12.7 Plotted Curves. An auxiliary view is often needed to draw the regular views, as, for example, the object in Fig. 12–9 (a). The top and auxiliary views and part of the front view can be readily drawn, as shown at I. Note that in the top view the hidden lines for the hole are projected from the hole in the auxiliary view.

To complete the front view, II, select any points, as A, B, and C, on the curve in the auxil-

iary view, and project them to the top view and then down to the front view, as shown. Transfer the heights a, b, and c from the auxiliary view to the front view with dividers. Note that each height dimension can be used to transfer two points on opposite sides of the semicircle in the auxiliary view. In addition to points A, B, and C, as many additional points should be selected as necessary to define the curve accurately. Sketch smooth curves through the points in the front view and apply the irregular curve, Sec. 3.30.

Fig. 12–10 (a) shows a case where an auxiliary view is needed to plot the curve of intersection between a cylinder and a plane. At I, the front view, the incomplete top view, and a partial auxiliary view (see Sec. 12.8) are shown. Note that in the front view the hidden lines for the large inclined hole are projected from the auxiliary view.

Draw light construction lines across the

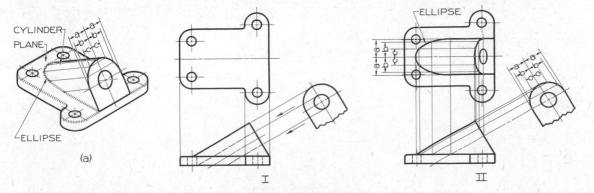

Fig. 12–10. Plotted Curves.

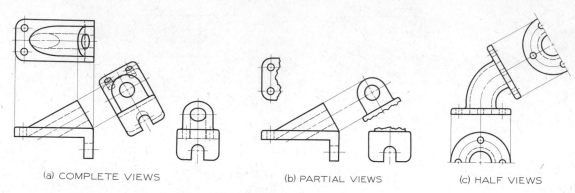

(a) COMPLETE VIEWS   (b) PARTIAL VIEWS   (c) HALF VIEWS

Fig. 12–11. Partial Views.

auxiliary view to locate points on the curve, as shown at II. Then project these points to the front view and then up to the top view, as shown. Transfer pairs of equal distances *a*, *b*, and *c* from the auxiliary view to the top view with dividers, as shown. Then sketch smooth curves through the points in the top view and apply the irregular curve, Sec. 3.30.

12.8 Partial Views. Fig. 12–11 (a) gives complete front, top, right-side, and auxiliary views of an Angle Bearing. You can give a complete shape description, while also omitting the drawing of difficult curves, by drawing *partial views,* as shown at (b). Note the use of break lines, Fig. 3–11, to limit the partial views.

For simple symmetrical objects, *half views* may be drawn, as shown at (c). Note the use

of center lines instead of break lines to limit the half views. See also Secs. 7.19 and 11.11.

12.9 Angles. As has been shown in Sec. 7.14, you can view an angle in its true size only when your line of sight is perpendicular to the plane of the angle. To see the true angle between two planes, you must look parallel to the line of intersection of the planes. For example, Fig. 12–12 (a), the true angle between the leaves of the book is seen if you look parallel to line AB. As shown at (b), line AB will appear as a point, and the two planes as lines.

At (c) a V-block drawing shows how the two inclined surfaces intersect at line AB, and since this line shows as a point in the front view, the true 120° angle is shown.

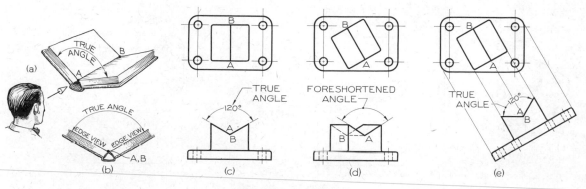

Fig. 12–12. Angles.

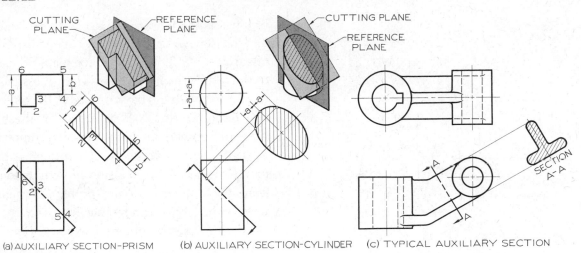

(a) AUXILIARY SECTION-PRISM   (b) AUXILIARY SECTION-CYLINDER   (c) TYPICAL AUXILIARY SECTION

Fig. 12–13. Auxiliary Sections.

At (d) the V-block is turned so that line AB does not appear as a point in the front view. Therefore the angle does not show true size and this drawing is unsatisfactory.

At (e) an auxiliary view is drawn so that the projection lines are parallel to line AB, the line shows as a point, and the true angle is shown. In industrial drafting, auxiliary views are used more for showing true angles than for showing true sizes of surfaces.

12.10 Auxiliary Sections. An auxiliary section is a section on an auxiliary plane—that is, a section viewed at an angle. The entire object may be shown in the auxiliary-section view, Fig. 12–13 (a), or the cross-hatched surface alone may be shown, (b). A typical example of an auxiliary section in machine drawing is shown at (c). Note that there is not enough space to show a revolved section of the arm (see Sec. 11.6), and also that visible lines behind the section are omitted.

12.11 True Length of Line. The true length of any oblique line can easily be found by means of an auxiliary view. For example, Fig.

12–14, let it be required to find the true length of the hip rafter 1–2. It is only necessary to draw an auxiliary view with the direction of sight perpendicular to the line in any view, as shown. The true length may also be found by revolution, Fig. 13–7 (c).

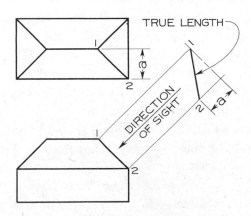

Fig. 12–14. True Length of Line.

12.12 Secondary Auxiliary Views. A *secondary auxiliary view* is any auxiliary view that is projected from a primary auxiliary view. A specific example of a secondary auxiliary view

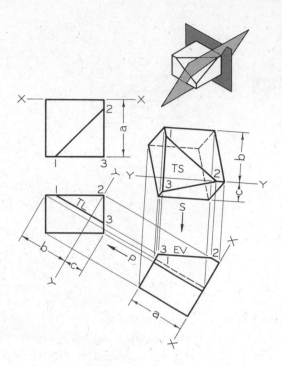

Fig. 12–15. Secondary Auxiliary Views.

produce a point view of line 1–3, as explained in Sec. 12.9, and therefore an edge view (EV) of surface 1–2–3. The reference plane X, used to draw the primary auxiliary view, is assumed to coincide with the back surface of the object, and reference line X–X is drawn in the top view and primary auxiliary view. To complete the primary auxiliary view, all depth measurements, such as a, are transferred with dividers from the top view to the primary auxiliary view with respect to reference line X–X.

To draw the secondary auxiliary view, a line of sight, arrow S, is selected that is perpendicular to the edge view of surface 1–2–3 in the primary auxiliary view. Reference plane Y is assumed cutting through the object to simplify the transfer of measurements, and reference line Y–Y is drawn in the front view and the secondary auxiliary view. Note carefully the transfer of measurements to either side of reference line Y–Y in both views and their relative locations with respect to the primary auxiliary view. For example, dimension b is transferred to the side of Y–Y *away* from the primary auxiliary view in both views.

12.13 Auxiliary-View Problems. Elementary auxiliary-view problems are given in Figs. 12–16 to 12–18, to be drawn either freehand or mechanically, with or without dimensions, as assigned. The problems that follow are practical drafting-room problems such as would be encountered by the industrial draftsman. All problems are to be drawn on Layout C (Appendix, page 474) unless otherwise indicated.

is shown in Fig. 12–15, in which it is required to draw an auxiliary view that will show the true size and shape of an oblique surface, such as surface (plane) 1–2–3. Since the true size (TS) of a plane will appear only in a view that has a line of sight normal (perpendicular) to that plane, it is first necessary to draw a primary auxiliary view that will show surface 1–2–3 as a line (edge view). In order to do this, a line of sight, arrow P, is selected that is parallel to a true length (TL) line of surface 1–2–3 in the front view. Thus, arrow P is drawn parallel to line 1–3 of the front view, since the resulting primary auxiliary view will

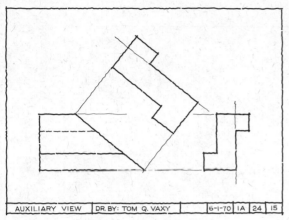

Fig. 12–16. Sketch.

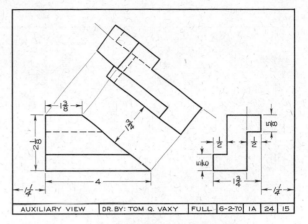

Fig. 12–17. Mechanical Drawing.

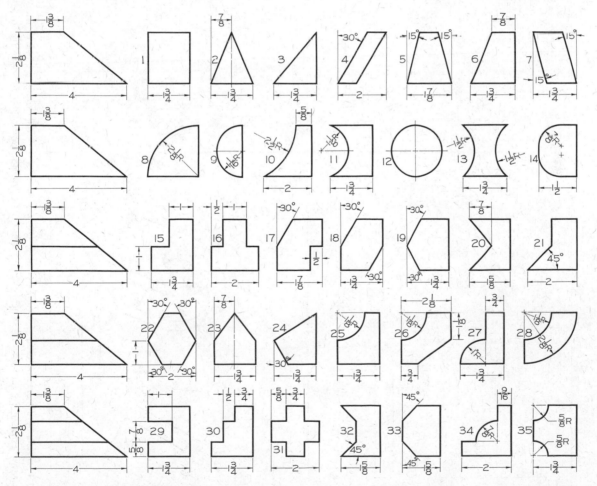

Fig. 12–18. Auxiliary-View Problems. Using Layout A (Appendix, page 474), make sketch of assigned problem as shown in Fig. 12–16, or mechanical drawing as in Fig. 12–17. In either case, draw complete auxiliary view, or partial auxiliary view of inclined face only, as assigned.

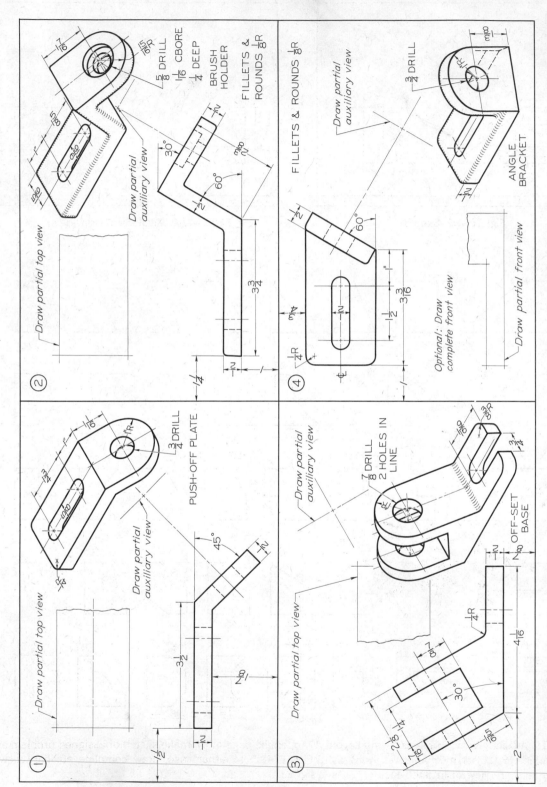

Fig. 12–19. Auxiliary-View Problems. Using Layout C (Appendix, page 474), make complete mechanical drawing of problem assigned. Omit pictorial drawings, spacing dimensions, and instructional notes. Include complete dimensions, moving dimensions from pictorials to the views.

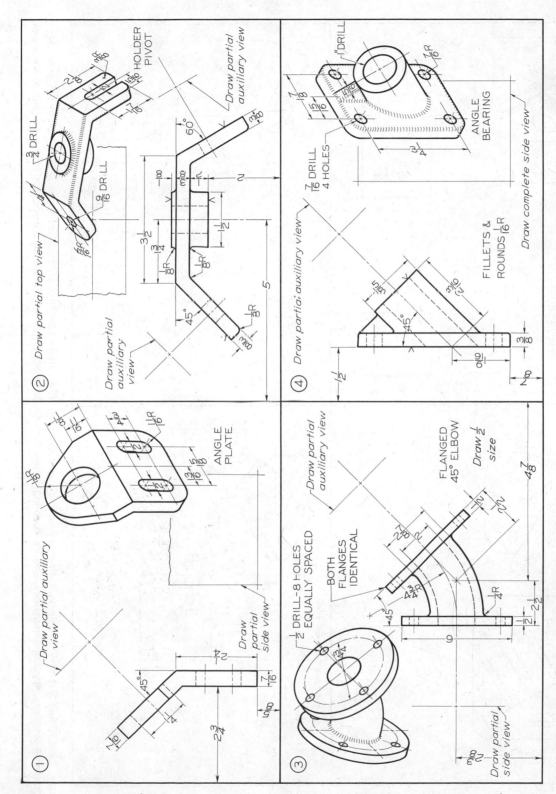

Fig. 12–20. Auxiliary-View Problems. Using Layout C (Appendix, page 474), make complete mechanical drawing of problem assigned. Omit pictorial drawings, spacing dimensions, and instructional notes. Include complete dimensions, moving dimensions from pictorials to the views.

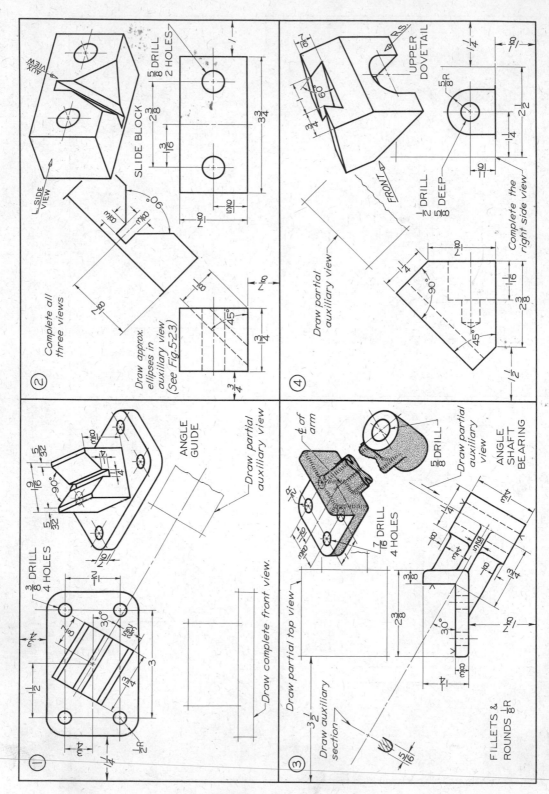

Fig. 12-21. Auxiliary-View Problems. Using Layout C (Appendix, page 474), make complete mechanical drawing of problem assigned. Omit pictorial drawings, spacing dimensions, and instructional notes. Include complete dimensions, moving dimensions from pictorials to the views.

Draw top and auxiliary views.

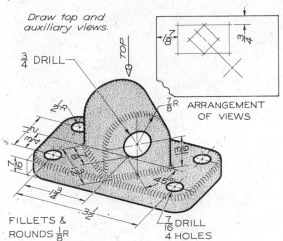

ARRANGEMENT OF VIEWS

FILLETS & ROUNDS ⅛R

Fig. 12-22. Bearing (Layout C).

Draw front, top, and partial auxiliary views.

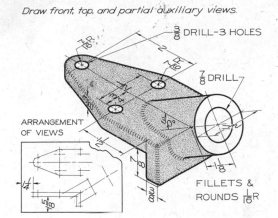

ARRANGEMENT OF VIEWS

FILLETS & ROUNDS ⅟₁₆R

Fig. 12-23. Holder Bracket (Layout C).

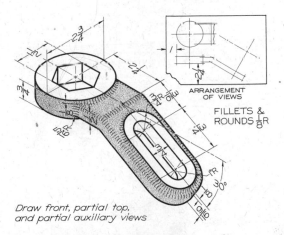

ARRANGEMENT OF VIEWS

FILLETS & ROUNDS ⅛R

Draw front, partial top, and partial auxiliary views

Fig. 12-24. Angle Arm (Layout C).

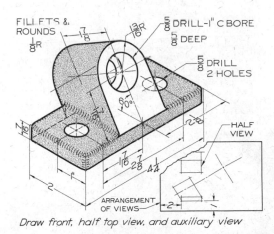

HALF VIEW

ARRANGEMENT OF VIEWS

Draw front, half top view, and auxiliary view

Fig. 12-25. End Bearing (Layout C).

Draw top, partial front, and auxiliary view.

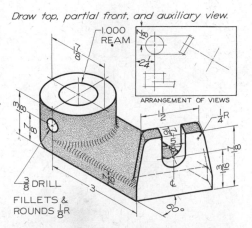

1.000 REAM

ARRANGEMENT OF VIEWS

⅜ DRILL

FILLETS & ROUNDS ⅛R

Fig. 12-26. Contact Arm (Layout C).

Draw front, & partial side & auxiliary views

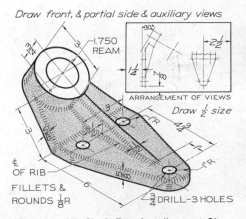

1.750 REAM

ARRANGEMENT OF VIEWS

Draw ½ size

₵ OF RIB

FILLETS & ROUNDS ⅛R

¾ DRILL-3 HOLES

Fig. 12-27. Shaft Bracket (Layout C).

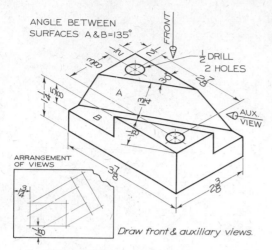

Draw front & auxiliary views.

Fig. 12–28. Angle Base (Layout C).

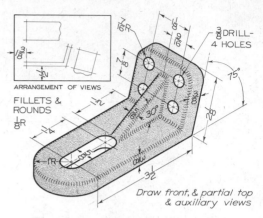

Draw front, & partial top
& auxiliary views

Fig. 12–29. Spar Bracket (Layout C).

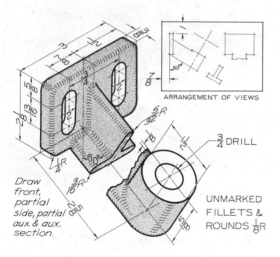

Draw front, partial side, partial aux. & aux. section.

UNMARKED FILLETS & ROUNDS $\frac{1}{8}$R

Fig. 12–30. Rod Guide (Layout C).

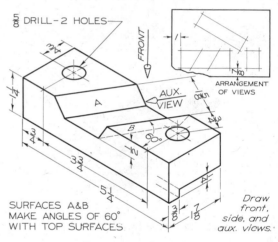

SURFACES A&B MAKE ANGLES OF 60° WITH TOP SURFACES

Draw front, side, and aux. views.

Fig. 12–31. Slotted Support (Layout C).

Draw front, left-side, & partial top & auxiliary views

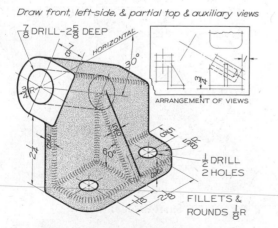

FILLETS & ROUNDS $\frac{1}{8}$R

Fig. 12–32. Angle Bearing (Layout C).

Draw front, & partial top & auxiliary views.

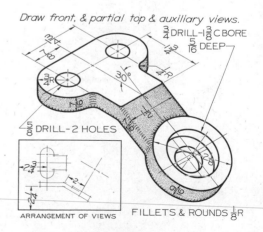

FILLETS & ROUNDS $\frac{1}{8}$R

Fig. 12–33. Spacing Lever (Layout C).

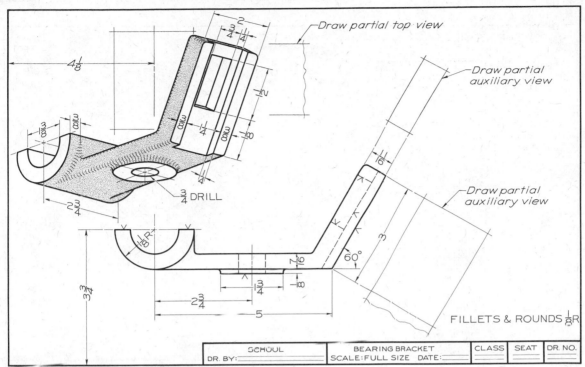

Fig. 12–34. Using Layout D (Appendix, page 475), make complete working drawing. Omit pictorial drawing, instructional notes, and spacing dimensions. Move dimensions from pictorial to views.

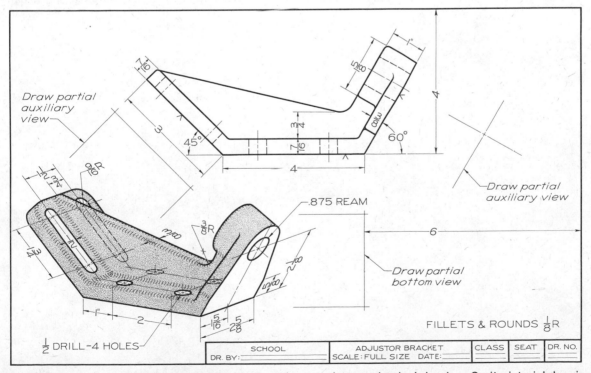

Fig. 12–35. Using Layout D (Appendix, page 475), make complete mechanical drawing. Omit pictorial drawing, instructional notes, and spacing dimensions. Move dimensions from pictorial to views.

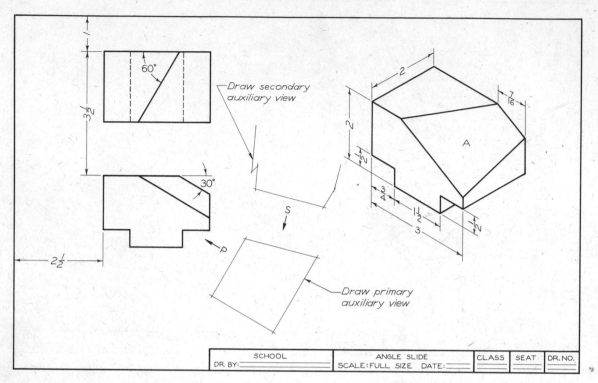

Fig. 12–36. Using Layout D (Appendix, page 475), draw complete primary and secondary auxiliary views so that the latter shows the true shape of oblique surface A. Omit pictorial drawing, instructional notes, and all dimensions.

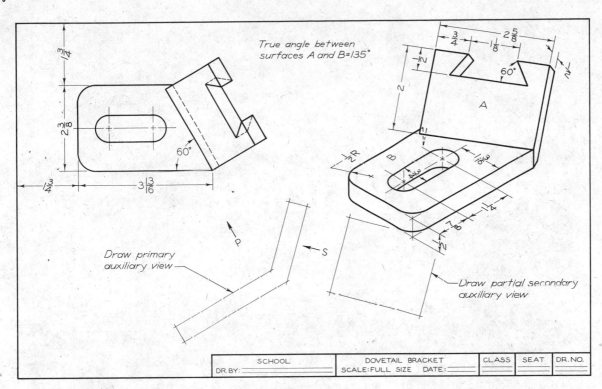

True angle between surfaces A and B=135°

Draw primary auxiliary view

Draw partial secondary auxiliary view

SCHOOL	DOVETAIL BRACKET	CLASS	SEAT	DR.NO.
DR.BY:	SCALE: FULL SIZE DATE:			

Fig. 12–37. Using Layout D (Appendix, page 475), draw a complete primary auxiliary view, and a partial secondary auxiliary view that shows the true shape of the oblique surface. Omit pictorial drawing, instructional notes, and all dimensions.

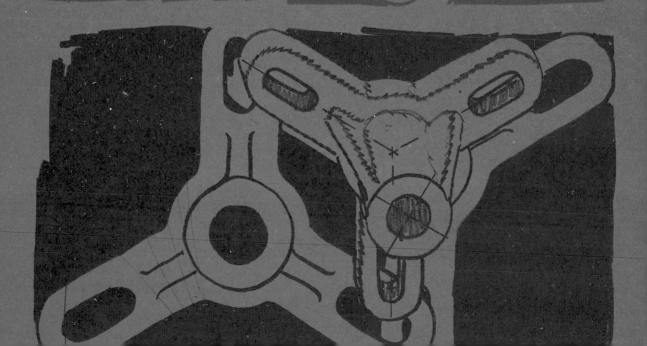

CHAPTER 13

REVOLUTIONS

13.1 Revolution. The normal position of an object for its three regular views is shown in Fig. 13–1 (a). If the object is revolved so as to bring the right end nearer to the observer in the front view, as shown at (b), the front and side views will change greatly in appearance, although the *height* will remain the same as before. The top view will be the same size and shape, but will be changed in position only. The amount of rotation, in this case 30°, shows in the top view. The revolution shown is said to be *clockwise;* in the opposite direction, it would be *counterclockwise.*

13.2 Auxiliary Views and Revolution. As shown in the previous chapter, an auxiliary view is obtained when you *view the object at an angle,* Fig. 13–2 (a). In other words, for an auxiliary view you shift your position with respect to the object until you can view it in the direction you wish. In this case the purpose is to obtain a view that shows the true size and shape of surface A.

You can obtain exactly the same view by *revolving the object* until the side view shows the true size and shape of surface A, as shown at (b). In this case the *axis of revolution* is assumed perpendicular to the front plane of the imaginary glass box, (c), and shows as a point in the front view, as shown at (b). The *direction of revolution,* in this case, is said to be *clockwise,* as shown by the small clock face. The axis may be assumed through any convenient point on the object.

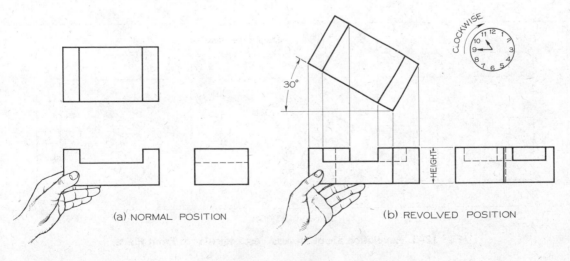

(a) NORMAL POSITION   (b) REVOLVED POSITION

Fig. 13–1. Revolution.

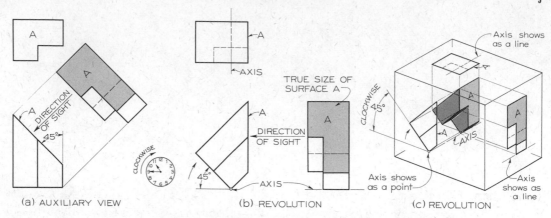

Fig. 13–2. Auxiliary Views and Revolution.

13.3 Revolution about Axis Perpendicular to Front Plane—Fig. 13–3. In this case, as shown at I, the axis AB is taken perpendicular to the front plane so that the axis shows as a point in the front view. As shown at II, the front view is copied in the revolved position, in this case 30° *counterclockwise*. Notice that the front view is exactly the same size and shape as before, but in a revolved position; hence this rule: *The view that is revolved is always the one where the axis is shown as a point, and this view is not changed in shape and size.* The top and side views will be

changed in shape and size, but the depth dimension remains the same; hence this rule: *In the views where the axis shows as a line, the dimension parallel to the axis remains unchanged.* Completed views are shown at III.

13.4 Revolution about Axis Perpendicular to Top Plane—Fig. 13–4. In this case, as shown at I, the axis is taken perpendicular to the top plane so that the axis shows as a point in the top view. As shown at II, the top view is copied in the revolved position, in this case 45° *clockwise*. The top view is exactly

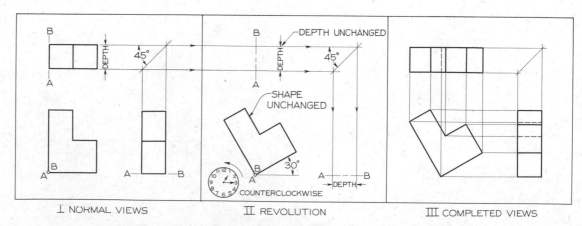

Fig. 13–3. Revolution about an Axis Perpendicular to Front Plane.

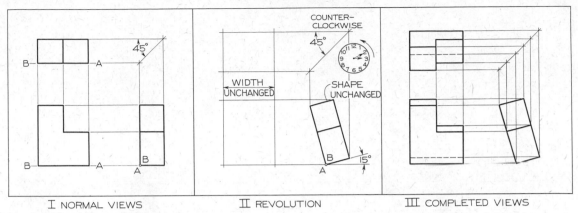

Fig. 13-4. Revolution about an Axis Perpendicular to Top Plane.

the same size and shape as before, since the axis shows as a point in this view; but the view is revolved.

In the front and side views, the axis shows as a line; therefore, the dimension of the object that is parallel to the axis—in this case the *height*—remains unchanged and can be projected from I or transferred with dividers or scale. The completed views are shown at III.

13.5 Revolution about Axis Perpendicular to Side Plane—Fig. 13-5. In this case, as shown at I, the axis is taken perpendicular

to the side plane so that the axis shows as a point in the side view. As shown at II, the side view is copied in the revolved position, in this case 15° *counterclockwise*. The side view is exactly the same size and shape as before, since the axis shows as a point in this view; but the view is revolved. In the front and top views, the axis shows as a line; therefore, the dimension of the object that is parallel to the axis—in this case the *width*—remains unchanged and can be transferred from I with dividers or scale. The completed views are shown at III.

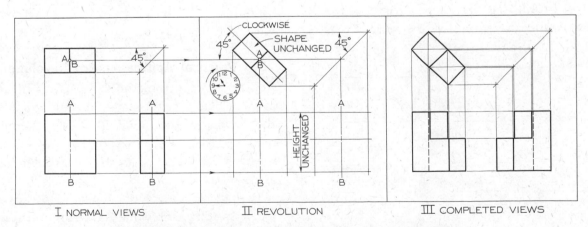

Fig. 13-5. Revolution about an Axis Perpendicular to Side Plane.

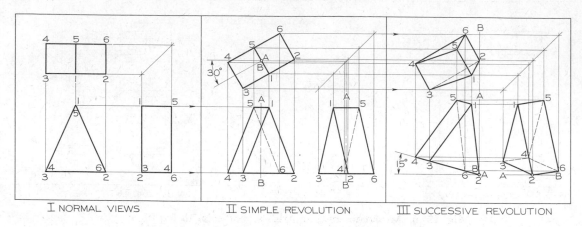

I NORMAL VIEWS   II SIMPLE REVOLUTION   III SUCCESSIVE REVOLUTION

Fig. 13-6. Successive Revolution.

13.6 Successive Revolutions—Fig. 13-6.

After one revolution, the object can be revolved further through as many stages as desired. The given views in this case are shown at I. As shown at II, the object is first revolved 30° counterclockwise about an axis perpendicular to the top plane. Then, III, the object is further revolved 15° clockwise about an axis perpendicular to the front plane. This can be continued indefinitely. Such drawings furnish excellent practice in projection of views.

Important Rule: Lines that are parallel on the object will be parallel in any view. Therefore, in drawing the lines in Fig. 13-6 at II and III, make sure that the various sets of parallel lines are truly parallel by using the triangle and T-square as shown in Fig. 3-20.

13.7 True Length of Line—Fig. 13-7.

The true length of any oblique line can be easily found by means of revolution. For example, (a), the element of the cone 1-2 is revolved to the position 1-2′ where the line shows true length in the front view. At (b) the same method is applied to the oblique edge of a pyramid. At (c) the true length of a hip rafter 1-2 is found in the same manner.

The true length of a line can also be found by means of an auxiliary view, as shown previously in Fig. 12-14.

13.8 Practical Applications of Revolution.

Revolution may sometimes be used instead of an auxiliary view to save time and simplify the drawing, as shown in Fig. 13-8 (a). A common application of revolution often occurs where a part is to be bent after machining, as shown at (b).

In some cases revolution is helpful in clarifying a drawing that would otherwise be confusing and difficult to draw. For example, Fig.

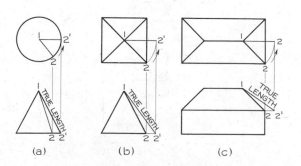

(a)   (b)   (c)

Fig. 13-7. True Length of Line.

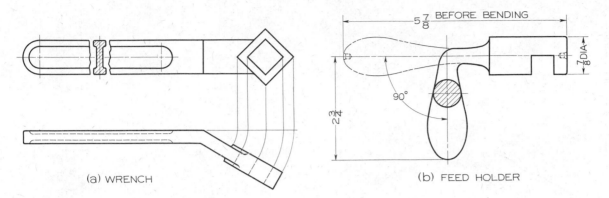

Fig. 13-8. Practical Applications of Revolution.

13-9 (a), a rib in an oblique position produces a confusing view, (b). Also, the holes in the base do not appear in their true distance from the rim and are easily confused with other lines. If the rib and the hole are revolved, a clearer drawing results, as shown at (c). See also Sec. 11.10.

Another example is shown at (d) and (e), in which the slotted arm is foreshortened and distorted; in addition, the view at (e) appears to be "amputated." Furthermore, the curves are difficult and time-consuming to draw. If the arm is revolved, (f), the resulting view is much simpler and more understandable.

13.9 Revolution Problems.
Simple revolution problems are given in Fig. 13–10, and some more difficult applications in Fig. 13–11. All are to be drawn with instruments. Those in Fig. 13–10 need not be dimensioned, but dimensions should be shown on the problems in Fig. 13–11.

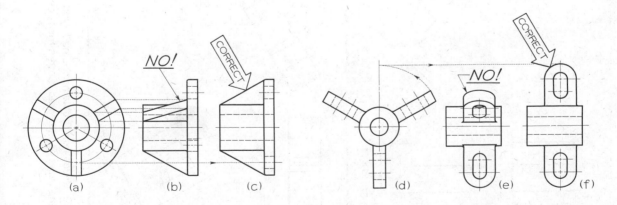

Fig. 13-9. Practical Applications of Revolution.

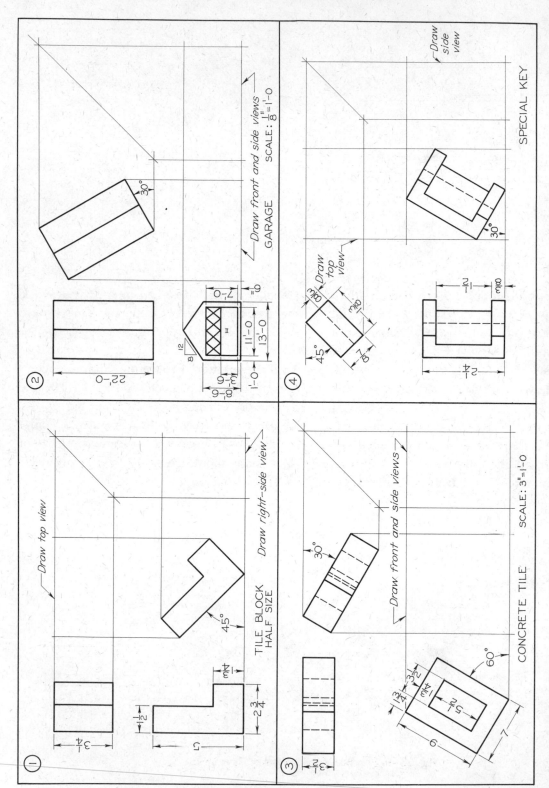

Fig. 13–10. Revolution Problems. Using Layout C (Appendix, page 474) for each problem, draw given views of assigned problems as shown, and add missing views as indicated. Omit dimensions and instructional notes. Move titles and scales to title strip.

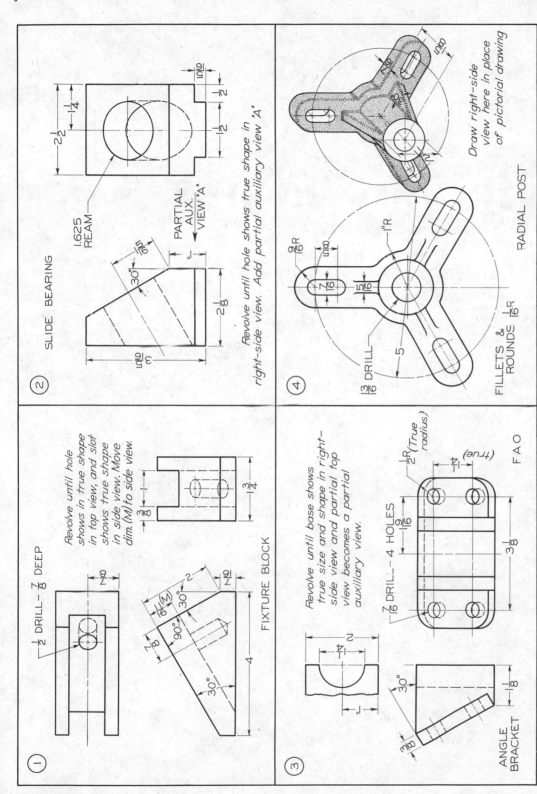

Fig. 13-11. Revolution Problems. Using Layout C (Appendix, page 474) for each problem, draw problem assigned. Omit instructional notes. Dimension drawings completely. Move titles to title strip.

CHAPTER 14

THREADS AND FASTENERS

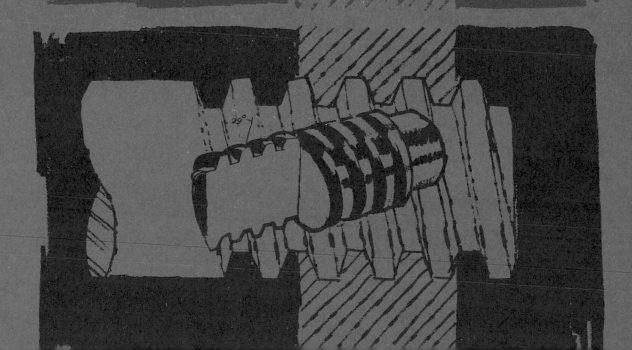

14.1 The Helix. All screw threads wind around a shaft in a curve called a *helix* (pronounced hē′ liks). If you wind a string around a pencil or any cylinder, the string will take the general form of a helix, Fig. 14–1 (a). Other applications are found in the spring, (b), the threads on a bolt, (c), or on a screw, (d), and the stripes on a barber pole, (e). If a right triangle is wrapped around a cylinder, as at (f), the hypotenuse will form a helix on the cylinder. The *lead* (pronounced leed) of the helix is the distance, parallel to the axis, from a point on one turn of the helix to the corresponding point on the next turn, as shown.

To draw a helix, draw two views of the cylinder, (g). Divide the circle into any number of equal parts, say 12. Then, on the rectangular view, assume a lead and divide it into the same number of equal parts; then draw parallel lines and number the divisions as shown. Project point 1 on the circle up to line 1, point 2 up to line 2, etc. Draw the helix through the points with the aid of the irregular curve, Sec. 3.30.

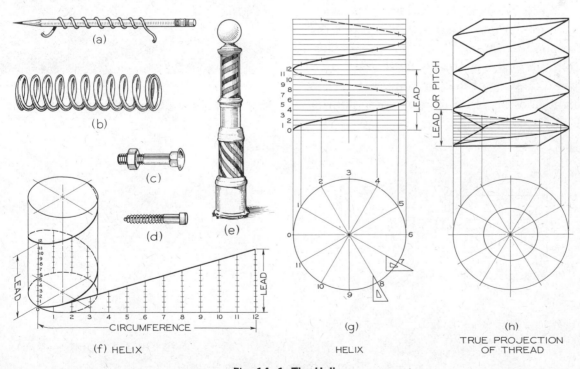

Fig. 14–1. The Helix.

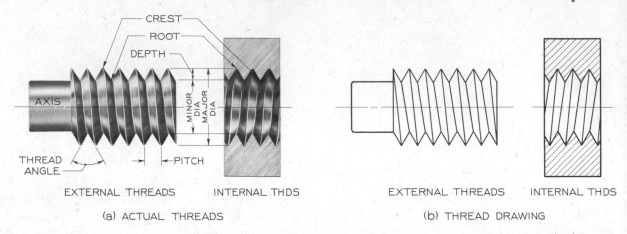

Fig. 14–2. Screw-Thread Terms.

The use of the helix in drawing the true projections of a screw thread is shown at (h). While the draftsman should understand the helix and how to draw it, a true helical thread drawing takes so much time that simpler approximations are used, as explained in the paragraphs that follow.

14.2 Screw Threads. In Fig. 14–2 (a) is shown a photograph of an *external thread* and of the corresponding *internal thread* in section, illustrating the various thread terms. At (b) is shown a drawing of the same threads in which straight lines are used to replace the difficult helical curves.

Courtesy Chicago Public Schools

Fig. 14–3. Cutting External Threads on a Lathe.

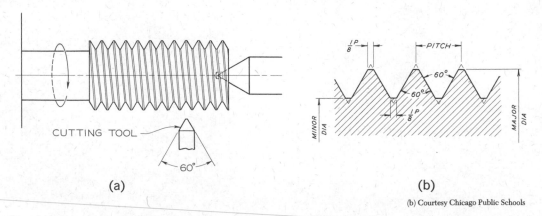

(a)   (b)

(b) Courtesy Chicago Public Schools

Fig. 14–4. Form of Thread (American National).

Screw threads are used principally to (1) *hold* parts together, as on a bolt or screw, (2) *adjust* parts with respect to each other, as on the adjusting screw on a bow compass, or (3) *transmit power,* as on a vise screw or a valve stem.

14.3 Cutting Threads on a Lathe. Threads are often cut on a lathe, particularly the larger threads, Fig. 14–3. The threads are formed by cutting helical grooves around the shaft, leaving the threads standing as helical ridges. As the shaft rotates in the lathe, the cutting tool moves slowly to the left. Several cuts are made, each a little deeper, until the correct depth is reached. For a top view of this operation, see Fig. 14–4 (a).

Large internal threads are cut in a manner similar to boring, Figs. 10–17 and 10–20 (g). Small internal and external threads are usually cut with taps and dies, Sec. 14.12.

14.4 Thread Forms. The *form* of thread is determined by the shape of the cutting tool, Fig. 14–4. Some of the basic thread forms are illustrated in Fig. 14–5. The *American National thread* is a general-purpose thread still widely used but now being replaced by the new *Unified thread,* (a). The *square thread,* (b), and the *Acme thread,* (c), are used (also in modified forms) for transmission of power, as on a vise.

Additional thread forms are illustrated in Fig. 14–6. The *sharp-V thread,* (a), is now used only to a limited extent where friction holding or adjusting is important. The *Whitworth thread,* (b), has been commonly used in Great Britain for general purposes corresponding to the *American National thread.* The new *Unified thread,* (c), is a compromise form of the Whitworth and American National forms jointly agreed upon by the two

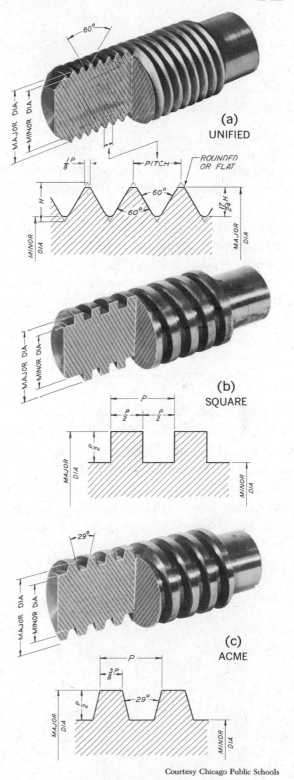

Courtesy Chicago Public Schools

Fig. 14–5. Thread Forms.

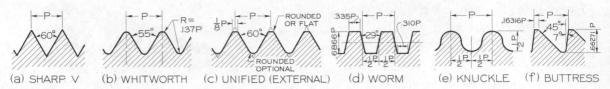

(a) SHARP V   (b) WHITWORTH   (c) UNIFIED (EXTERNAL)   (d) WORM   (e) KNUCKLE   (f) BUTTRESS

Fig. 14–6. Other Thread Forms.

countries largely in order to standardize military equipment, and is now the new USA Standard thread. The *worm thread*, (d), is used on shafts to carry power to worm gears. The *knuckle thread*, (e), is usually rolled from sheet metal, but is sometimes cast, and is used in various modified forms on bottle tops, light sockets and globes, and such items. The *buttress thread*, (f), is used to transmit great power in one direction, as on jacks and on breech locks of large guns. In addition to these, many different *special* forms of threads are used in industry for unusual requirements.

14.5 Thread Pitch. The *pitch* of a thread is the distance, parallel to the axis, from a point on one thread to the corresponding point on the next adjoining thread, Figs. 14–2 (a), 14–5, and 14–6. The pitch, P, is usually expressed in tables (see Appendix, pages 454–455) in terms of the *number of threads per inch*, which means the number of pitches per inch. To find the pitch, you simply divide 1 by the number of threads per inch. Thus, if a thread has 4

threads per inch, Fig. 14–7 (a), the pitch is $\frac{1}{4}''$, and the threads are relatively large. If a thread has a larger number of threads per inch, say 16, as at (b), the threads are relatively small. In measuring "threads per inch" the scale, (c), or a thread pitch gage, (d), can be used. For square or Acme threads, (e) and (f), each pitch includes a thread and a space.

14.6 Right-Hand (RH) and Left-Hand (LH) Threads. A right-hand, or RH, thread advances into a nut when turned clockwise, Fig. 14–8 (a). A left-hand, or LH, thread advances into a nut when turned counterclockwise, (b). If your right or left hand is placed alongside the threads, the direction of the thumb indicates whether the thread is RH or LH, as shown. All threads are understood to be RH unless designated specifically LH in the thread note, Sec. 14.18.

14.7 Multiple Threads. A *single thread* is composed of a single ridge around a shaft, with each turn next to the previous turn, Fig. 14–9

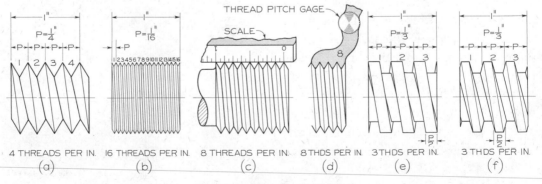

4 THREADS PER IN.   16 THREADS PER IN.   8 THREADS PER IN.   8 THDS PER IN.   3 THDS PER IN.   3 THDS PER IN.
(a)   (b)   (c)   (d)   (e)   (f)

Fig. 14–7. Threads Per Inch.

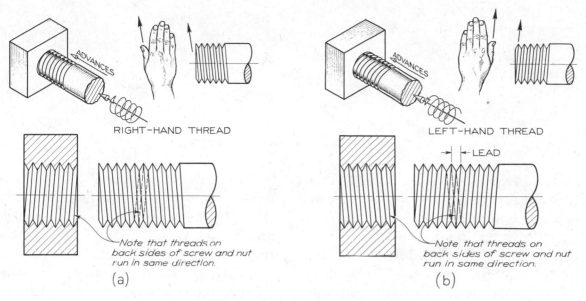

Fig. 14-8. Right-Hand and Left-Hand Threads.

(a). This can be likened to a single cord wound around a rod, each turn packed tightly against the previous one, as shown. *Multiple threads* are composed of two or more ridges side by side, (b) to (d), which may be likened to two or more cords wound around a rod. Note that the *slope* (or slant of the crest lines) of a single thread is determined by the offset measurement ½P, a double thread by the offset measurement P, a triple thread by the offset measurement 1½P, etc.

The *lead* of a thread is the distance a threaded shaft advances into the nut in one revolution. As shown in Fig. 14-9, in a single thread the lead is equal to the pitch; in a double thread, it is twice the pitch; in a triple thread, it is three times the pitch.

Note that in a drawing of a single or a triple

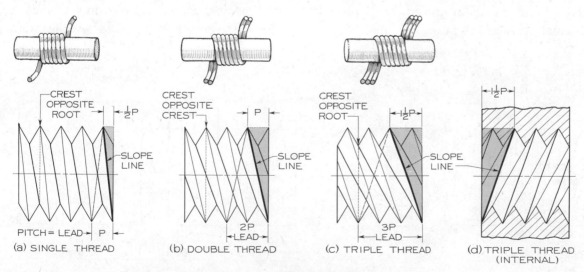

Fig. 14-9. Multiple Threads.

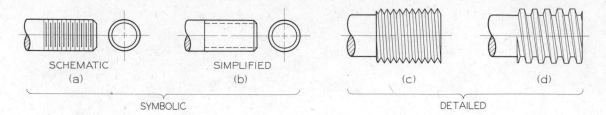

Fig. 14-10. Thread Representation (External).

thread a crest is opposite a root, and in the case of a double or a quadruple thread a crest is opposite a crest. This holds for all forms of threads.

Multiple threads are used wherever quick action but not great power is required, as on valve stems, fountain pen caps, toothpaste tube caps, and such applications.

14.8 Symbolic and Detailed Threads. In drawing threads, the idea should be to represent them as simply as possible. The simplest method of drawing threads is the *symbolic method*, Fig. 14-10 (a) and (b). This method is used for small diameters (under approximately 1″ on the drawing) where more complete thread pictures would be difficult or impractical to draw. The symbolic method is used for *all forms of threads*, as American National, Unified, square, or Acme.

The nearest approximation to the true thread picture is the *detailed* thread representation, Fig. 14-10 (c) and (d), in which straight lines replace the helical curves. This method may be used for diameters of approximately 1″, or over, on the drawing. Thus, a $2\frac{1}{2}″$ diameter thread to half scale will be $1\frac{1}{4}″$ diameter on the drawing and may be drawn by the detailed method.

14.9 Detailed Sharp-V, Unified, and American National Threads. As shown in Figs. 14-4 to 14-6, the sharp-V, Unified, and American National thread forms are the same except for rounds or flats on the crests and roots. These small features are ignored, and all three threads are drawn in the same way by the detailed method, Fig. 14-11. Note how the external threads appear on the shaft, (a), and in the end view, (b). Note how the internal

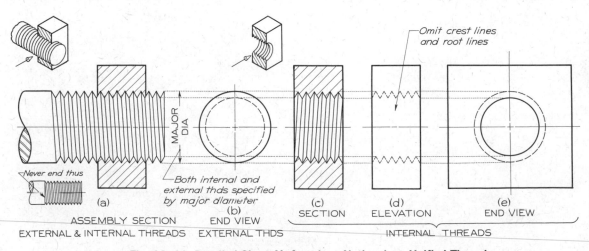

Fig. 14-11. Detailed Sharp-V, American National, or Unified Threads.

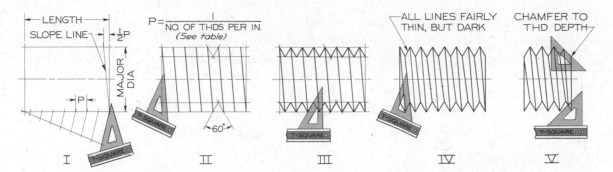

Fig. 14–12. **Steps in Drawing External American National, Unified, or Sharp-V Threads.**

threads appear in section, (c), in elevation, (d), and in the end view, (e).

The steps in drawing a single RH external thread are shown in Fig. 14–12.

I. Lay out the major diameter and the thread length with light construction lines. Then find the pitch, P, for this diameter by using Table 1 (Appendix, pages 454–455). See also Fig. 14–7 (a). For example, if this is a 3″ diameter thread, there are 4 threads per inch, and $P = \frac{1}{4}''$. Establish the slope of the crest lines by the offset measurement $\frac{1}{2}P$, or $\frac{1}{8}''$. Then set off distances P along the lower line, as shown. In this case $P = \frac{1}{4}''$, and can easily be set off with the scale. If the pitches cannot be set off with the scale, use the parallel-line method, as shown (see Fig. 5–2 or 5–3).

II. Draw fine parallel crest lines, using triangle and T-square, as shown. These lines can be drawn finished weight at once. Draw two 60° V's and then guide lines for the root diameter of the thread, as shown.

III. Draw all V's finished weight.

IV. Draw root lines finished weight.

V. Construct chamfer, if required. Note that this will cause the last crest line to change position slightly.

In the final drawing, all lines should be fairly thin and dark. Note that the root lines will not be parallel to the crest lines.

14.10 Detailed Square Threads. Detailed square threads are shown in Fig. 14–13. Note carefully the differences between the internal

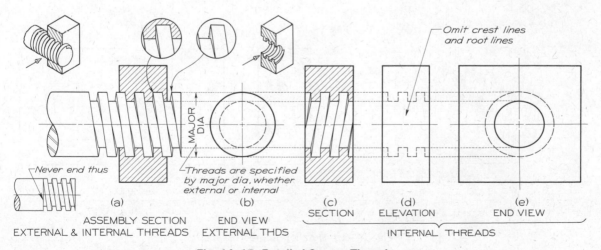

Fig. 14–13. **Detailed Square Threads.**

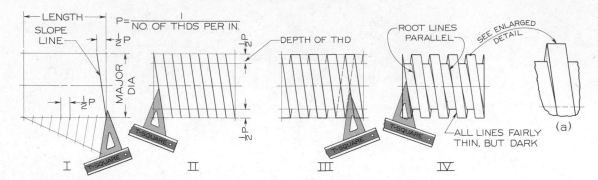

Fig. 14–14. Steps in Drawing External Square Threads.

and external threads at (a) and (c), and the differences between the external thread alone and when mated with the internal thread in section, (a), as shown in the small circles.

The steps in drawing an external square thread are shown in Fig. 14–14.

I. Lay out the major diameter and the thread length with light construction lines. Then find the pitch, P, for this diameter by using Table 3 (Appendix, page 456). See Fig. 14–7 (e). For example, if this is a $1\frac{1}{2}''$ diameter thread, there are 3 threads per inch, and P $= \frac{1}{3}''$. Along the lower line, set off distances

$\frac{1}{2}P$ ($\frac{1}{6}''$). You can do this by using the parallel-line method, Fig. 5–2 or 5–3, as shown, or by dividing $1''$ into 6 parts with the bow dividers. Then establish the slope of the threads by the offset measurement $\frac{1}{2}P$ ($\frac{1}{6}''$).

II. Draw parallel crest lines, using the triangle and T-square. These can be drawn finished weight at once. Draw light guide lines for the depth of the thread, as shown, using measurements $\frac{1}{2}P$.

III. Heavy-in tops of threads. Draw visible back crest lines, finished weight.

IV. Draw parallel visible root lines finished

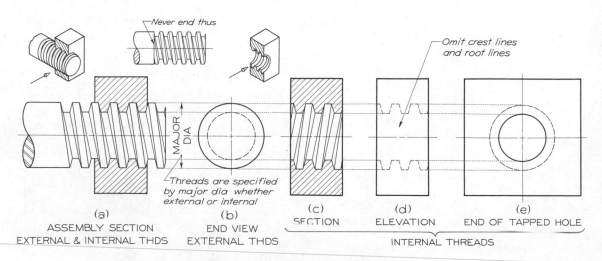

Fig. 14–15. Detailed Acme Threads.

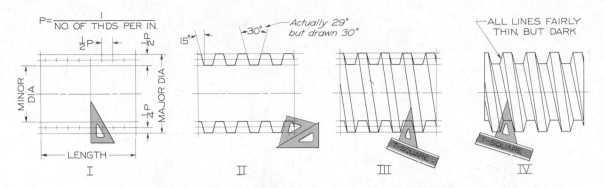

Fig. 14–16. Steps in Drawing External Acme Threads.

weight, and heavy-in bottoms of thread spaces. See detail at (a).

All final lines should be fairly thin and dark.

14.11 Detailed Acme Threads.
Detailed Acme threads are shown in Fig. 14–15, and the steps in drawing an external Acme thread in Fig. 14–16.

I. Lay out the major diameter and the thread length with light construction lines. Then find the pitch, P, for this diameter by using Table 3 (Appendix, page 456). See Fig. 14–7 (f). For example, if this is a $1\frac{1}{2}''$ diameter thread, there are 3 threads per inch, and $P = \frac{1}{3}''$. Draw root-diameter guide lines, making the thread depth $\frac{1}{2}P$ or $\frac{1}{6}''$. Divide $1''$ into sixths by trial with the bow dividers, or use the parallel-line method, Fig. 5–2 or 5–3. Then draw construction lines halfway between these lines and the outside lines, and set off $\frac{1}{2}P$ ($\frac{1}{6}''$) distances along these construction lines.

II. Draw sides of threads 15° with vertical through the $\frac{1}{2}P$ points. These can be drawn finished weight at once. Then heavy-in tops of threads and bottoms of thread spaces.

III. Draw parallel crest lines, finished weight, using the triangle and T-square.

IV. Draw parallel root lines, finished weight, as shown. All final lines should be fairly thin and dark.

14.12 Cutting Threads with Taps and Dies.
Small external threads are often cut by hand with a *die*, Fig. 14–17. An enlarged view of a die is shown at (b). The die fits in a die holder, and the thread is cut merely by turning the tool until the proper thread length is obtained. For production work, the die may be power driven on a lathe or in a drill press.

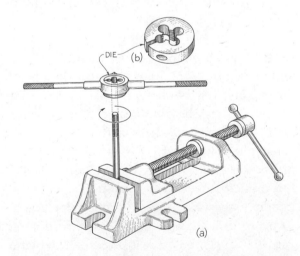

Fig. 14–17. Cutting a Small External Thread with a Die.

Small internal threads are usually cut with a *tap*, such holes being called *tapped holes*, Fig. 14–18 (a). The tap is a small fluted cutting tool with cutting teeth shaped to cut the threads that are desired. First, the original hole is drilled, (b). The depth of the hole must be

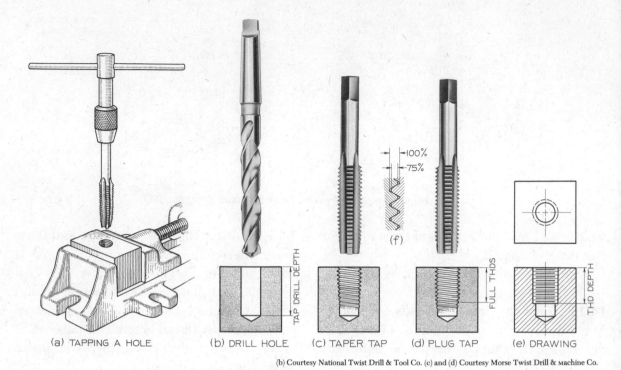

(a) TAPPING A HOLE   (b) DRILL HOLE   (c) TAPER TAP   (d) PLUG TAP   (e) DRAWING

(b) Courtesy National Twist Drill & Tool Co. (c) and (d) Courtesy Morse Twist Drill & Machine Co.

Fig. 14–18. Tapping a Small Hole.

several thread pitches deeper than the thread length required, to allow space for the end of the tap and for the metal chips. The diameter of the drill is slightly greater than the root diameter of the thread so that the actual thread engagement will be, for best results, about 75% of the full depth of the thread. Tap drill sizes for Unified threads are given in Table 1 (Appendix, pages 454–455). It is good practice to give tap drill information in the thread note, Fig. 14–22 (I), but this is often omitted, and the shop man consults tap drill tables.

Taps are available in sets of three, the *taper tap*, the *plug tap*, and the *bottoming tap*. The taper tap, Fig. 14–18 (c), is always used first and has a considerable tapered portion at the end to guide the tap, but does not have full threads near the end. To extend the threads

deeper in the hole, the plug tap, (d), is then used. Although this still leaves several imperfect threads in the bottom of the hole, this tap is usually the last one used. The tapped hole as it would appear on a drawing is shown at (e). If necessary to tap practically to the bottom of the hole, the bottoming tap (not shown in Fig. 14–18) is used, producing a hole as shown in Fig. 14–20 (f), (g), (p), and (r).

A "blind" tapped hole is one that does not go through a piece, Fig. 14–18 (e). A "through" tapped hole goes entirely through. In general, blind tapped holes should be avoided where possible, because chips in the bottom may cause jamming.

For production work, tapping is done in a lathe or drill press, in which a tapholder is used to hold and drive the tap.

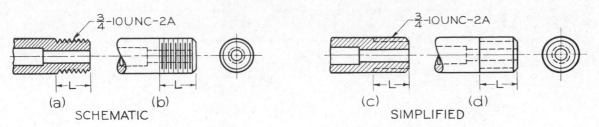

Fig. 14–19. Symbolic External Threads.

14.13 Symbolic Threads. *Symbolic thread symbols* are used for small diameters, under approximately 1″ on the drawing, to represent *all forms of threads*. Two forms are approved as USA Standard, the *schematic* and the *simplified*. External thread symbols are shown in Fig. 14–19. Note that when the schematic form is drawn in section, (a), the actual thread forms are drawn; otherwise, the presence of threads would not be evident. In elevation, (b), the crest lines are represented by long thin lines and the roots by short thick lines, both at right angles to the shaft. At (c) and (d) are shown the simplified forms in which hidden

lines are drawn parallel to the axis at the approximate depth of the thread.

Symbolic internal threads are shown in Fig. 14–20. Note that schematic and simplified symbols are alike except in sectional views.

14.14 To Draw Schematic External Symbolic Threads. The steps in drawing external symbolic threads are shown in Fig. 14–21. Since the representation is strictly symbolic, no attempt is made to use actual thread depth and pitch. Values for depth and pitch, which can be easily laid off with the scale and which result in a pleasing approximation of the

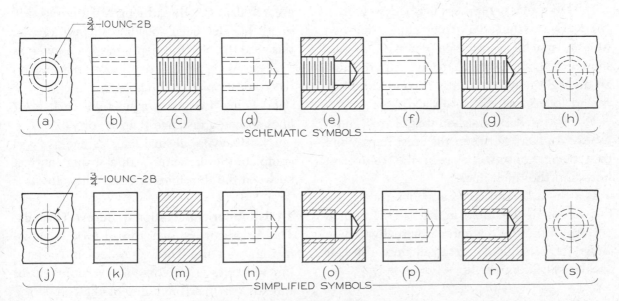

Fig. 14–20. Conventional Internal Threads.

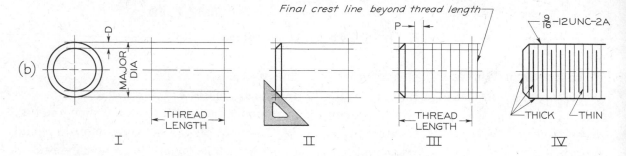

	MAJOR DIAMETER	#5 (125) TO #12 (216)	$\frac{1}{4}$	$\frac{5}{16}$	$\frac{3}{8}$	$\frac{7}{16}$	$\frac{1}{2}$	$\frac{9}{16}$	$\frac{5}{8}$	$\frac{11}{16}$	$\frac{3}{4}$	$\frac{13}{16}$	$\frac{7}{8}$	$\frac{15}{16}$	1
(a)	DEPTH, D	$\frac{1}{32}$	$\frac{1}{32}$	$\frac{1}{32}$	$\frac{3}{64}$	$\frac{3}{64}$	$\frac{1}{16}$	$\frac{1}{16}$	$\frac{1}{16}$	$\frac{1}{16}$	$\frac{5}{64}$	$\frac{3}{32}$	$\frac{3}{32}$	$\frac{3}{32}$	$\frac{3}{32}$
	PITCH, P	$\frac{3}{64}$	$\frac{1}{16}$	$\frac{1}{16}$	$\frac{1}{16}$	$\frac{1}{16}$	$\frac{3}{32}$	$\frac{3}{32}$	$\frac{3}{32}$	$\frac{3}{32}$	$\frac{1}{8}$	$\frac{1}{8}$	$\frac{1}{8}$	$\frac{1}{8}$	$\frac{1}{8}$

Fig. 14–21. Schematic External Threads (Full Size).

threads, are shown in the table, Fig. 14–21 (a).

I. Lay out the major diameter (in this case $\frac{9}{16}''$) and the desired thread length with light construction lines. In the table find the depth D for a $\frac{9}{16}''$ diameter thread. This is $\frac{1}{16}''$. Draw the inner circle in the end view $\frac{1}{16}''$ smaller in radius than the large circle, and draw guide lines for the depth of the thread, as shown.

II. Draw 45° chamfer, finished weight.

III. Draw thin dark crest lines, finished weight, spaced a distance P apart ($\frac{3}{32}''$, as shown in table). Note that the final crest line on the right may fall slightly beyond the actual thread length, which is satisfactory.

IV. Draw dark finished-weight root lines by eye midway between the crest lines. Note the strong contrast between the thin crest lines and the thick lines.

If a thread is drawn to a reduced scale, use the values in the table in Fig. 14–21, which correspond to the diameter as actually drawn. Thus, for a 1″ diameter thread drawn to half scale, use the values for $\frac{1}{2}''$ diameter.

14.15 To Draw Schematic Internal Symbolic Threads. The steps in drawing "blind" tapped holes are shown in Fig. 14–22. Values for depth and pitch are shown in the table in Fig. 14–21.

I. Draw hidden circle equal to the major diameter (say $\frac{3}{8}''$) and solid circle a distance D ($\frac{3}{64}''$) less in radius than the hidden circle. Lay out rectangular view with light construction lines, making the tap drill depth several pitches greater than the thread length if the tap drill depth has not been specified in the note. In this case the tap drill depth has been given as $\frac{7}{8}''$. Note that the drill depth does not include the conical point of the hole.

II. Draw finished-weight thin dark crest lines spaced a distance P apart, or $\frac{1}{16}''$.

III. Heavy-in all final lines as necessary to complete the drawing, with a strong contrast between the thin lines and the thick lines.

14.16 American National Screw Threads. The American National thread *form* is shown in Fig. 14–4 (b). The old American National thread tables list five *series* of numbers of threads per inch for different diameters: the *Coarse Thread Series* (NC), the *Fine Thread Series* (NF), and the *8-Pitch Series* (8N), *12-*

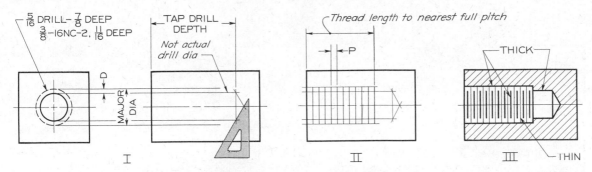

Fig. 14–22. Schematic Internal Threads (Full Size).

Pitch Series (12N), and *16-Pitch Series* (16N). The 8-, 12-, and 16-Pitch series all have the same number of threads per inch for all diameters. The *Coarse* and *Fine* series are given in Table 1 (Appendix, pages 454–455).

Also standardized are four *classes of fit*. The term *fit* refers to how closely the screw fits in the threaded hole—that is, to the amount of "play" between the two parts. The four fits are the Loose Fit (Class 1), used for rough work; the Free Fit (Class 2), used for the great bulk of screw thread work; the Medium Fit (Class 3), used for the better grades of work, as in automobile engines; and the Close Fit (Class 4), used where a very snug fit is required, as in certain aircraft engine parts.

14.17 Unified Threads. The new Unified thread *form* is shown in Fig. 14–6 (c). The USA Standard "Unified Screw Threads" (USAS B1.1—1960) lists six different *series* of numbers of threads per inch for different diameters, and selected combinations of special diameter and special pitch. The six series are the *Coarse Thread Series* (UNC or NC) recommended for general use, the *Fine Thread Series* (UNF or NF) for automotive and aircraft work, the *Extra Fine Thread Series* (UNEF or NEF) for aircraft work and other uses where a very

fine thread is required, and the *8-Thread Series* (8N), *12-Thread Series* (12UN or 12N), and *16-Thread Series* (16UN or 16N). The 8-, 12-, and 16 Thread series all have the same number of threads per inch for all diameters. For numbers of threads per inch and tap drill sizes, see Table 1 (Appendix, pages 454–455).

Exact fits are controlled by dimensions given in tables, and the draftsman simply indicates by a symbol the fit desired. In these symbols the letter A refers to external threads, and B refers to internal threads. See Fig. 14–23 (b) and (c). Classes 1A and 1B take the place of and are similar to Class 1 of the old American National; Classes 2A and 2B are intended for the normal production of screws, bolts, and nuts and for a variety of other uses; and Classes 3A and 3B provide for highly accurate and close fitting requirements. Classes 2 and 3 have been retained from the old American National.

14.18 Thread Notes. Typical thread notes are shown in Fig. 14–23. Notes for internal threads are attached to the circular views, (a) and (b). Notes for external threads are attached to the "side" views of the threads, (c) to (f). Threads are always understood to be single threads, Sec. 14.7, unless the note indicates a double thread, (e), or some other multiple.

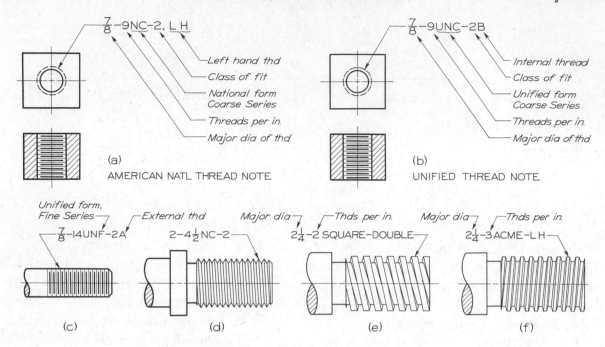

Fig. 14–23. **Thread Notes.** (See also Fig. 10–2.)

Also, threads are understood to be RH, Sec. 14.6, unless LH is given in the note, as at (a) and (f).

14.19 USA Standard Pipe Threads. USA Standard pipe threads* are either *tapered* or *straight,* and have a 60° angle of thread. As shown in Fig. 14–24, pipe threads are drawn by the symbolic method, the detailed method seldom being used. The actual taper of a taper pipe thread is $\frac{1}{16}''$ per inch *on diameter,* but this is exaggerated to $\frac{1}{16}''$ per inch *on radius* to make the taper show up more clearly, as shown at (a). Straight pipe threads are drawn as in Fig. 14–24, but without taper.

The *nominal size* of pipe does not correspond to the outside diameter (OD) except in sizes 14″ and larger. For smaller sizes, the outside diameter is larger than the nominal size. Thus, a 1″ pipe has an outside diameter of

1.315″, as shown in Table 6 (Appendix, page 457).

To draw the threads as in Fig. 14–24, lay out the *outside* diameter of the pipe as given in the table. For drawing purposes, make the thread length approximately equal to the outside diameter. Space crest and root lines as in Fig. 14–21, using the outside diameter of the pipe as "major diameter" in the table.

For further information on piping drawings, see Sec. 15.14.

14.20 USA Standard Bolts and Nuts. Bolts and nuts, Fig. 14–25, are used to hold parts together, usually so that they can be taken apart later for repair or replacement. Two standard forms are used, the *hexagon* and the *square.* Square heads and nuts are chamfered at 25°; hexagon heads and nuts are chamfered at 30°. Both are drawn at 30° for simplicity.

Regular Series bolts and nuts are for general use, while the *Heavy Series* are slightly larger

---

* USAS B2.1—1960.

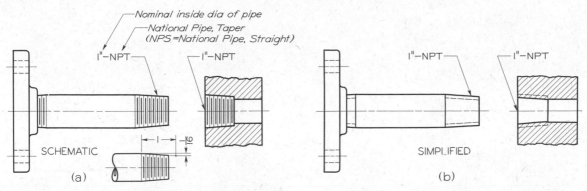

Fig. 14–24. Symbolic Representation of Pipe Threads.

and intended for heavier use. Square head bolts appear only in the Regular Series, while hexagon head bolts, hexagon nuts, and square nuts appear in both series.

Square heads and nuts are rough, or *unfinished*. Hexagon heads and nuts may be *unfinished* or *finished*. A *washer face* is machined on finished hexagon bolts and nuts. This is actually $\frac{1}{64}''$ thick, but is drawn $\frac{1}{32}''$ to show up more clearly. The threaded end of a bolt may be rounded or chamfered, but is usually drawn with a 45° chamfer to the thread depth.

The proportions of bolts and nuts, based on the diameter D of the bolt, which are either exact sizes or close approximations for drawing purposes, are as follows, where W = width across flats, H = height of head, and T = thickness of nut:

Regular Hexagon and Square Bolts and Nuts

$$W = 1\tfrac{1}{2}D \qquad H = \tfrac{2}{3}D \qquad T = \tfrac{7}{8}D$$

Heavy Hexagon Bolts and Nuts and Heavy Square Nuts*

$$W = 1\tfrac{1}{2}D + \tfrac{1}{8} \qquad H = \tfrac{2}{3}D \qquad T = D$$

Unfinished bolts have coarse threads, Class 2A, while finished bolts have coarse, fine, or 8-pitch threads, Class 2A. Unfinished nuts have coarse threads, Class 2B, while finished nuts have coarse, fine, or 8-pitch threads, Class 2B. Minimum thread lengths are 2D plus $\frac{1}{4}''$ for bolts up to 6″ in length, and 2D plus $\frac{1}{2}''$ for bolts over 6″ in length. Bolts too short for these formulas are threaded up to the head.

* There are no heavy square bolts.

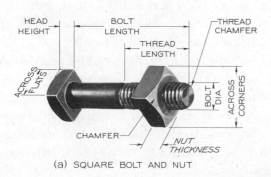

(a) SQUARE BOLT AND NUT

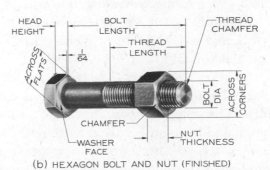

(b) HEXAGON BOLT AND NUT (FINISHED)

Courtesy National Screw & Mfg. Co.

Fig. 14–25. Bolt and Nut Terms.

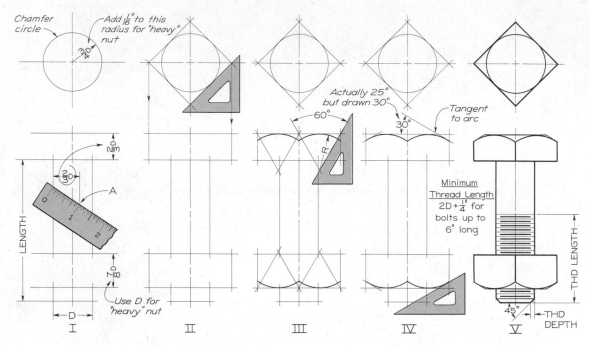

Fig. 14–26. Regular Square Bolt and Nut (Unfinished).

Bolt lengths have not been standardized because of the endless variety required by industry. The following increments (steps in lengths) are compiled from manufacturer's catalogs:

Square Head Bolts
Lengths $\frac{3}{4}''$ to $1\frac{1}{2}''$: $\frac{1}{4}''$ increments
Lengths $2''$ to $10''$: $\frac{1}{2}''$ increments
Lengths $11''$ to $30''$: $1''$ increments

Hexagon Head Bolts
Lengths $\frac{3}{4}''$ to $8''$: $\frac{1}{4}''$ increments
Lengths $8\frac{1}{2}''$ to $20''$: $\frac{1}{2}''$ increments
Lengths $21''$ to $30''$: $1''$ increments

14.21 To Draw USA Standard Bolts and Nuts. The steps in drawing USA Standard square and hexagon bolts and nuts are shown in Figs. 14–26 and 14–27. Proportions are based on diameter D and closely conform to the actual dimensions. Occasionally, where unusual accuracy is required, bolts and nuts are drawn to actual dimensions, as given in Table 7 (Appendix, pages 458–459). In Figs. 14–26 and 14–27:

I. Lay out bolt diameter D, the bolt length (under side of head to tip end), the height of the head and thickness of the nut, and the chamfer circle. For finished hexagon bolts and nuts, draw a washer face (actually $\frac{1}{64}''$, but drawn $\frac{1}{32}''$ thick), Fig. 14–27, *included in* the head height and nut thickness. Note that square bolts and nuts are always unfinished, and hence have no washer faces.

An easy way to find the various proportions based on D is to use the scale as shown in both figures. For example, Fig. 14–26 (I), make a small dot at A, then set the dividers between this dot and the left side of the bolt, and transfer the $\frac{2}{3}$D distance to sct off the height of the head.

II. Draw a square or a hexagon about the

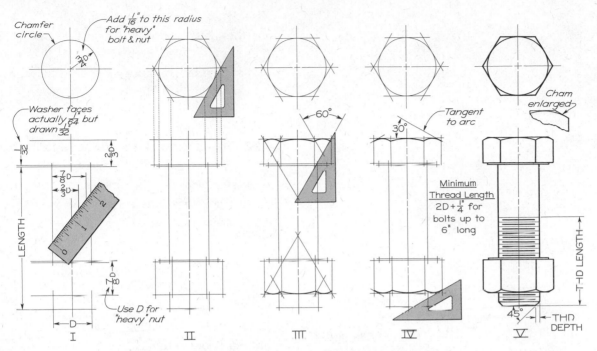

Fig. 14–27. Regular Hexagon Bolt and Nut (Finished).

Note: Unfinished hexagon bolt has this same head height and nut height, but washer face is omitted.

chamfer circle, and project down to the head and nut to establish the corners.

III. Locate centers for arcs with 30° × 60° triangle as shown, and draw final dark arcs.

IV. Draw 30° chamfers tangent to arcs, as shown.

V. Lay off thread length (see Sec. 14.20) from tip end of bolt, and draw threads. Construct 45° chamfer on end of bolt to thread

depth, as shown. Darken all lines as necessary to complete the drawing.

Bolts and nuts are preferably drawn "across corners" so as to show the maximum number of faces in each view, as shown in Figs. 14–26 and 14–27. In Fig. 14–28 (a) the top view is correctly placed to show the hexagon bolt head "across corners" in the front view. However, the true projection of the side view

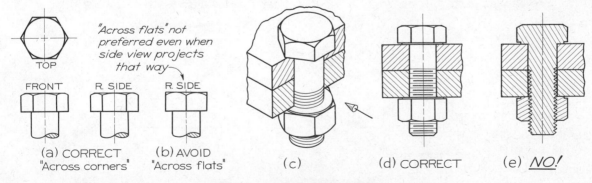

Fig. 14–28. Drawing Bolts and Nuts.

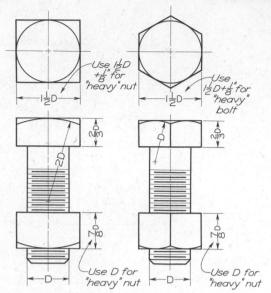

Fig. 14–29. Bolts "Across Flats."

would show "across flats," (b), which is not desirable because it looks like a square head "across corners." The conventional and correct method is to draw the head "across corners" in the side view, as shown at (a). Occasionally, a head or a nut may be in a restricted position in an assembly drawing and must be represented "across flats." In such cases, use the proportions shown in Fig. 14–29.

It is customary not to section bolts and nuts when drawn in assembly, Fig. 14–28 (c) and (d), because they do not in themselves require sections for clearness. Incorrect sectioning is shown at (e). The same holds true for all screws and similar parts. See Sec. 15.13.

It is a good rule to make holes $\frac{1}{32}''$ larger than the bolts up to $\frac{3}{8}''$ diameter, and $\frac{1}{16}''$ larger above $\frac{3}{8}''$ diameter. However, clearance is not shown on the drawing unless necessary to make clear that the bolt threads are not engaged with internal threads in the hole.

14.22 Specifications for Bolts and Nuts.

Bolts and nuts are often listed in parts lists, on assembly drawings, and other records. Standard specifications are given, of which the following are typical:

Example (Complete):
$\frac{7}{16}$–14UNC–2A × $2\frac{1}{4}$ FIN HEX BOLT

Example (Abbreviated):
$\frac{7}{16}$ × $2\frac{1}{4}$ FIN HEX BOLT

Example (Complete):
$\frac{3}{4}$–10UNF–2A HEAVY SQ NUT

Example (Abbreviated):
$\frac{3}{4}$ HEAVY SQ NUT

14.23 Studs.

A stud is a rod threaded on both ends, one end of which is screwed tightly in a permanent position into a main casting. The other end passes through a clearance hole in the part to be held down, such as a cylinder head, and a nut is screwed on the free end, Fig. 14–30 (a). Studs are not standardized, and hence must be drawn and dimensioned on detail drawings. See Fig. 14–30 (b).

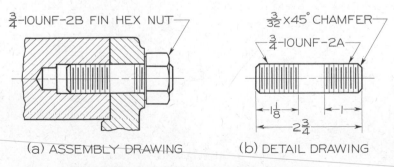

(a) ASSEMBLY DRAWING   (b) DETAIL DRAWING

Fig. 14–30. Stud Application and Dimensioning.

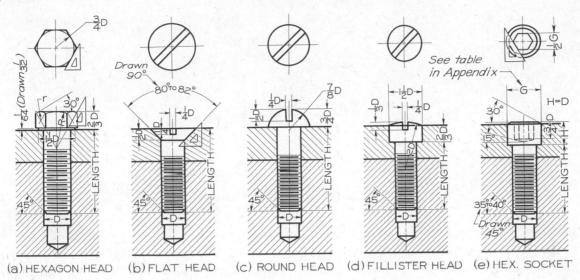

(a) HEXAGON HEAD   (b) FLAT HEAD   (c) ROUND HEAD   (d) FILLISTER HEAD   (e) HEX. SOCKET

Fig. 14–31. USA Standard Cap Screws.

14.24 USA Standard Cap Screws. Cap screws ordinarily pass through a clearance hole in one member and screw into the other member. The clearance hole is usually drilled $\frac{1}{32}''$ larger than the screw, up to $\frac{3}{8}''$ diameter, and $\frac{1}{16}''$ larger for diameters over $\frac{3}{8}''$. However, the clearance is not shown on the drawing unless necessary for clearness, in which case it is drawn $\frac{1}{16}''$ larger in diameter. Similarly, the counterbores for the fillister heads and socket heads are usually made $\frac{1}{32}''$ or $\frac{1}{16}''$ larger in diameter than the heads, and this clearance is usually not shown on the drawing. Cap screws are regularly produced with finished heads and chamfered points, and are used on machines where accuracy and appearance are important.

There are five types of heads, as shown in Fig. 14–31. Exact dimensions for the hexagon head cap screw are given in Table 7 (Appendix, page 458), while exact dimensions for the slotted and socket head types are given in Table 8 (Appendix, page 460). These dimensions may be used where the drawing requires exact sizes, but this seldom occurs. In Fig. 14–31 proportions, given in terms of diameter

D, closely conform to the actual dimensions, and produce almost exact drawings with relatively little effort. The hexagon head cap screw, (a), is drawn in the same manner as the finished hexagon bolts, Fig. 14–27. The minimum length of thread on hexagon head cap screws* is $2D + \frac{1}{4}''$ for lengths up to 6″, and $2D + \frac{1}{2}''$ for lengths over 6″. Hexagon head cap screws too short for these formulas are threaded to within $2\frac{1}{2}$ threads of the head for lengths up to 1″, and $3\frac{1}{2}$ threads for lengths greater than 1″. The minimum length of thread on slotted cap screws* is $2D + \frac{1}{4}''$; and when the screws are too short for this formula, they are threaded to within $2\frac{1}{2}$ threads of the head. See tables in USAS B18.3—1961 for thread lengths of hexagon socket cap screws. Lengths of screws and thread data are given in notes with Table 8 (Appendix, page 460).

In drawing tapped holes for cap screws, allow one or two threads beyond the end of the screw for clearance, and allow a drill depth beyond the threads equal to several more threads, Fig. 14–31. See also Secs. 14.12 and

---

* USAS B18.2.1—1965 and USAS B18.6.2—1956.

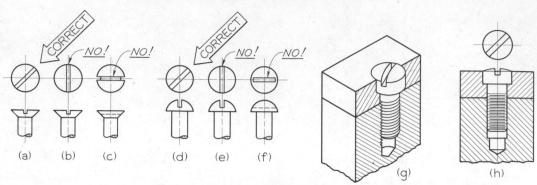

Fig. 14–32. Drawing Cap Screws.

14.15. Note that in the bottom of all the tapped holes in Fig. 14–31 the threads are omitted so as to show the ends of the screws more clearly. For information on drilled, countersunk, or counterbored holes, see Secs. 9.24, 10.11, and 10.12.

Correct and incorrect methods of drawing slotted screw heads are shown in Fig. 14–32 (a) to (f). It is customary not to section screws when drawing them in assembly, (g) and (h), because they do not in themselves require sectioning for clearness. See also Fig. 14–28 for similar treatment of bolts.

Typical cap screw notes are as follows:

Example (Complete):

$\frac{7}{16}$–14UNC–3 × $2\frac{1}{4}$ HEXAGON CAP SCREW

Example (Abbreviated):

$\frac{7}{16}$ × $2\frac{1}{4}$ HEX CAP SCR

Example (Complete):

$\frac{1}{2}$–13NF–3 × $2\frac{1}{2}$ FILLISTER HEAD CAP SCREW

Example (Abbreviated):

$\frac{1}{2}$ × $2\frac{1}{2}$ FILL HD CAP SCR

14.25 USA Standard Machine Screws. Machine screws are in general smaller than cap screws and are available in a greater variety of heads. Eight styles of head are standardized; four of the most common are shown in Fig. 14–33. Machine screws are particularly adapted to screwing into thin materials, and

on all lengths under 2″ are threaded to the head.* Machine screws over 2″ in length have a minimum thread length of $1\frac{3}{4}$″. Machine screws are used extensively in firearms, jigs, fixtures, and other small mechanisms.

In Fig. 14–33 proportions, given in terms of diameter D, conform closely to the actual dimensions, and produce almost exact drawings. Actual dimensions are given in Table 9 (Appendix, page 461). Threads are National Coarse or National Fine, Class 2 fit. Clearance holes and counterbores should be made larger than the screws, as explained for cap screws in Sec. 14.24.

Typical machine screw notes are as follows:

Example (Complete):

NO. 5 (.125)–40NC–2 × $1\frac{1}{2}$ FILLISTER HEAD MACHINE SCREW

Example (Abbreviated):

NO. 5 (.125) × $1\frac{1}{2}$ FILL HD MACH SCR

Example (Complete):

$\frac{5}{16}$–24NF–2 × $2\frac{1}{4}$ OVAL HEAD MACHINE SCREW

Example (Abbreviated):

$\frac{5}{16}$ × $2\frac{1}{4}$ OVAL HD MACH SCR

14.26 USA Standard Set Screws. Set screws are used to prevent relative motion, usually rotary, between two parts, such as the motion

* USAS B18.6.3—1962.

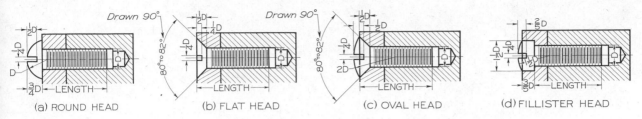

Fig. 14–33. USA Standard Machine Screws.

of a pulley hub on a shaft. A set screw is screwed into one part so that its point bears firmly against another part, Fig. 14–34 (a). If the point of the set screw is cupped, (e), or if a flat is milled on the shaft, (a), the screw will hold more firmly. There are four types of head, (a) to (d), and six styles of points, (e) to (k). The headless set screws are coming into greater use because they have no projecting heads to catch clothing and cause accidents.

Most of the dimensions in Fig. 14–34 are USA Standard* formula dimensions, and the resulting drawings are almost exact representations. Length L is the overall length for headless set screws, and is the distance from the

* USAS B18.6.2—1956 and USAS B18.3—1961.

underside of the head to the point for square head set screws.

Set screws are usually made of steel and are casehardened. Headless set screws are threaded with either National Coarse or National Fine threads, Class 2, while square head set screws may have Coarse, Fine, or 8-Pitch threads, Class 2A. The Coarse thread is usually applied to the square head set screw, which is generally used in the rougher grades of work. Typical set screw notes are as follows:

Example (Complete):

$\frac{1}{2}$–13UNC–2A × 1$\frac{1}{2}$ SQUARE HEAD FLAT POINT SET SCREW

Examples (Abbreviated):

$\frac{1}{2}$ × 1$\frac{1}{2}$ SQ HD FLAT PT SET SCR

$\frac{3}{8}$ × 1$\frac{1}{4}$ HEX SOCK CUP PT SET SCR

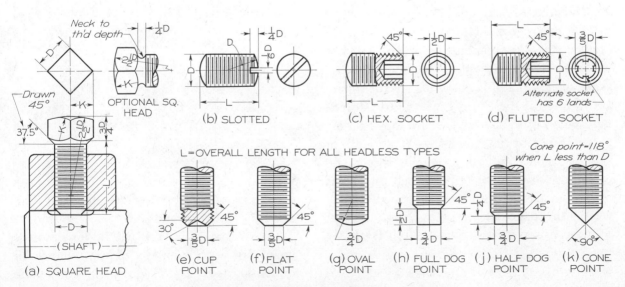

Fig. 14–34. USA Standard Set Screws.

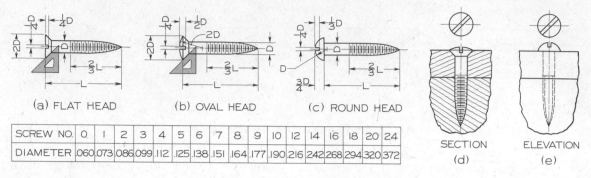

SCREW NO.	0	1	2	3	4	5	6	7	8	9	10	12	14	16	18	20	24
DIAMETER	.060	.073	.086	.099	.112	.125	.138	.151	.164	.177	.190	.216	.242	.268	.294	.320	.372

Fig. 14–35. USA Standard Wood Screws.

14.27 USA Standard Wood Screws. Wood screws with three types of heads have been standardized,* Fig. 14–35. Any of them may be plain-slotted or cross-recessed. The proportions, based on diameter D, closely conform to actual dimensions and are more than sufficiently accurate for use on drawings. Applications are shown at (d) and (e). Note that the threads are drawn by the symbolic method, Sec. 14.13. Crest lines and root lines are spaced by eye to appear approximately as in Fig. 14–35.

A typical wood screw note is:

NO. 5 × 1¼ FILL HD WOOD SCR

14.28 Keys. *Keys* are used to prevent relative motion between shafts and wheels, couplings, cranks, and other machine parts. A square key is shown in Fig. 14–36 (a). A flat key is the same, except that it is not as high as the square key. Either of these may have the top surface tapered ⅛″ in 12″, in which case they become square taper or flat taper keys. The width of square and flat keys is generally about one-fourth the shaft diameter. One-half the height of the key is sunk into the shaft. See Table 10 (Appendix, page 462).

A gib head key is shown at (b). It is exactly

the same as the square taper or flat taper key, except that a gib head is added. The head provides for easy removal. See Table 10 (Appendix, page 462).

The Pratt & Whitney key, (c), has rounded ends and is seated in a keyseat of the same shape. Two-thirds of a P & W key is sunk into the shaft.

The Woodruff key, (d), is semicircular in shape and fits in a semicircular keyseat in the shaft. The top of the key fits into a plain rectangular keyway. Sizes of keys for given shafts are not standardized, but a good rule is to select a key whose diameter is about equal to the shaft diameter. See Tables 11 and 12 (Appendix, page 463) for dimensions.

Typical notes for keys are as follows:

¼ × 1½ SQ KEY
¼ × 3/16 × 1½ FLAT KEY
NO. 204 WOODRUFF KEY
NO. 10 PRATT & WHITNEY KEY

A *keyseat* is in the shaft; a *keyway* is in the hub or surrounding part. Typical notes for keyseats and keyways are shown in Fig. 9–30 (C), (D), (J), and (N).

In keyway notes, the width dimension is given first, and then the depth. The depth dimension is measured along the side of the

---

* USAS B18.6.1—1961.

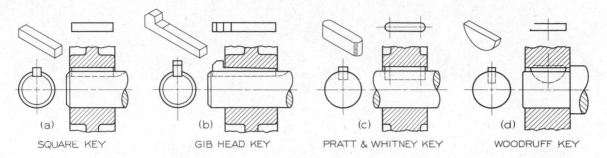

Fig. 14-36. Keys.

keyway and not at the center. Typical notes for keyways are as follows:

$$\frac{3}{8} \text{ WIDE} \times \frac{3}{16} \text{ DEEP KEYWAY}$$
$$\text{or} \quad \frac{3}{8} \times \frac{3}{16} \text{ KEYWAY}$$

14.29 Locking Devices. Many types of devices to prevent nuts from unscrewing too easily are available, the most common of which are shown in Fig. 14–37. USA Standard *jam nuts*, (a) and (g), are somewhat thinner than the regular nuts (see USAS B18.2.2—1965). The finished and semifinished nuts have a $\frac{1}{64}''$ washer face or are chamfered on both sides. Jam nuts are drawn in the same manner as ordinary hexagon nuts, Fig. 14–27, but with a thickness of $\frac{1}{2}$D (approximate). The USA Standard lock washer, (b), has a slot whose edges tend to "dig in" when the nut is loosened. Cotter pins are used with slotted nuts, (c), castle nuts, (d), or with ordinary nuts,

(e). There are many patented nuts of various kinds, of which the Esna stop nut, (f), is representative. A set screw may be kept from loosening by means of a jam nut, as shown at (g).

14–30 Rivets. *Rivets* are considered permanent fastenings, as distinguished from removable fastenings, such as bolts or screws. They are generally used to hold sheet metal or rolled steel shapes together, and are made of wrought iron, soft steel, copper, or other materials.

USA Standard large rivets are used in the structural work of bridges, buildings, and in ship and boiler construction. They are shown in their exact formula proportions in Fig. 14–38. The button head and cone head rivets are commonly used in tank and boiler construction.

A typical riveted joint is shown at (g). Note

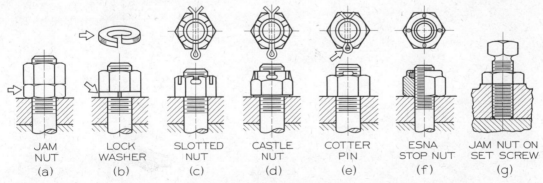

JAM NUT (a)   LOCK WASHER (b)   SLOTTED NUT (c)   CASTLE NUT (d)   COTTER PIN (e)   ESNA STOP NUT (f)   JAM NUT ON SET SCREW (g)

Fig. 14-37. Locking Devices.

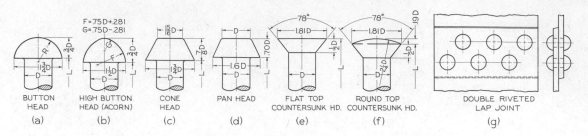

Fig. 14–38. USA Standard Large Rivets (adapted from USAS B18.4—1960).

that the side view of each rivet shows the shank of the rivet with both heads made with circular arcs, and the circular view of each rivet is represented by only a visible circle for the head. Where there are many such circles to draw, the drop-spring bow compass is a favorite instrument.

14.31 Problems. The problems in Figs. 14–39 and 14–40 are given to provide practice in drawing detailed threads, symbolic threads, and some of the more common fasteners. These problems are all designed to be drawn with instruments on Layout C (Appendix, page 474).

A large number of practical problems involving threads and fasteners are given at the end of the following chapter.

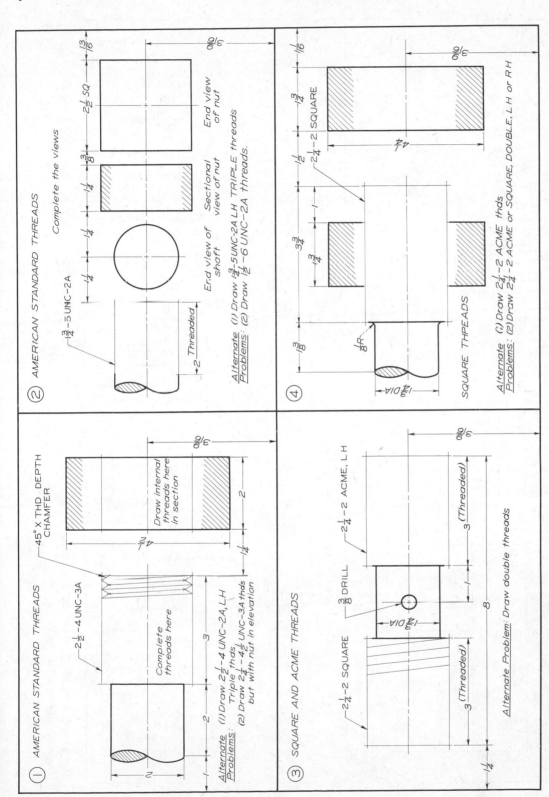

Fig. 14-39. Detailed Threads. Using Layout C (Appendix, page 474), draw problems assigned. Omit all inclined lettering. Transfer titles to title strip.

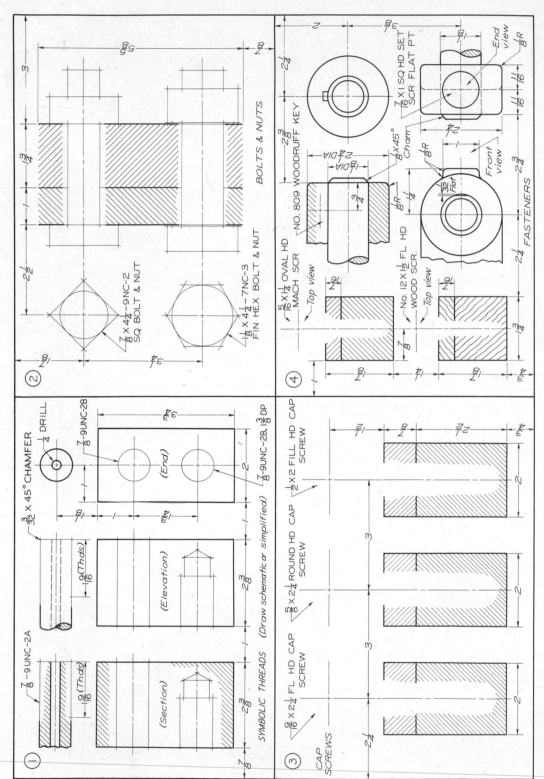

Fig. 14–40. Symbolic Threads and Fasteners. Using Layout C (Appendix, page 474), draw problem assigned. Omit all inclined lettering. Transfer titles to title strip.

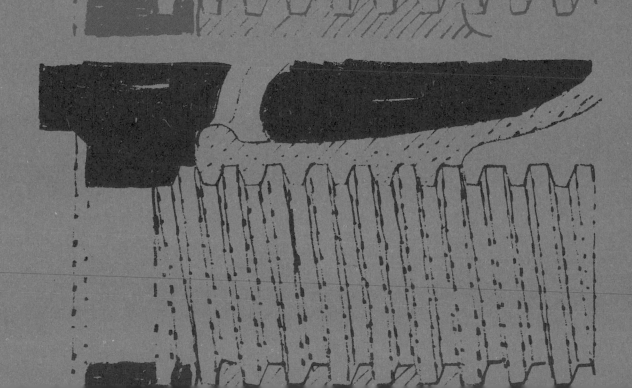

CHAPTER 15

WORKING DRAWINGS

15.1 Industrial Drafting. A typical drafting room in a large manufacturing company is shown in Fig. 15–1. The people who work here are engineers, designers, and draftsmen, each doing his part in planning the products to be manufactured. When you look at some complicated structure, such as a supersonic bomber, a full-color printing press, or a numerically controlled milling machine, do not be discouraged. Not one, but hundreds of persons have contributed their time, their "know-how," and their originality to these modern

Courtesy The Warner & Swasey Co.

Fig. 15–1. Industrial Drafting Room.

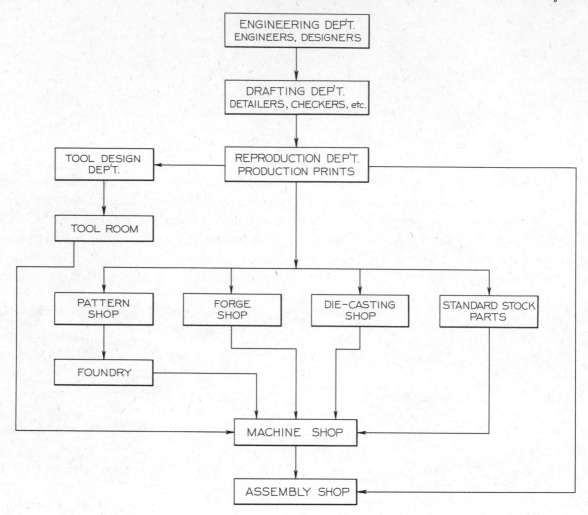

Fig. 15–2. Routing of Drawings.

wonders. Each person is more or less of a specialist who has learned how to handle problems in a relatively limited field.

A chart showing how drawings are routed from the engineering department to the drafting department and on to all the main divisions of a manufacturing plant is shown in Fig. 15–2. For a discussion of the various shops, see Chapter 10. Usually a design engineer conceives the basic idea or an idea for improvement, but often the engineering department gets valuable ideas from draftsmen, mechanics, or salesmen. In any case, it is only natural at this point for the designer to set down the new idea in the form of a freehand sketch. For example, at the Brown & Sharpe Manufacturing Company, a designing engineer had an idea for improving the adjustment of the arm on a milling machine, and he set this down rapidly in a freehand sketch, Fig. 15–3. The device that he had in mind, which was eventually developed, was an "Arm Adjusting Mechanism," shown on the machine in the upper part of Fig. 15–4.

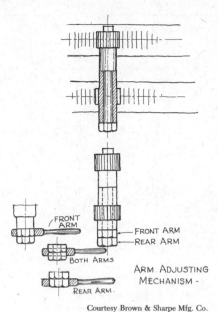

Courtesy Brown & Sharpe Mfg. Co.

Fig. 15–3. Idea Sketch for Improvement on Milling Machine.

Next, the same designer who made the sketch, or someone equally experienced, made a *layout drawing* or *design assembly,* drawn accurately to scale in pencil and without dimensions, Fig. 15–5. This is an assembly drawing showing the mechanism in position in the milling machine and including the views necessary to show the size and shape of each part of the mechanism, but dimensions are omitted. Layouts are drawn full size, if possible, to enable the designer to visualize more clearly the actual sizes of the parts as he draws them. When drawing to half size or less, the designer may tend to design large, heavy, or clumsy parts. On the other hand, in drawing double size or larger, he may tend to get an exaggerated idea of sizes and draw them too small.

Special attention is given to clearances of moving parts so that they will not interfere with other parts, to ease of assembly, and to serviceability. The designer makes use of all the sources of information available to him,

including physics, chemistry, mathematics, mechanics, strength of materials, and other subjects that the engineering student learns in college. He adopts the conclusions and recommendations obtained from experimental tests and laboratory studies. He also uses the sound experience that he and his company have obtained from success and failure of past machines, following carefully the performance and maintenance records of machines in use. Much design information is compiled in handbooks, to which he refers frequently.

Some mechanisms are relatively simple in form and operation, so that most of the design effort consists in determining functional shapes and arrangement. However, the parts may be subject to heavy loads, in which event careful computation of strength is necessary in order that the parts may be designed to correct sizes. A good example of this type of design problem would be a simple hoist. Other mechanisms, such as typewriters and adding machines, are not subject to large stresses. With them, the chief problem is one of the arrangement and shape of parts for effective operation and low-cost production.

Courtesy Brown & Sharpe Mfg. Co.

Fig. 15–4. "Arm Adjusting Mechanism" on the Milling Machine.

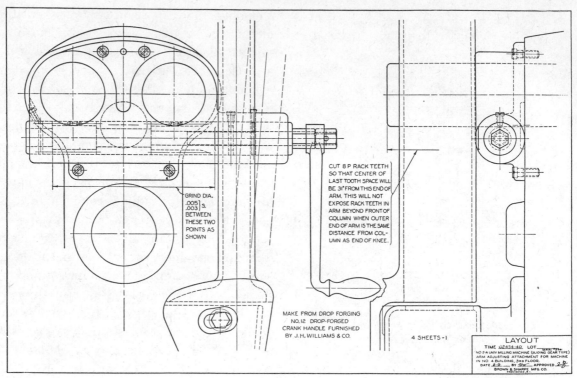

Fig. 15-5. Design Layout of "Arm Adjusting Mechanism."

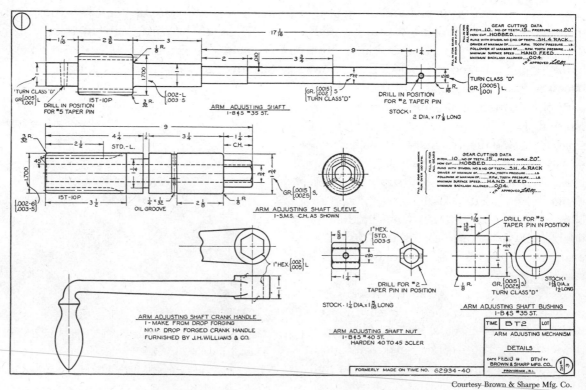

Fig. 15-6. Details of "Arm Adjusting Mechanism."

After the layout has been approved by the chief engineer or others delegated by him, it is turned over to the draftsmen to make the *production drawings*. In relatively small firms, the engineering and drafting may be concentrated in a single department, whereas in larger firms they are separated, as shown in Fig. 15–2. In any case, the *draftsmen* or *detailers* use the layout as a guide and make the detail drawings, Fig. 15–6. The draftsman "picks off" the sizes directly from the layout with the scale or the dividers. The detail drawings show the necessary views, dimensions, and notes required in the shop to make the parts without additional instructions. The details may all be drawn on one sheet, as in Fig. 15–6, or each part on a separate sheet.

Finally, an *assembly drawing*, Fig. 15–7, will be needed to guide workers in assembling parts properly and for general reference

throughout the shops. Since the original layout, Fig. 15–5, is an assembly drawing, it can often be traced, omitting unnecessary adjacent parts and making any other desirable changes to suit the purpose.

If ink tracings are required, these are made by *tracers* in ink, usually on tracing cloth, as discussed in Chapter 8. A tracer is a draftsman who is skilled in doing ink work and who usually is not sufficiently experienced to do detailing. In these modern times ink tracings are made less and less since pencil drawings are satisfactory and easier and quicker to make.

Before blueprints are made, the production drawings must be carefully checked by a *checker*, Sec. 15.6, to make sure they are correct before releasing prints to the shops.

If the mechanism contains new ideas, a *patent drawing* is made, Fig. 15–8, and sent

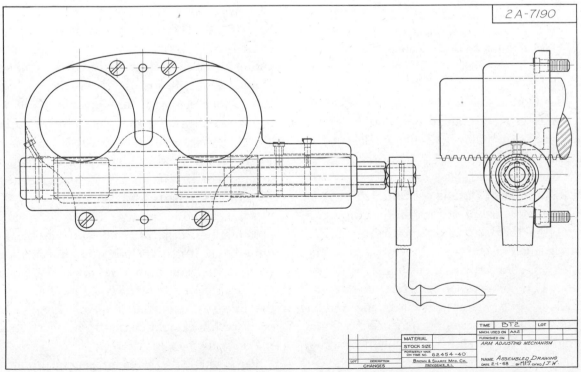

Courtesy Brown & Sharpe Mfg. Co.

Fig. 15–7. Assembly of "Arm Adjusting Mechanism."

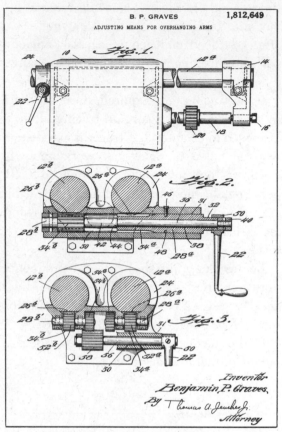

Courtesy Brown & Sharpe Mfg. Co.

Fig. 15–8. Patent Office Drawing for "Arm Adjusting Mechanism."

to the United States Patent Office. Patent drawings are line-shaded, lettered in script, and otherwise follow the rigid rules of the Patent Office.

15.2 Number of Details per Sheet.

Two general methods are used regarding grouping of details on sheets. If the structure or mechanism is small or composed of few parts, all the details may be drawn on one sheet, as in Fig. 15–6. This is common practice in drawing details of jigs, fixtures, valves, and similar mechanisms. The assembly may also be shown along with the details, but usually it is drawn on a separate sheet.

When several details are drawn on one sheet, first attention should be given to spacing. Before beginning to draw any one view or part, block in all details with construction lines, as shown in Fig. 15–9. Ample spacing should be allowed for dimensions and notes. A good method to determine spacing is to cut out rectangular scraps of paper roughly equal to the sizes of the views, and to place these on the sheet in position. Locations are then marked lightly on the sheet, and the scraps of paper are removed.

If possible, the same scale should be used for all details; otherwise, the different scales should be clearly noted under each detail. Each detail should be represented by the regular views, sections, or auxiliary views needed to describe the part clearly, as shown in Chapters 6, 7, 11, and 12, and should have all necessary dimensions and notes, as described in Chapter 9.

One common method of identifying the parts is to letter a title note under each detail, Fig. 15–10. The detail number may be encircled, and the title underlined for emphasis. Under the title the material and the number of pieces required in the assembly are indicated, together with any other general information regarding the piece, for example, f ALL OVER, HARDEN & GRIND.

Another method of identifying the parts is to give a *parts list*, or bill of materials, Fig. 15–11. Such a parts list takes the place of the title notes shown in Fig. 15–10, as regards part number, title, material, and number required; and it may include other information, such as pattern numbers, stock sizes, and weights. The parts list may be located above the title block, *reading upward*, Fig. 15–11, or in the upper right corner of the sheet, *reading downward*.

Parts should be listed in general order of

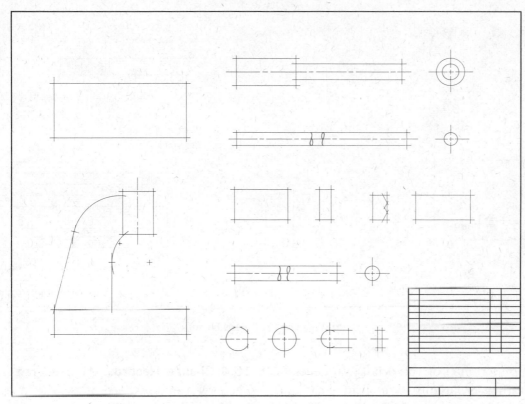

Fig. 15–9. "Blocking in" the Views. (See Fig. 15–45.)

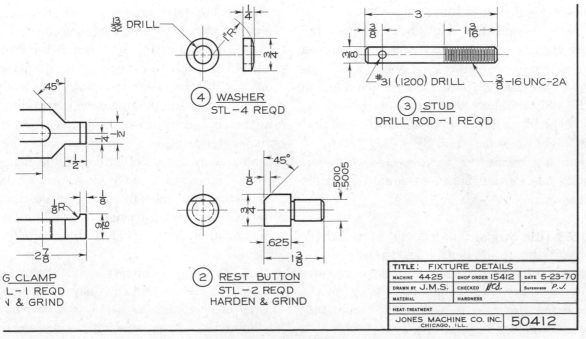

Fig. 15–10. Parts Notes.

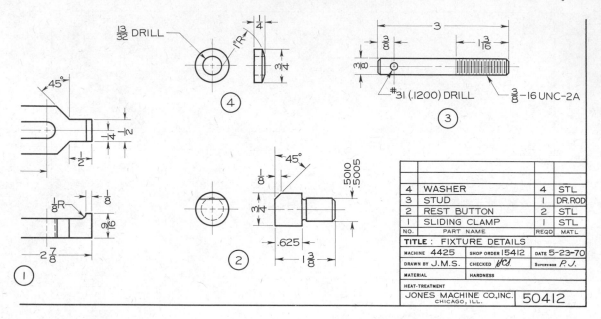

Fig. 15–11. Parts List.

size or importance of the details. In general, main castings or large parts are listed first, and the standard parts or small details last. Standard parts are listed but not drawn unless they are to be altered in some way.

Many companies have now adopted the practice, for convenience in filing, of drawing each detail, no matter how small, on a separate sheet, Fig. 15–12. Many have adopted the USA Standard sheet sizes, based on $8\frac{1}{2}''$ × $11''$, or $9''$ × $12''$, and multiples thereof, Sec. 3.5. The sizes based on $8\frac{1}{2}''$ × $11''$ are most common because of the convenience in correspondence and filing in standard letter file drawers.

15.3 Title Strips. Every drawing should have a *title strip* or title block, Figs. 15–10 to 15–12, whose purpose is to show in an organized way all necessary information not shown on the drawing itself. In industry it is almost universal practice to use printed sheets on which the border and title strip are already printed, so that the draftsman merely fills in the blanks.

15.4 Change Records. After a drawing has been released and used, it is often necessary to make changes, and it is important to keep a complete and accurate record of every change. For this purpose a *change strip*, or *revision strip*, is included at some convenient place on the drawing, as shown at the lower left of Fig. 15–12. An encircled letter is placed on the drawing near the place where the change was made, and the same letter is given in the change strip. Some companies use triangles or squares around the letters, and others use numbers instead of letters. In the change strip, the change is briefly described, usually by stating what the affected part of the drawing *was* before the change. In addition, the draftsman gives his initials and the date.

15.5 Drawing Numbers. Every drawing should be numbered. This may be a simple serial number, such as 30418, or it may have a letter after it or before it to indicate the sheet size, such as 30418A or A30418. Thus, a size A sheet may be $8\frac{1}{2}''$ × $11''$ and size B sheet $11'' \times 17''$. See Sec. 3.5. All sorts of numbering schemes are in use in which the various parts of the number are used to indicate such things as model number of the machine, the general nature of the part, the use of the part, and so on. In general, it is advisable to use simple serial numbers and to avoid using the drawing number to convey other information, as such practice can lead to confusion.

The drawing number is also the number of the part itself. It should be lettered very boldly, at least $\frac{1}{4}''$ high, in the lower right corner and also in the upper left corner of the sheet, Fig. 15–12. For this purpose a Leroy pen, Sec. 4.16, or a Speedball pen, Sec. 4.18, is recommended.

15.6 Checking. The young draftsman in industry soon finds that *correctness* is very important. His initials on the drawing identify him with it and with any errors he has made. When the drawing is completed, it is turned over to another person to check it. In small offices, checking may be done by the original designer or by an experienced draftsman. In large offices, drawings are inspected by a *checker*, who follows a rigid procedure so as to be sure not to overlook anything. In general, the checker covers the following:

1. Soundness of the Design—function, strength, economy, manufacturability, serviceability, ease of assembly and disassembly, lubrication, repairs, and like considerations.

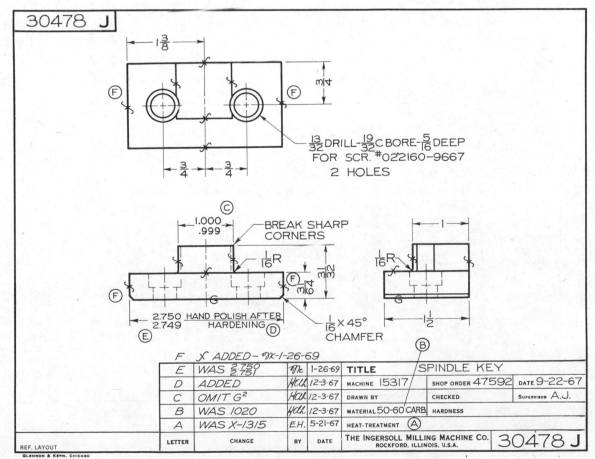

Fig. 15–12. Detail Drawing.

2. Views—scale, sections, partial views, auxiliary views, fillets and rounds.

3. Dimensions and Notes—errors, omissions, duplications, tolerances, shop operations, finish marks.

4. Legibility—linework and lettering for clearest reproduction.

5. Clearances—no interference of moving parts.

6. Materials—best choice, stock sizes, heat treatment, other considerations.

7. Standard Parts—specified wherever possible.

8. Title Block Information.

Throughout this procedure, the checker follows carefully the company engineering and drafting standards, Sec. 15.7. Checking is usually done on a print so as not to mark up the original. Ordinary lead pencils may be used on Ozalid prints or others that have white backgrounds. On blueprints, colored pencils are most satisfactory. Often, a "check-assembly" is drawn to make sure that all parts fit and function in the assembly in a proper manner.

15.7 Standards. Each firm has its own engineering standards, made available to all engineers and draftsmen in loose-leaf form. These include, in addition to engineering data, definite standards that all draftsmen are expected to follow regarding the items listed in Sec. 15.6. Various groups of industries also have standards that sum up the recommended practices in each type of industry—as, for example, the standards of the Society of Automotive Engineers covering the entire automotive industry. Above these are the many publications of the United States of America Standards Institute, including the *USA Standard Drafting*

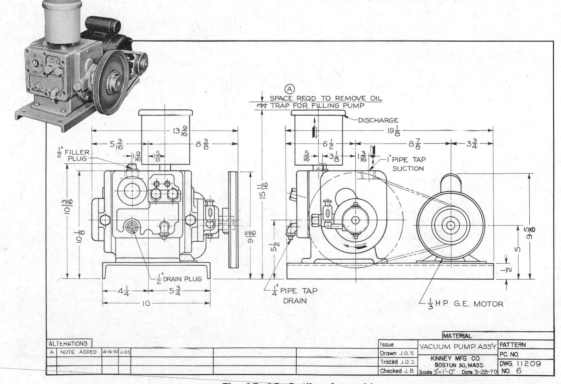

Fig. 15–13. Outline Assembly.

Manual, Y-14. It is the "last word" in drafting standards in this country, and all drawings in this book are drawn in conformity with it.

15.8 Assembly Drawings. An *assembly drawing* shows the entire assembled machine or structure, or the assembly of some unit such as the carburetor or the transmission of an automobile. Assemblies vary in character according to their uses. In general, they are: (1) *design assemblies* or *layouts*, (2) *outline* or *installation assemblies*, (3) *working-drawing assemblies*, and (4) *general assemblies*.

15.9 Outline Assembly. An *outline assembly*, or *installation assembly*, Fig. 15–13, shows one or more views of an assembly "in outline,"

the purpose being to give general information as to character and size of the unit and how it fits in its environment. Little or no sectioning is generally needed, but it may be used if necessary. Small, relatively unimportant details are omitted—as, for example, many screw heads in Fig. 15–13. The dimensions given are only the principal overall and center-to-center distances needed to clarify questions of installation. The outline assembly is widely used in catalogs and other sales literature. When there are several sizes of one machine, dimensions are usually indicated by capital letters and values for each letter given in a table.

15.10 Working-Drawing Assembly. A *working-drawing assembly*, Fig. 15–14, is a com-

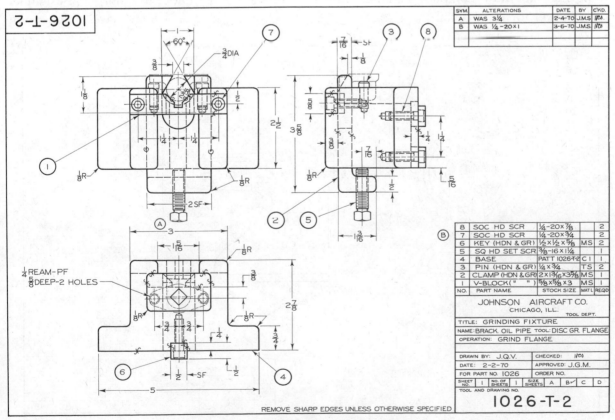

Fig. 15–14. Working-Drawing Assembly.

bined detail and assembly drawing giving complete dimensions and notes for all parts. This method is used in place of separate detail and assembly drawings when the mechanism is relatively simple and all parts can be adequately represented in a single assembly drawing. It often happens that, while most parts can be clearly shown and dimensioned in the assembly, some cannot. In this case these parts are detailed separately on the same sheet, and the drawing becomes a combination of working-drawing assembly and detail drawing.

Since working-drawing assemblies eliminate the cost of separate detail drawings, they are widely used in classes of work where this is possible. Examples are drawings of jigs, fixtures, valves, locomotive sub-assemblies, aircraft sub-assemblies, and certain work not requiring the most complete manufacturing information for each detail.

15.11 General Assembly. A *general assembly*, Fig. 15–15, shows *how the parts fit together and how the assembly functions*. Its chief use is in the assembly shop where all the finished parts are received and put together. The assembly man does not need to learn the shape of any part from the assembly drawing, as he has all the actual parts, ready to be assembled. Consequently, the views used on an assembly drawing are simply those that show clearly how all the parts fit together. These views may be one or more regular views, sections of any kind, auxiliary views, and partial views. Sectioning in assembly drawing is discussed in Sec. 15.13.

As a rule, no dimensions are given on a general assembly drawing, but occasionally some special dimension is given that is related to the function of the entire assembly. For example, a drawing of a jack may have a dimension

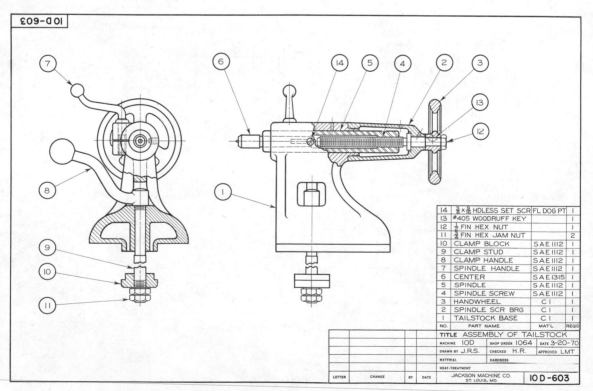

14	$\frac{3}{8}$ X $\frac{7}{16}$ HDLESS SET SCR	FL DOG PT	1
13	#405 WOODRUFF KEY		1
12	$\frac{1}{2}$ FIN HEX NUT		1
11	$\frac{1}{2}$ FIN HEX JAM NUT		2
10	CLAMP BLOCK	SAE 1112	1
9	CLAMP STUD	SAE 1112	1
8	CLAMP HANDLE	SAE 1112	1
7	SPINDLE HANDLE	SAE 1112	1
6	CENTER	SAE 1315	1
5	SPINDLE	SAE 1112	1
4	SPINDLE SCREW	SAE 1112	1
3	HANDWHEEL	C 1	1
2	SPINDLE SCR BRG	C 1	1
1	TAILSTOCK BASE	C 1	1
NO.	PART NAME	MAT'L	REQD

TITLE ASSEMBLY OF TAILSTOCK

MACHINE IOD	SHOP ORDER 1064	DATE 3-20-70
DRAWN BY J.R.S.	CHECKED H.R.	APPROVED LMT
MATERIAL	HARDNESS	
HEAT-TREATMENT		

| LETTER | CHANGE | BY | DATE | JACKSON MACHINE CO. ST. LOUIS, MO. | IOD-603 |

Fig. 15–15. General Assembly Drawing.

showing the maximum height when open and the minimum height when closed, or a drawing of a vise may have a dimension showing the maximum opening of the jaws.

Frequently, with large or complicated machines it is not possible to show all parts in one assembly. In such situations a separate drawing is made showing a group of related parts that form a unit of the whole machine and make what is called a *sub-assembly* or *unit assembly*. See Fig. 15–7.

15.12 Title Strips and Parts Lists on Assemblies. Title strips on assembly drawings are the same as on detail drawings, Sec. 15.3, except that the title includes the word "assembly," as in the following examples:

CONNECTING ROD ASSEMBLY
ASSEMBLY OF GRINDER VISE

Parts lists are similar to those on detail drawings where a number of details are shown on the same sheet, Fig. 15–11. The parts list may be placed in any convenient open corner on the drawing, but preferably is located to read up from the title block, Fig. 15–15, or down from the upper right-hand corner of the sheet. Ordinarily the information given for each part includes the part number, part name, material, and number of pieces required in the assembly. Part No. 1 is usually the main base or casting and is followed by the other parts

in general order of decreasing size, and finally by the standard parts, such as bolts, screws, bearings, and pins.

In order to save drafting time in lettering and to facilitate filing and record-keeping, it is quite common, particularly in large plants, to give parts lists in typewritten form on separate parts list sheets.

On the assembly drawing, each part usually is identified with its description in the parts list. This is done by lettering the part numbers in $\frac{7}{16}''$ or $\frac{1}{2}''$ diameter circles near the assembly, and drawing leaders to each part where it is most clearly shown, Figs. 15–14 and 15–15. The circles should be arranged in groups in vertical or horizontal rows, and not scattered in disorder on the sheet. Leaders should not cross, and none should be drawn vertically or horizontally on the sheet.

15.13 Assembly Sections. In assembly drawings where several adjacent parts are sectioned, it is necessary to draw the section lines in different directions in order to distinguish the pieces clearly, as shown in Fig. 15–16. The first large area, (a), is section-lined at 45°. The next large area, (b), is then section-lined at 45° in the opposite direction. Additional areas are section-lined at 30° or 60° with horizontal, (c) and (d). If necessary, in order to make any area contrast with the others, any

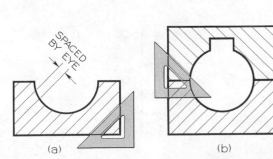
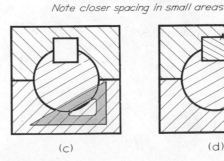
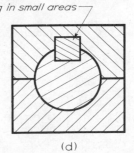

Fig. 15–16. Section Lining (Full Size).

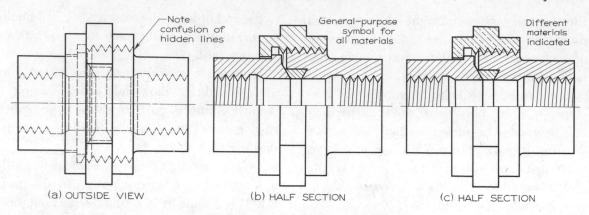

(a) OUTSIDE VIEW   (b) HALF SECTION   (c) HALF SECTION

Fig. 15–17. Assembly Sectioning.

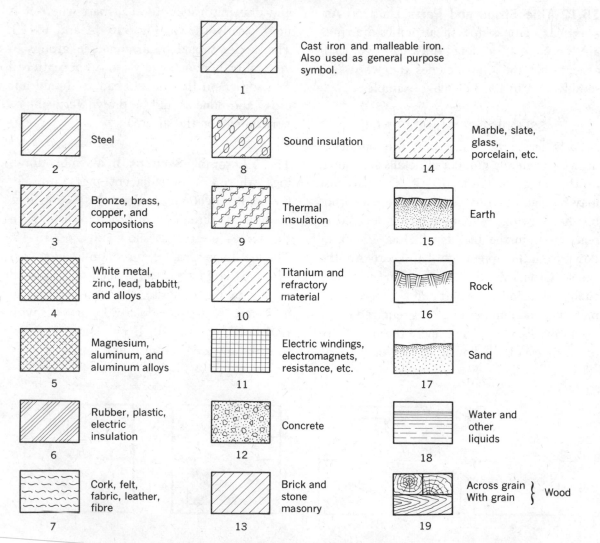

1 — Cast iron and malleable iron. Also used as general purpose symbol.

2 — Steel

3 — Bronze, brass, copper, and compositions

4 — White metal, zinc, lead, babbitt, and alloys

5 — Magnesium, aluminum, and aluminum alloys

6 — Rubber, plastic, electric insulation

7 — Cork, felt, fabric, leather, fibre

8 — Sound insulation

9 — Thermal insulation

10 — Titanium and refractory material

11 — Electric windings, electromagnets, resistance, etc.

12 — Concrete

13 — Brick and stone masonry

14 — Marble, slate, glass, porcelain, etc.

15 — Earth

16 — Rock

17 — Sand

18 — Water and other liquids

19 — Across grain / With grain } Wood

Fig. 15–18. USA Standard Symbols for Section Lining (USAS Y14.2—1965 Tentative).

other angle may be used. Note that section lines do not meet at the visible lines separating the areas, and that for small areas the lines are drawn closer together.

In Fig. 15–17 (a) is shown a regular outside view of an assembled unit, illustrating how interior shapes are not made clear by means of hidden lines. The separate pieces overlap and it is impossible to distinguish between them. A half section of the same unit is shown at (b) in which the four pieces are clearly shown by section lines in opposite directions.

It is sometimes desirable on assembly sections to indicate the different materials by means of symbolic section lining, as shown at (c). The USA Standard symbols for section lining are shown in Fig. 15–18. On detail drawings of individual parts, it is recommended that the symbol for cast iron be used for all materials and that the exact specification of materials be given in a note under the views or in the title strip.

In sectioning very thin parts, such as sheet metal, gaskets, and like pieces, when there is not enough space for section-lining, the sec-

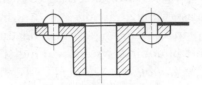

Fig. 15–19. Sectioning Thin Parts.

tioned parts may be shown solid, as in Fig. 15–19.

In assembly sections, there may lie in the path of the cutting plane some solid parts that themselves do not require sectioning and where the use of sectioning would make the drawing less clear and would lengthen the time required to draw it. In such cases it is customary *not to section* such solid parts. A pictorial drawing of a flange coupling is shown in Fig. 15–20 (a), in which the bolts, nuts, shafts, and keys are not sectioned but are left in full form. The corresponding assembly drawing is shown at (b). Other parts that are usually not sectioned are ribs, gear teeth, spokes, screws, nails, ball and roller bearings, and pins. See Figs. 14–28 and 14–32.

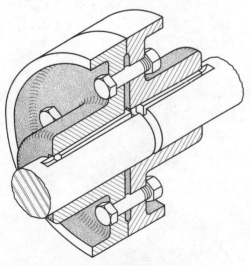

(a) PICTORIAL ASSEMBLY

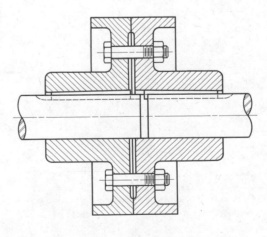

(b) ASSEMBLY

Fig. 15–20. Solid Parts in Assembly Sections.

15.14 Piping Drawings. Piping drawings are actually assembly drawings. In drawings for power plants, pumping plants, heating systems, and plumbing systems, piping is represented by *double-line drawings*, Fig. 15–21 (a), or *single-line drawings,* (b), depending upon how complete a picture is necessary. For the most part, the single-line drawings are used because of the saving in drafting time. Since piping usually runs in many directions, ordinary top, front, and side views may be very confusing, if not impossible to read, because of many overlapping pipes. For this reason it is common to draw piping systems pictorially, either in isometric, Fig. 15–22 (a), or in oblique, (b). For complete information on pictorial drawing methods, refer to Chapter 16.

15.15 Computer Graphics. The use of electronic computers today in nearly every phase of engineering, science, business, and industry is well known. Today's high-speed computers are capable of carrying out long and complex sequences of operations, manipulating information, and making logical comparisons and routine decisions, if properly programmed, all at tremendous speed and without human intervention.

In general, computers may be classified as one of two kinds—*digital* and *analog*. A digital computer counts by digits, going from one to two to three, and so on, in distinct steps, whereas the analog computer measures continuously, without steps. A slide rule, an electric kitchen clock, and the speedometer on a

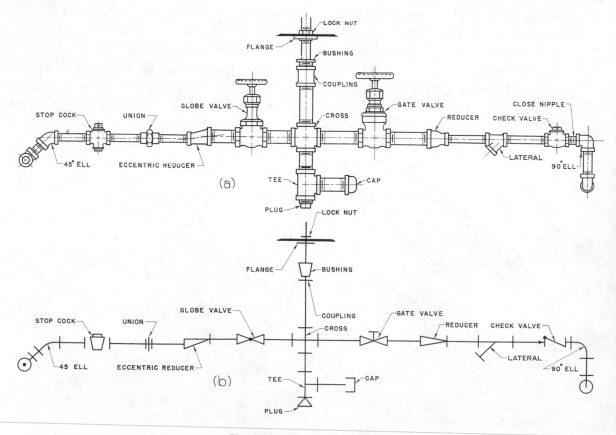

Fig. 15–21. Piping Drawings.

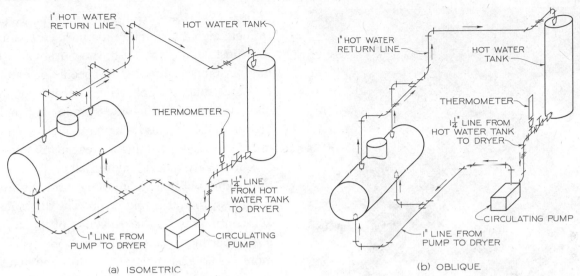

Fig. 15–22. Pictorial Piping Drawings.

car are all examples of an analog device. An abacus or a wrist watch, with the second hand advancing in regular steps, are examples of digital devices. Digital computers are more widely used than analog computers because they are more flexible and can do a greater variety of jobs.

In recent years, research in computer graphics—that is, the graphical use of digital computers by design engineers—has brought about great changes in the methods and processes of producing drawings, graphs, diagrams, etc. Several of the systems developed use a computer in conjunction with a cathode-ray tube and light pen (a photo cell in a pen-like tube) or with an *X–Y recorder* or plotter. One of the earliest developments was the MIT "Sketchpad" system, which uses a cathode-ray tube and a light pen for graphic input to a computer. The light pen is connected to the computer in a manner such that it will send an electric signal to the computer whenever it sees a spot of light on the oscilloscope face. Used like a pen, it imitates the drawing ability of a real pen by having the computer program track

its motion over the face of the tube, thus permitting the engineer to draw shapes on the face of the tube that can be stored numerically in the computer's memory unit. He can also revise or erase the shapes appearing on the tube much as he would on paper.

Digital computers can also be used with several makes of X–Y plotters to produce drawings, graphs, charts, and other graphical data. The plotting is produced by the incremental movement of a pen relative to the surface of the graph paper along the X- and Y-axes.

One of the newest computer-aided design systems is the *DAC-I system* (Design Augmented by Computers), Fig. 15–23, developed by the General Motors Research Laboratories. The system consists of three parts: a large IBM-7094 computer with extra memory unit, a console with a TV-like screen through which the design engineer and computer communicate with each other, and an image processor that reproduces on film or paper the designs produced on the console. The design engineer begins with his sketches and writes out a description of his design problem in computer

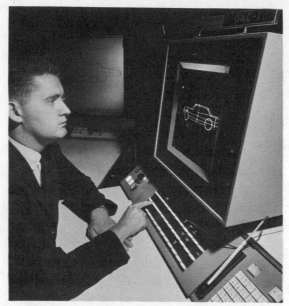

Courtesy General Motors Corp.

Fig. 15–23. General Motors Corp. DAC–I System.

language (programming) that is then stored in the computer memory section. The design engineer then sits at the console and instructs the computer to work out the problem he has described to it, communicating through a typewriter-like keyboard. When the computer has finished, the results show up on the screen as a drawing. The designer can get any view or a perspective of, for example, a fender, by simply touching a special pencil to the screen. He can change the sketch by altering the description in the computer's memory section. When he obtains a design that he likes, he can get a permanent copy of the sketch and a mathematical model defining the design in all its dimensions in the form of equations.

Although many of these systems are still in the experimental stage, they will undoubtedly become a reality in the not too distant future. It should not be assumed, however, that all drawings in the future will be made by computers, and that engineers, designers, and draftsmen will no longer be needed. The cost

of many of these computer systems is extremely high, thus prohibiting most companies from installing them. Also, their usefulness in most instances is limited to specific or repetitive and routine applications because of the high cost of operating the system. It should be remembered, also, that although a computer is capable of doing a great many things, it is in reality a giant moron since it cannot think, and will not do anything more, or anything less, than it is told to do. The draftsman will continue to be the backbone of the drafting department, although his work will be more creative as some of the tiresome details are automated.

Although computer graphics will be an important part of engineering work in the future, a knowledge of the principles of technical drawing will still be a prime requisite for engineers, designers, and draftsmen engaged in this activity, and, if anything, additional training in this subject will be necessary.

15.16 Welding Drawings. A welding drawing is, in one sense, an assembly drawing because it shows a number of individual pieces joined together by welding to form a single unit.

In earlier days, it was common practice to indicate welding information on a drawing by simply lettering a note such as "To be completely welded" or other similar instructions. This practice resulted in costly shop mistakes, with not enough or too many welds being made, due primarily to a lack of specific welding information on the drawing. Since it would be impractical to show the welds themselves on a drawing, graphical welding symbols were developed by the American Welding Society, and are completely described in their publication AWS A2.0-58, "Welding Symbols." These symbols have also been adopted by the United States of America Standards

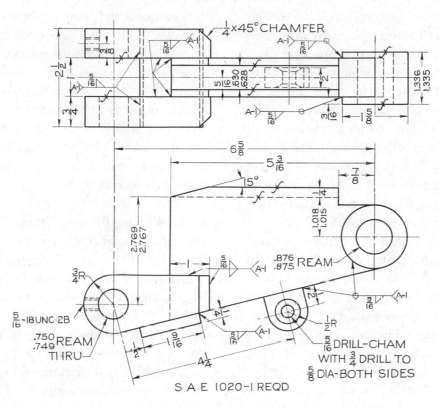

Fig. 15–24. Welding Drawing.

Institute, and published as USAS Y32.3—1959, *Graphical Symbols for Welding*. Information pertaining to standard welding symbols is shown in Table 18 (Appendix, pages 470–71). A completely dimensioned welding drawing is shown in Fig. 15–24.

The complete welding symbol, shown in Table 18 (Appendix, pages 470–71), consists of eight elements, or as many of these elements as are necessary. These are (1) reference line, (2) arrow, (3) basic weld symbols, (4) dimensions and other data, (5) supplementary symbols, (6) finish symbols, (7) tail, and (8) specification, process, or other references. A welding symbol may be drawn mechanically or freehand, but it is important that the elements of a symbol maintain standard locations with respect to each other. For complete in-

formation and procedures regarding the specification and application of welding symbols, the AWS and USA Standards should be consulted.

15.17 Aerospace Drafting. No other industry in modern times has experienced such a rapid growth as the aerospace industry, a vast industrial complex that designs and builds aircraft, missiles, and space vehicles. The name aerospace is used because it generally is understood to include all means of both air and space travel. What a generation ago was considered science fiction is today scientific fact. The aerospace industry has developed modern aircraft, many of which are capable of flying at supersonic speeds, for both civilian and military use. It has also, through cooperation with government agencies such as the National Aeronautics and Space Administration (NASA), conducted successful space programs that have resulted in astronauts being orbited into space and rockets sent to the moon.

The size and importance of the aerospace industry can readily be appreciated in the development of a supersonic jet fighter or a guided missile. The development of each requires the efforts of over 2100 engineers and draftsmen expending several million man-hours in the preparation of 18,000 drawings, which contain the designs and information for 100,000 different parts. Aerospace drafting, Fig. 15–25, is therefore a composite of many specialized types of drafting such as mechanical, electrical, structural, sheet-metal, etc., and includes many different kinds of drawings such as layout drawings, working drawings, detail drawings, assembly drawings, pictorial drawings, installation drawings, and many others. A competent aerospace draftsman will possess a working knowledge of the principles and techniques of technical drawing, and have a basic understanding of the terminology, methods, and procedures used in the aerospace industry.

Because of the many complex and varied activities in the aerospace industry, it would be impossible to cover the subject in any detail in a few short paragraphs. Standards have been developed, however, with which the draftsman should be familiar and which will serve as a source of reference. Among these are the SAE *Aerospace Drafting Manual*, government standards and publications, and individual company drafting standards.

15.18 Working Drawing Problems. The problems on the following pages have been taken from blueprints of actual machines or devices manufactured and used today. These range from comparatively simple problems to the more complicated problems that may be assigned to the superior student. The student, with the advice of his instructor, is to select his own sheet size (see Appendix, pages 474–75) for the problem assigned. The first step should be to make "thumbnail sketches" of the views of each part, and obtain the instructor's approval. After that, the instructor may assign more complete sketches, fully dimensioned, followed by the mechanically drawn detail and assembly drawings.

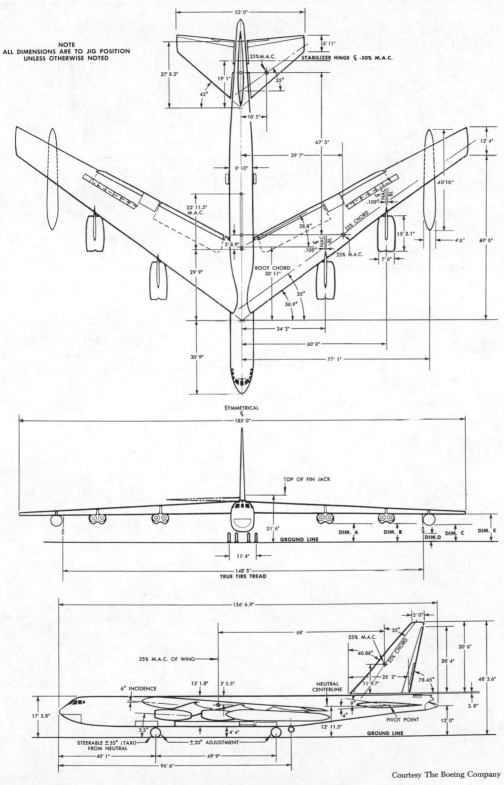

Courtesy The Boeing Company

Fig. 15–25. Typical Aerospace Drawing.

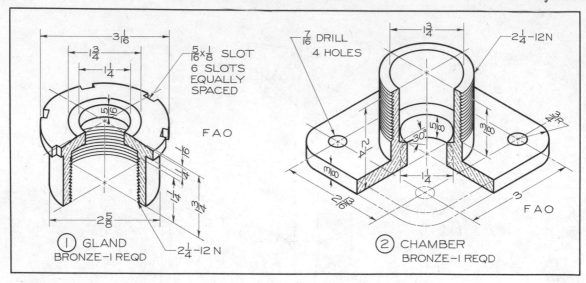

Fig. 15–26. Stuffing Box.

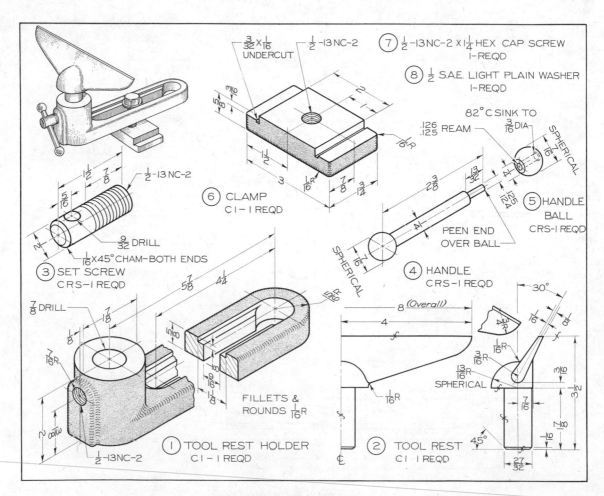

Fig. 15–27. Tool Post.

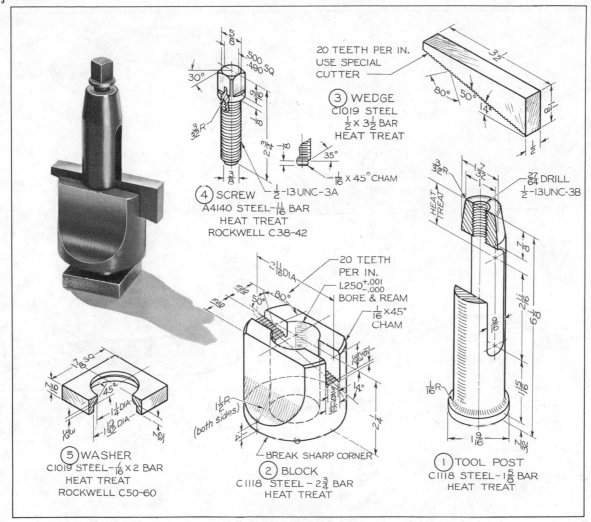

Fig. 15–28. Tool Post.

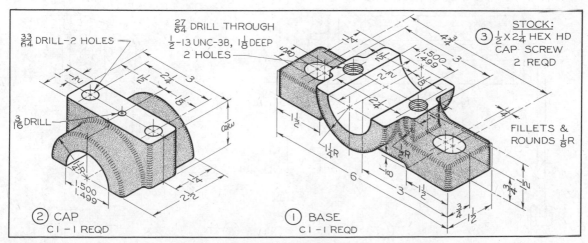

Fig. 15–29. Pillow Block.

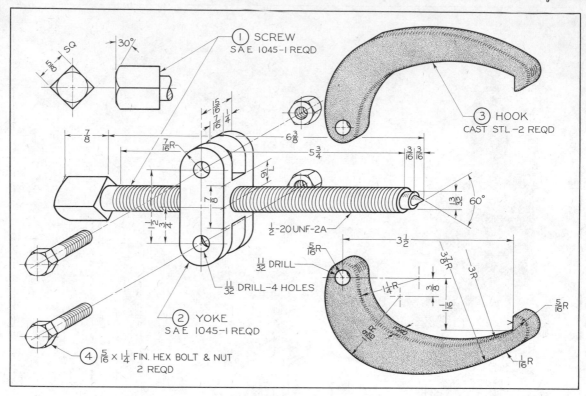

Fig. 15–30. Puller.

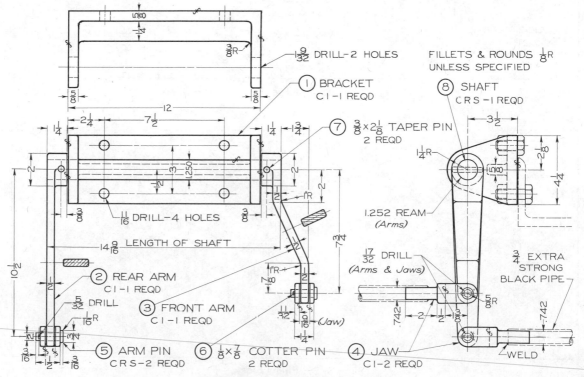

Fig. 15–31. Cylinder Cock Controls.

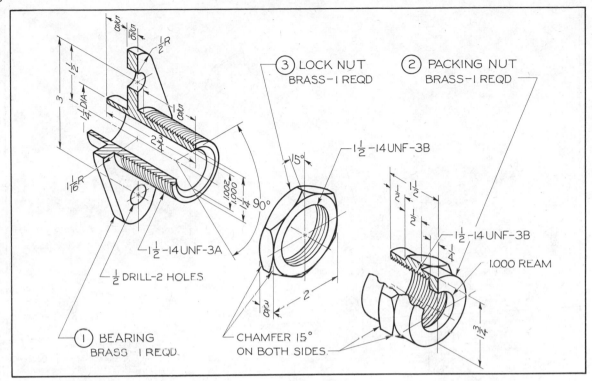

Fig. 15–32. Stuffing Box.

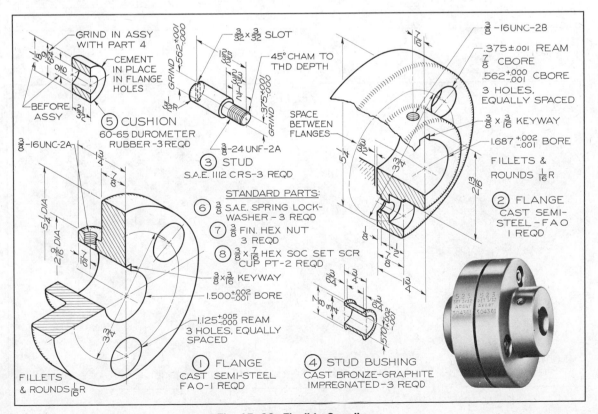

Fig. 15–33. Flexible Coupling.

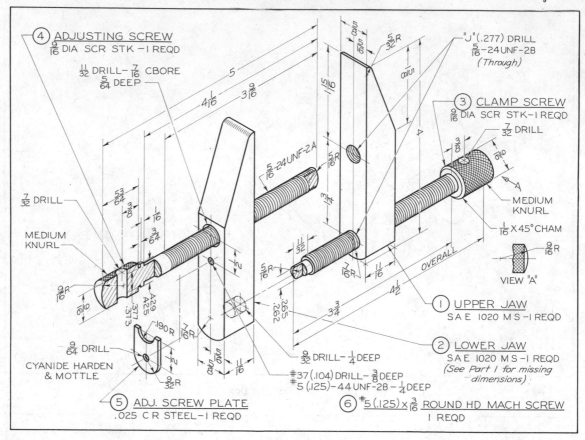

Fig. 15–34. Toolmaker's Clamp.

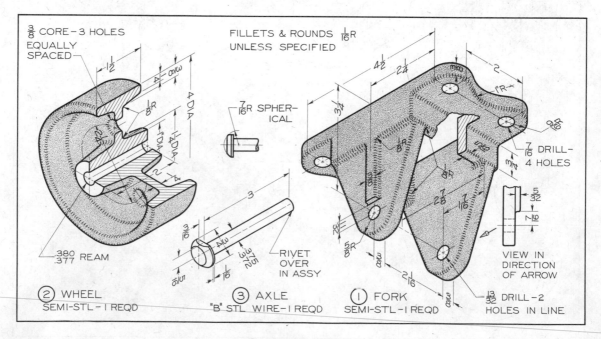

Fig. 15–35. Caster.

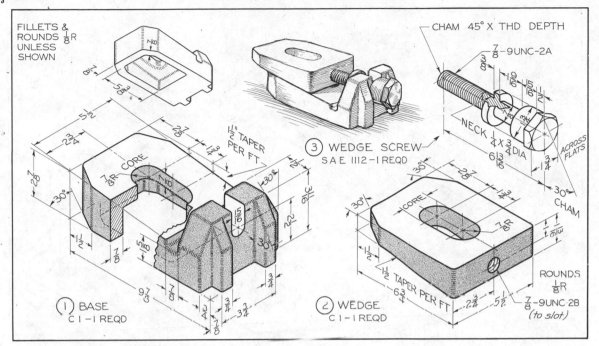

Fig. 15-36. Leveling Wedge.

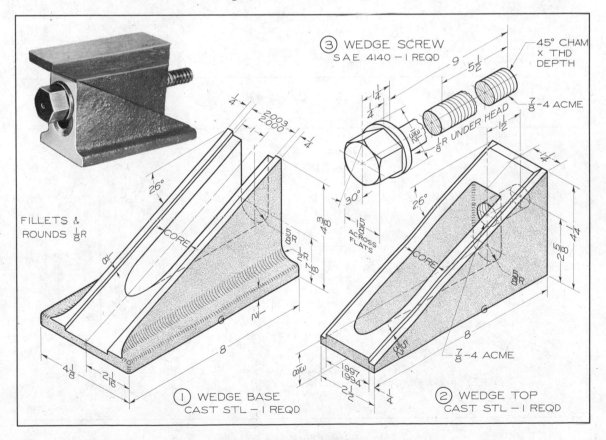

Fig. 15-37. Raising Block.

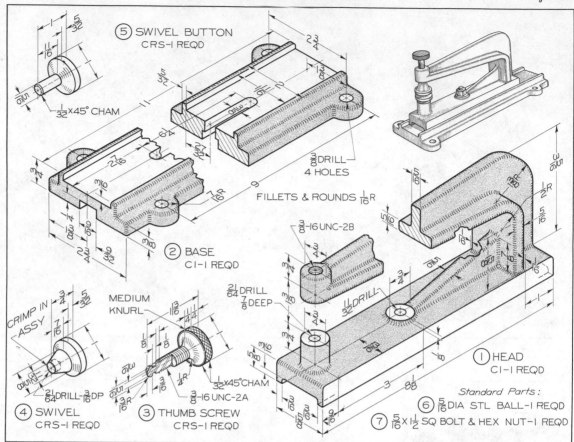

Fig. 15–38. Clamping Head.

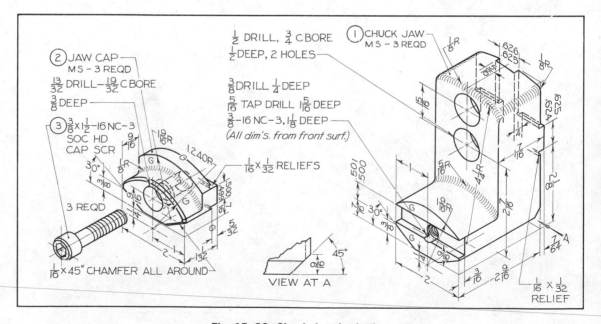

Fig. 15–39. Chuck Jaw for Lathe.

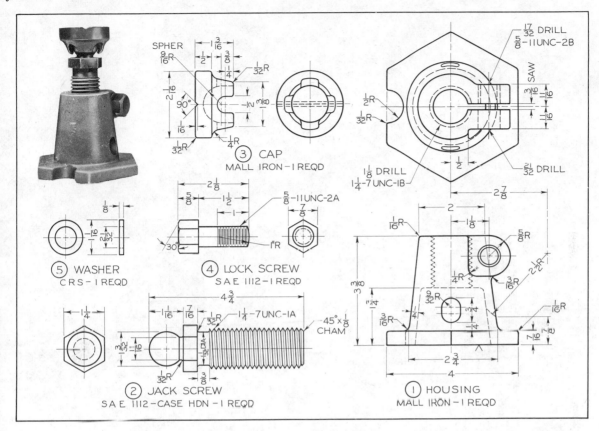

Fig. 15-40. Planer Jack.

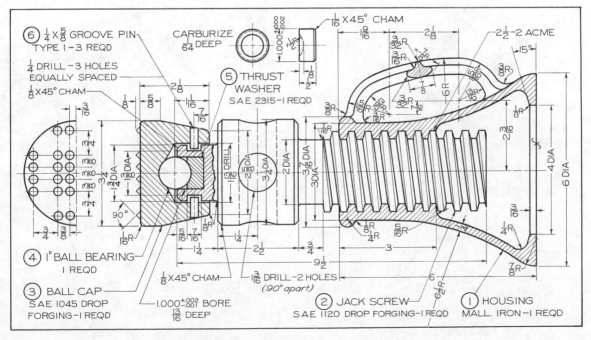

Fig. 15-41. Loco Screw Jack.

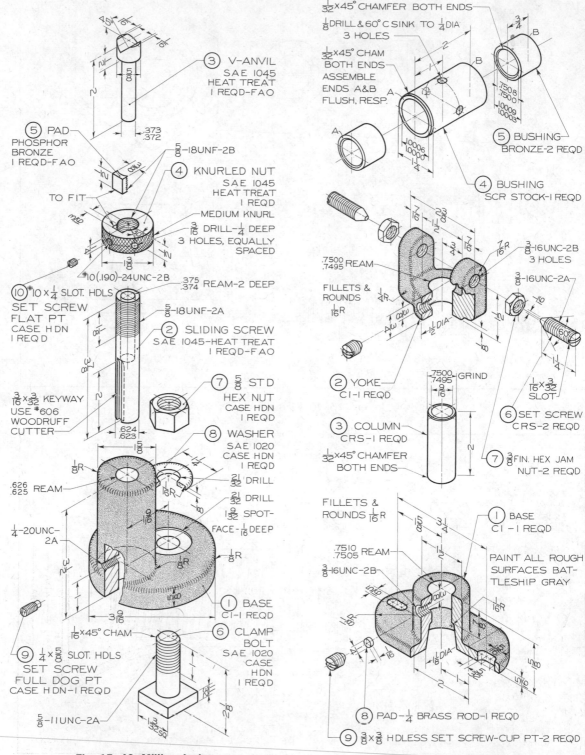

Fig. 15–42. Milling Jack.

Fig. 15–43. Shaft Support.

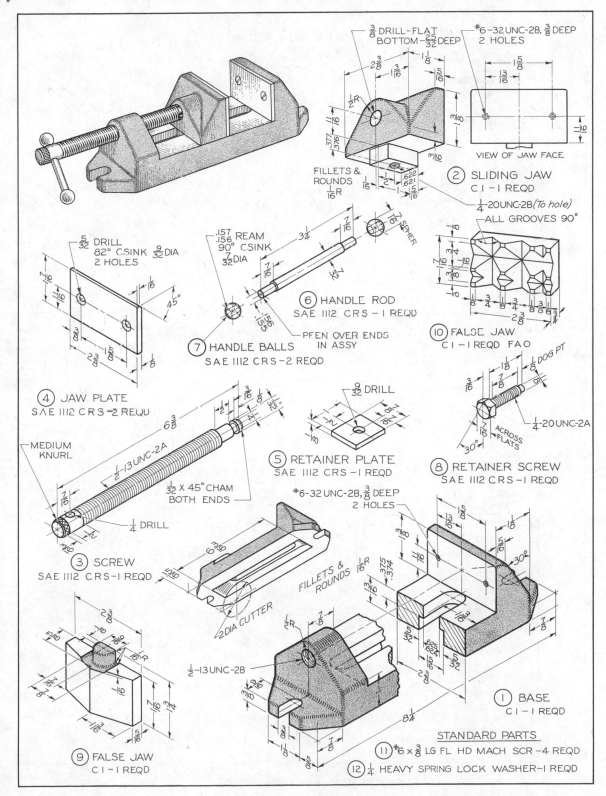

Fig. 15–44. Drill Press Vise.

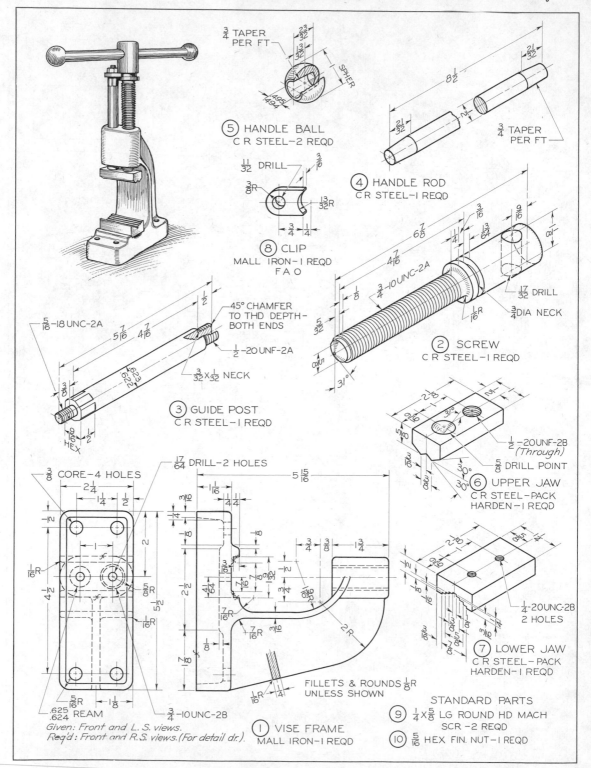

¾ TAPER PER FT

⑤ HANDLE BALL
C R STEEL – 2 REQD

⑧ CLIP
MALL IRON – 1 REQD
F A O

⑪ DRILL

④ HANDLE ROD
C R STEEL – 1 REQD

¾ TAPER PER FT

② SCREW
C R STEEL – 1 REQD

¾–10UNC–2A

45° CHAMFER TO THD DEPTH– BOTH ENDS

½–20UNF–2A

³⁄₃₂ × ¹⁄₃₂ NECK

⑤⁄₁₆–18UNC–2A

③ GUIDE POST
C R STEEL – 1 REQD

HEX

½–20UNF–2B
(Through)
⅝ DRILL POINT

⑥ UPPER JAW
C R STEEL – PACK HARDEN – 1 REQD

³⁄₈ CORE – 4 HOLES

¹⁷⁄₆₄ DRILL – 2 HOLES

¼–20UNC–2B
2 HOLES

⑦ LOWER JAW
C R STEEL – PACK HARDEN – 1 REQD

FILLETS & ROUNDS ⅛R UNLESS SHOWN

.625 / .624 REAM

¾–10UNC–2B

Given: Front and L. S. views.
Req'd: Front and R.S. views.(For detail dr.).

① VISE FRAME
MALL IRON – 1 REQD

STANDARD PARTS

⑨ ¼ × ⅝ LG ROUND HD MACH SCR – 2 REQD

⑩ ⁵⁄₁₆ HEX FIN. NUT – 1 REQD

Fig. 15–45. Pipe Vise. (See Fig. 15–9.)

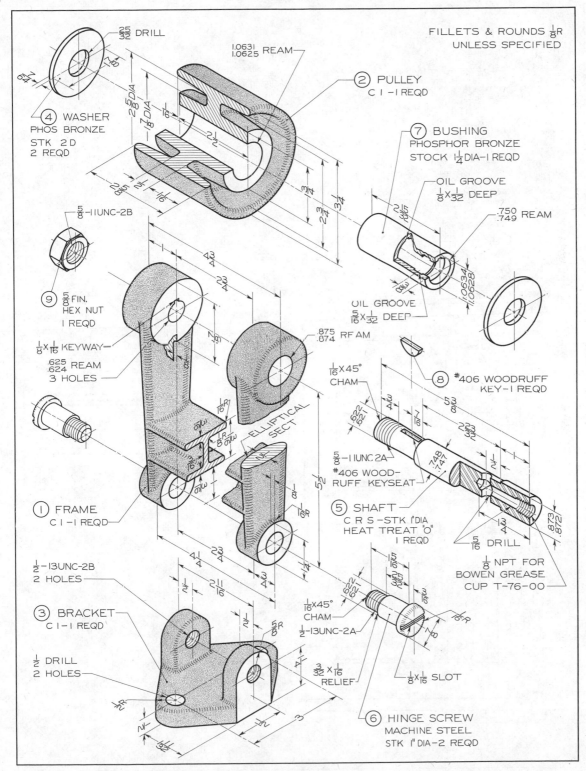

Fig. 15-46. Belt Tightener.

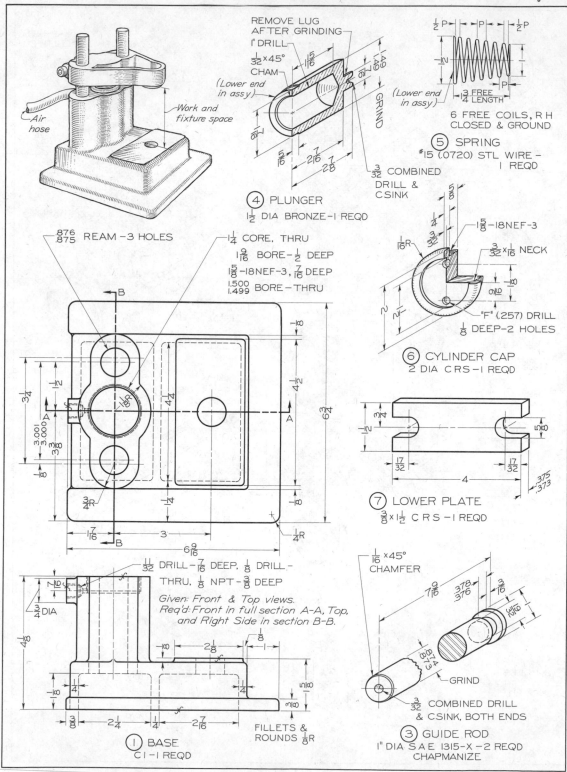

Fig. 15–47. Drilling and Tapping Jig. (See also Fig. 15–48.)

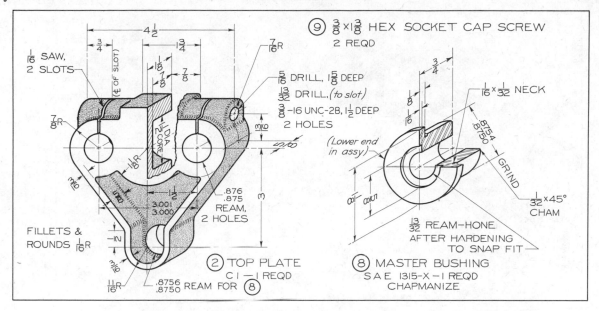

Fig. 15–48. Drilling and Tapping Jig. (See also Fig. 15–47.)

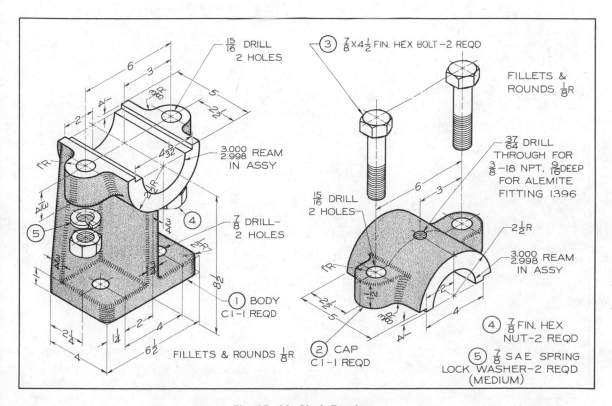

Fig. 15–49. Shaft Bearing.

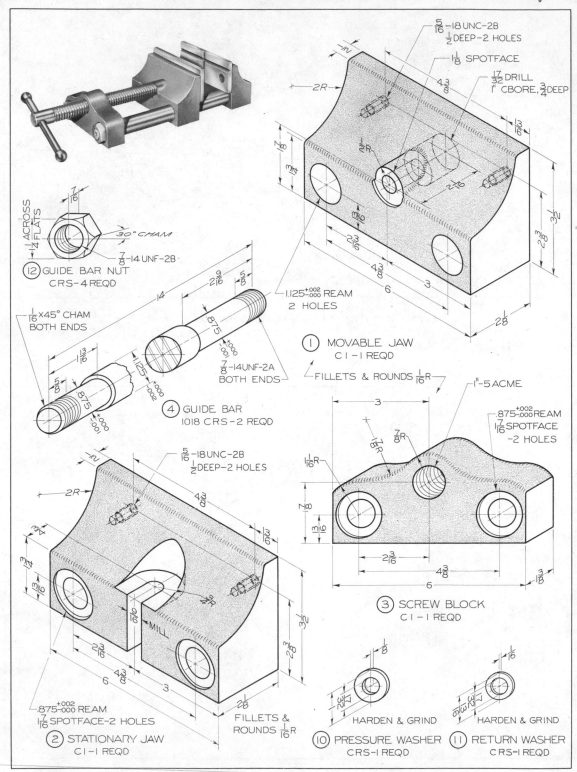

Parts annotations

① MOVABLE JAW
C1 – 1 REQD

$\frac{5}{16}$ –18 UNC–2B
$\frac{1}{2}$ DEEP–2 HOLES

$1\frac{1}{8}$ SPOTFACE

$\frac{17}{32}$ DRILL
1" CBORE, $\frac{3}{4}$ DEEP

1.125^{+002}_{-000} REAM
2 HOLES

FILLETS & ROUNDS $\frac{1}{16}$ R

⑫ GUIDE BAR NUT
CRS – 4 REQD

$\frac{7}{16}$
$\frac{1}{4}$ ACROSS FLATS
30° CHAM
$\frac{7}{8}$ –14 UNF–2B

④ GUIDE BAR
1018 CRS – 2 REQD

$\frac{1}{16}$ × 45° CHAM BOTH ENDS
$\frac{7}{8}$ –14 UNF–2A BOTH ENDS

③ SCREW BLOCK
C1 – 1 REQD

1"–5 ACME

$.875^{+002}_{-000}$ REAM
$1\frac{7}{16}$ SPOTFACE –2 HOLES

② STATIONARY JAW
C1 – 1 REQD

$\frac{5}{16}$ –18 UNC–2B
$\frac{1}{2}$ DEEP–2 HOLES

MILL

$.875^{+002}_{-000}$ REAM
$1\frac{7}{16}$ SPOTFACE–2 HOLES

FILLETS & ROUNDS $\frac{1}{16}$ R

⑩ PRESSURE WASHER
CRS–1 REQD
HARDEN & GRIND

⑪ RETURN WASHER
CRS–1 REQD
HARDEN & GRIND

Fig. 15–50. Drill Press Vise. (See also Fig. 15–51.)

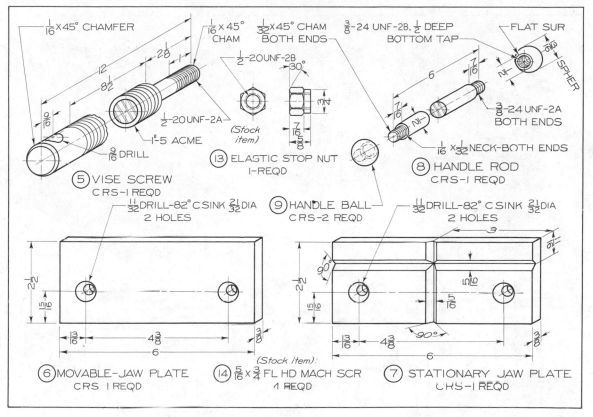

Fig. 15–51. Drill Press Vise. (See also Fig. 15–50.)

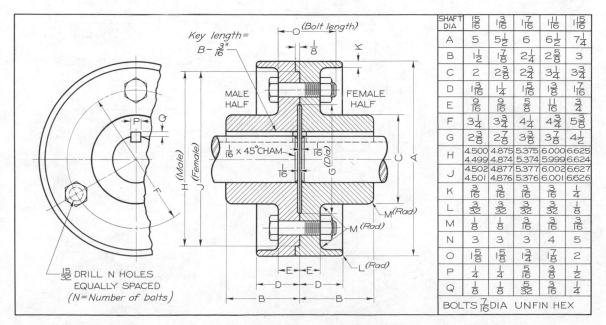

Fig. 15–52. Flange Coupling.

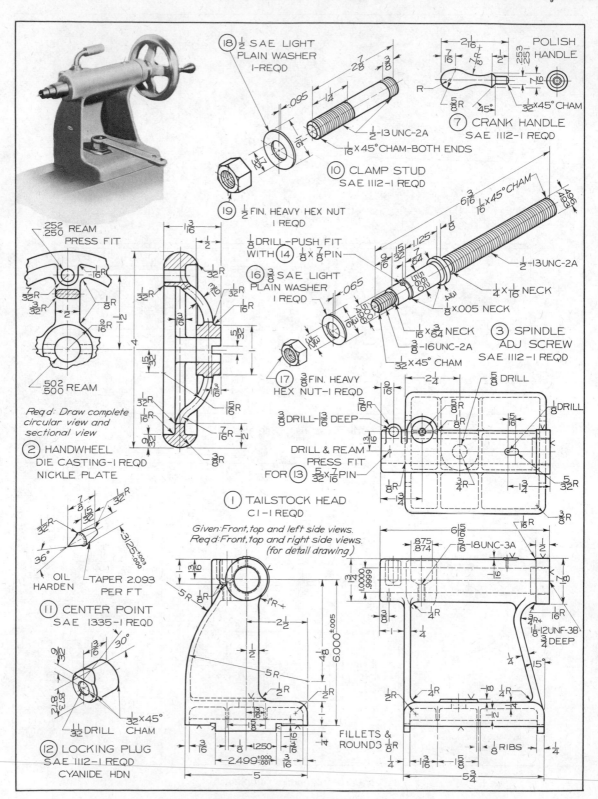

Fig. 15–53. Lathe Tailstock. (See also Fig. 15–54.)

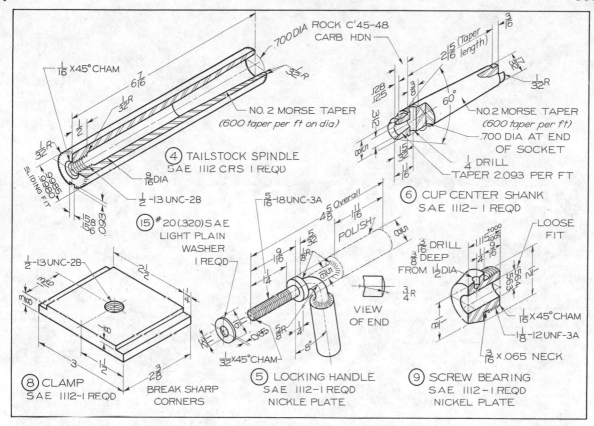

Fig. 15-54. Lathe Tailstock. (See also Fig. 15-53.)

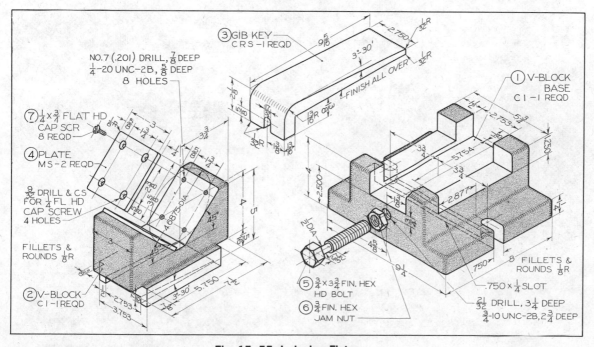

Fig. 15-55. Indexing Fixture.

CHAPTER 16

PICTORIAL DRAWINGS

16.1 Pictorial Drawing. A *pictorial drawing* is one in which the object is viewed in such a position that several faces appear in a single view. Pictorial drawings are excellent for showing the *appearance* of objects, and are often used to supplement multiview drawings—especially to show untrained people the appearance of objects that they cannot visualize from the views on the blueprints.

Pictorial drawings are used extensively in catalogs and sales literature; in technical books;* in Patent Office drawings; in piping and wiring diagrams; in machine, structural, and architectural drawing; and in furniture illustrations. Although only an artist or an

* Most of the pictorial illustrations in this book were drawn by the methods discussed in this chapter.

Courtesy Wettlaufer Mfg. Co.

Fig. 16–1. Automobile Stylist at Work.

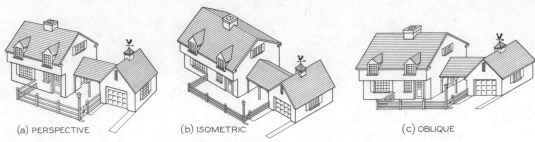

Fig. 16–2. Types of Pictorial Drawings.

architect may prepare pictures of complicated objects, anyone who is interested can learn to make ordinary pictorial drawings. A professional automobile stylist, or pictorial draftsman, is shown at work in Fig. 16–1.

16.2 Types of Pictorial Drawings. The three most common types of pictorial drawings are shown in Fig. 16–2. *Perspective,* (a), is the most natural representation, being geometrically the same as a photograph. However, perspectives are comparatively difficult to draw. Other methods, which are usually quite satisfactory and are much easier to draw, are *isometric,* (b), and *oblique,* (c). All three types will be discussed below.

16.3 Isometric Sketching. One of the most effective ways to sketch an object pictorially is to sketch it *in isometric.* Take the object in your hand, and tilt it toward you approxi-

mately as shown in Fig. 16–3 (a). The front edge (and those edges parallel to it) will appear vertical, and the two lower edges (and those parallel to them) will appear about 30° with horizontal.

I. Start by sketching the enclosing box, sloping lines AC and AD at about 30° with horizontal. Make the height AB equal to the height of the block, the width AD equal to the width of the block, and the depth AC equal to the depth of the block.

II. Block in the right-angled notch.

III. Darken all final lines.

This same procedure is used in sketching any rectangular object, such as the television cabinet in Fig. 16–4. Do not be discouraged by a seemingly complicated object, for the method is the same—only there are more lines. Why not try a sketch of your drawing table, a chair, or something in your home such as a china cabinet?

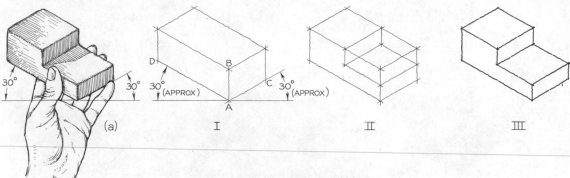

Fig. 16–3. Isometric Sketching.

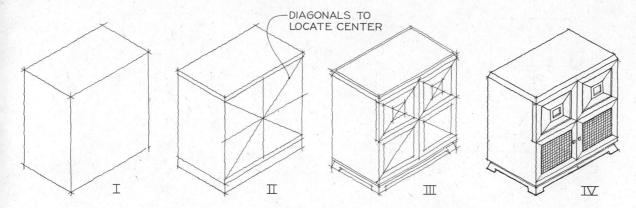

Fig. 16-4. Isometric Sketch of Television Cabinet.

16.4 Sketching Isometric Ellipses. In isometric, circles appear as ellipses. The steps in sketching a cylinder are shown in Fig. 16-5.

I. Sketch the enclosing box.

II. Sketch diagonals and center lines of the ends.

III. Sketch the ellipses and complete the cylinder. Note that the major axes of the ellipses are at right angles to the center line of the cylinder. Isometric sketches of other common objects showing the major axes of ellipses at right angles to center lines are shown at (a).

16.5 Sketching on Isometric Cross-Section Paper. One of the best aids in learning to sketch in isometric is cross-section paper, as shown in Fig. 16-6. Two given views are shown at (a).

I. Sketch the enclosing box. Count the isometric grid spaces equal to the corresponding

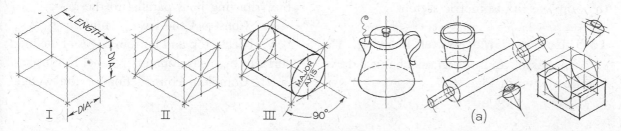

Fig. 16-5. Isometric Ellipses.

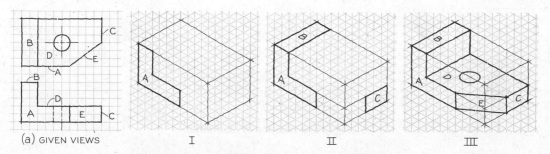

Fig. 16-6. Sketching Isometric from Given Views.

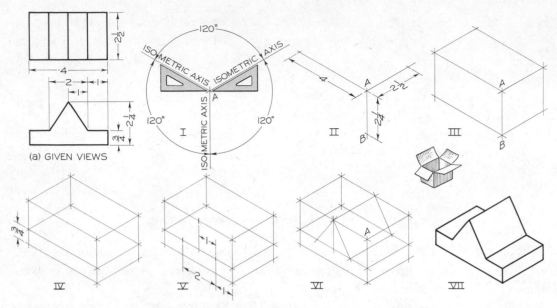

Fig. 16–7. Steps in Isometric Mechanical Drawing.

squares on the given views. Sketch surface A in isometric, as shown.

II. Sketch surfaces B and C in isometric.

III. Sketch surfaces D and E and the ellipse to complete the isometric sketch.

16.6 Isometric Mechanical Drawing. The steps in making an isometric mechanical drawing are shown in Fig. 16–7.

I. Draw the *isometric axes* 120° apart.

II. Set off along the axes the height ($2\frac{1}{4}''$), the width ($4''$), and the depth ($2\frac{1}{2}''$).

III. Complete the enclosing construction box, drawing lines parallel to the axes.

IV. Construct the base $\frac{3}{4}''$ high.

V. Locate top and bottom edges of inclined surfaces by measurements parallel to the axes.

VI. Complete construction of the wedge.

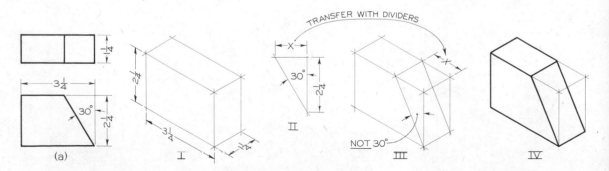

Fig. 16–8. Angles in Isometric.

VII. Darken all final lines.

Note, II and III, that the drawing could have been started at corner B instead of A.

16.7 Angles in Isometric.

As shown in Fig. 7–17, angles may appear either larger or smaller than true size, depending upon the direction in which they are viewed. Therefore, in isometric, angles cannot be set off directly with the protractor.

Angles are constructed by locating the endpoints of inclined lines by measurements parallel to the axes, Fig. 16–7 (V). Lines that are not parallel to the axes are called *non-isometric lines,* and are not true length. In isometric, all measurements must be made along *isometric lines*—that is, parallel to the axes.

If an angle is given *in degrees,* as in Fig. 16–8 (a), it is necessary to convert it into linear measurements.

I. Draw the construction box. Note that the actual 90° angles at all corners of the object are in no case 90° in the isometric drawing.

II. Since the 30° angle cannot be set off in degrees, it is necessary to draw the triangle full size to find the needed linear measurement X.

III. Transfer this dimension with dividers to the isometric drawing, and draw the parallel inclined lines. Remember: *Lines that are parallel on the object itself will be parallel in isometric.*

IV. Heavy-in all final lines.

16.8 Offset Measurements.

In Fig. 16–9 (a) is shown a triangular pyramid in which all surfaces are oblique except the base. Such an object is drawn by "box construction," I and II, in which all corners are located by offset measurements along isometric lines, or by means of "skeleton construction," III and IV, in which the base is drawn and then the vertex located on the vertical center line.

16.9 Other Positions of the Isometric Axes.

The isometric axes may be drawn in any desired position provided that the angle between them is held at 120°. Objects customarily viewed from below, Fig. 16–10 (a), may be drawn with *reversed axes*—that is, with two axes sloping downward. Also, reversed axes may be used to get a better view of an object, as shown at (b) and (c).

Long objects may be effectively drawn with the long axis horizontal, (d).

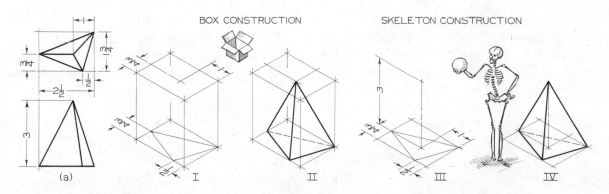

Fig. 16–9. Offset Measurements.

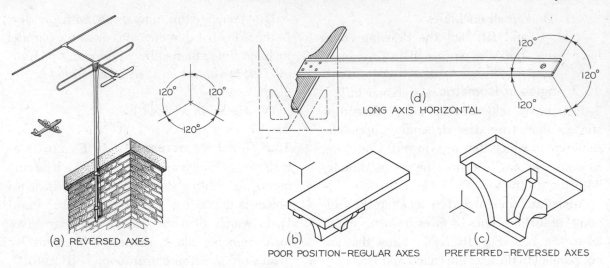

(a) REVERSED AXES

(d)
LONG AXIS HORIZONTAL

(b)
POOR POSITION–REGULAR AXES

(c)
PREFERRED–REVERSED AXES

Fig. 16–10. Other Positions of Axes.

16.10 Circles in Isometric. As shown in Secs. 2.9 and 16.4, circles appear as ellipses in isometric. An approximate ellipse, which can be easily drawn with the compass from four centers, is sufficiently accurate for nearly all isometric drawings. The steps in drawing this *four-center ellipse* are shown in Fig. 16–11.

Four-center ellipses, as they would be constructed on the four sides of a cube, are shown in Fig. 16–12 (a). All diagonals are horizontal or 60° with horizontal; hence the entire construction can be made with the T-square and the 30° × 60° triangle. Note also that *all diagonals are perpendicular bisectors of the sides of the parallelograms.*

An application of ellipses in drawing a pipe fitting is shown at (b). Note that the smaller ellipses require their own construction; that

is, the same centers cannot be used for two or more concentric ellipses.

To make an isometric drawing of a cylinder with its axis vertical, Fig. 16–12:

I. Draw the isometric ellipse for the upper end. Then drop centers A, B, and C down a distance equal to the height of the cylinder, as shown. Draw horizontal line A'C' and the lines B'A' and B'C' at 60° with horizontal.

II. Complete the cylinder by drawing two small arcs and one large arc corresponding to those in the upper ellipse.

16.11 Arcs in Isometric. Two views of an object with rounded corners are shown in Fig. 16–13 (a). The rounded corners could be drawn in isometric as shown at (b), in which the complete ellipses are constructed. How-

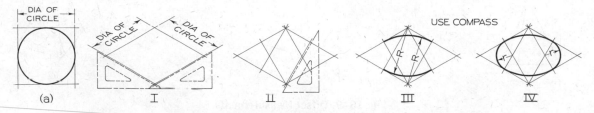

Fig. 16–11. Steps in Drawing Four-Center Ellipse.

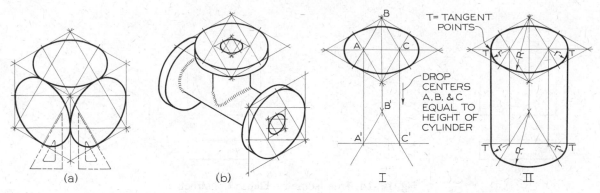

Fig. 16-12. Ellipse Construction.

ever, only one arc is actually required at each corner; hence, only a part of the construction is needed. It is only necessary, as shown at I, to set off the radius R from each corner and to draw perpendiculars that intersect at the required centers, as shown. Note that the compass arcs are not the same radius at both ends of the block, II, or equal to the actual radius, R.

16.12 True Isometric Ellipse. In cases where the four-center ellipse is not accurate enough, the true isometric ellipse can be drawn by plotting points and using the irregular curve, Fig. 16-14.

Around the given circle, (a), draw a square and diagonals, as shown. Where the diagonals cut the circle, draw lines parallel to the sides of the square. Draw this set of lines in iso-

metric, as shown at (b), transferring distances *a* and *b* with dividers. This method provides eight points on the ellipse. The ellipse is then drawn through the points with the irregular curve, Sec. 3.30. Use the curve as shown in Fig. 3-40.

A comparison of the true ellipse with the four-center ellipse is shown at (c). The four-center ellipse is slightly shorter and "fatter" than the true ellipse.

When more than eight points are needed for greater accuracy, draw as many parallel lines, spaced at random, across the given circle as desired, as shown at (d). Draw these lines in the isometric, (e), transferring distances *a*, *b*, *x*, and *y* with dividers.

To locate points on the bottom ellipse, drop points of the upper ellipse down a distance

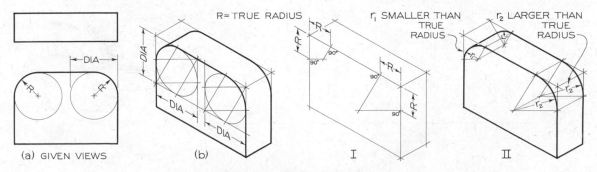

Fig. 16-13. Arcs in Isometric.

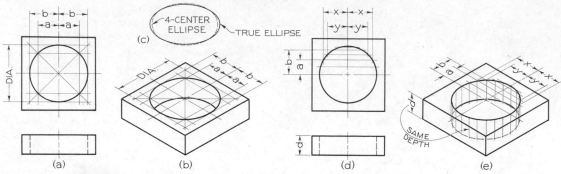

Fig. 16–14. True Isometric Ellipse Construction.

equal to the height d of the block and draw the ellipse, part of which will be hidden, through these points.

16.13 Ellipse Guides. One of the chief difficulties in pictorial drawing is the frequent necessity for drawing ellipses. In order to draw true ellipses in less time than required even for the four-center ellipse, many types of ellipse guides, or templates, are available. These are plastic sheets with various sizes of elliptical openings. Some of these, such as the Instrumaster Isometric Stencil,* also provide the angles needed to construct isometric drawings, as well as scales printed along the edges, Fig. 16–15 (a). The position for drawing an ellipse on top of a cube is shown at (b).

---

* Ridgways of New York, 110 W. 42nd St., New York, N.Y. 10036.

For ink work, the Leroy pens* and the technical fountain pens, Secs. 4.18 and 8.2, which are available in a variety of sizes, are recommended. Insert triangles under the stencil to separate it from the paper and prevent ink from running underneath, (c).

16.14 Curves in Isometric. Curves are drawn in isometric by plotting a series of points on the curve. In Fig. 16–16 (a) are the given views of a book end to be drawn in isometric. Draw a series of parallel construction lines across the view, as shown. It is best to space these equally with the scale or dividers so that they can be easily transferred to the isometric, I. Lines 1, 2, 3, 4, etc., are drawn the same length in isometric as in the given view.

---

* Keuffel & Esser Co., Hoboken, N.J.

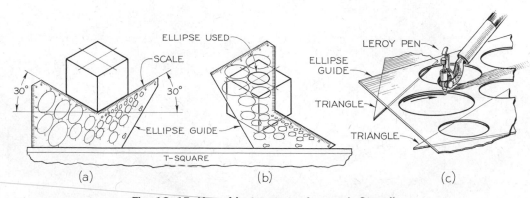

Fig. 16–15. Use of Instrumaster Isometric Stencil.

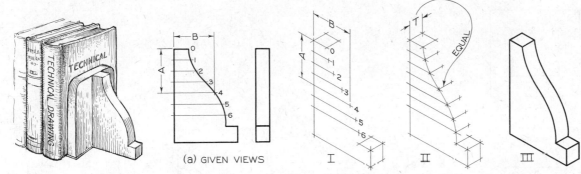

Fig. 16–16. Isometric Curves.

To draw the back curve, draw parallel construction lines equal in length to the thickness of the block, as shown at II. Finally, darken all required lines, III, using the irregular curve, Sec. 3.30.

16.15 Intersections. To draw the curve of intersection between a cylindrical hole and an oblique plane, Fig. 16–17, first draw the ellipse, representing the hole in isometric, in the top plane of the enclosing isometric construction box. A series of imaginary parallel cutting planes are then drawn in the isometric, corresponding to those that were used to obtain the elliptical intersection in the regular views. Points are then projected down from the top plane of the construction box to the oblique plane to obtain the desired curve of intersection, as shown at (b).

To draw the curve of intersection between two cylinders, Fig. 16–18, a series of imaginary cutting planes are passed through both cylinders and parallel to their axes, as shown. Each plane will cut lines (elements) from both cylin-

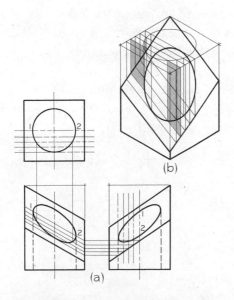

Fig. 16–17. Intersection of Oblique Plane and Cylinder.

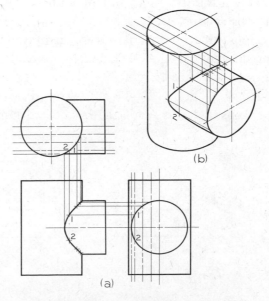

Fig. 16–18. Intersection of Cylinders.

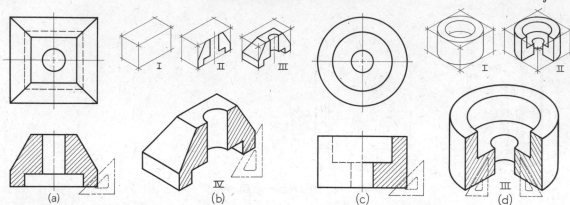

Fig. 16–19. Isometric Sections.

ders that intersect at points common to both cylinders, or, more specifically, will locate points on the curve of intersection, as shown at (b). In all problems involving a curve of intersection, as many points as are necessary should be plotted to assure a smooth curve. The curves of intersection may be drawn with ellipse guides or an irregular curve.

16.16 Isometric Sectioning. Interior shapes are exposed in isometric, as in multiview drawing, by means of sections. A full section is shown in Fig. 16–19 (a) and (b), with the steps in construction shown above. In drawing full sections in isometric, it is best to draw the cut surface first, and then add the remaining lines in the back half.

A half section is shown at (c) and (d), with the steps in construction shown above. In this case, it is best to block in the entire object, and then cut out the section.

Avoid drawing section lines parallel or perpendicular to any principal lines of the drawing. Generally, section lines at 60° with horizontal produce the best effect, but other angles are permissible. In a half section, slope the lines in opposite directions, (d).

16.17 Isometric Dimensioning. Isometric drawings may be dimensioned, if desired. In general, dimensions should be made to lie in the isometric planes (extended) of the object. Many examples are shown among the problems in this book, for example, on page 128.

16.18 Oblique Sketching. Another simple way to sketch a rectangular object pictorially is in oblique, Fig. 16–20.

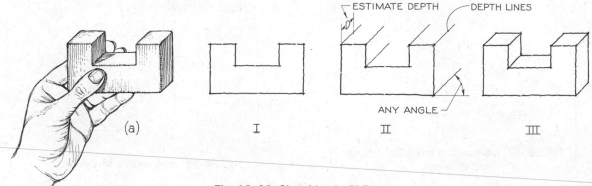

Fig. 16–20. Sketching in Oblique.

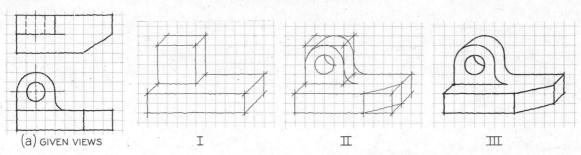

Fig. 16-21. Sketching in Oblique on Cross-Section Paper.

I. Sketch the front of the object in true size and shape.

II. Sketch the *depth lines* parallel to each other and at any convenient angle, say 30° or 45° with horizontal. Cut off the depth lines in such a way that the object will look natural. For simplicity in sketching, this may be full depth (actual depth of block), but more natural results are obtained if this depth is three-quarter or half size. If the sketch is at half depth, it is called a *cabinet sketch*.

III. Sketch remaining lines to complete the drawing.

16.19 Oblique Sketching on Cross-Section Paper.

Oblique sketches may be made easily on ordinary cross-section paper. In Fig. 16-21

(a), two views are given whose dimensions can be determined by counting the $\frac{1}{4}''$ squares.

I. Sketch the given object $2\frac{1}{2}''$ wide (10 squares) and $1\frac{1}{2}''$ high (6 squares). Sketch the depth lines at 45° diagonally through the squares as shown. Excellent results are obtained if the depth lines pass diagonally through half as many squares as the actual number given—in this case, two as compared to four.

II. Sketch the remaining features.

III. Darken all final lines.

16.20 Oblique Mechanical Drawing.

The steps in making an oblique mechanical drawing are shown in Fig. 16-22.

I. Construct the *oblique axes* lightly. The

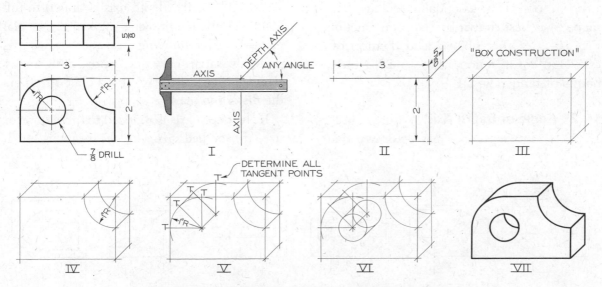

Fig. 16-22. Steps in Oblique Drawing—Box Construction.

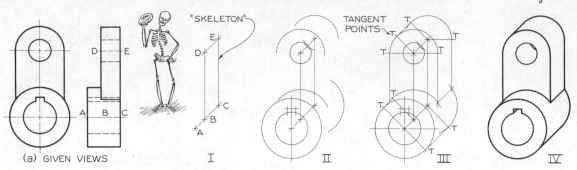

Fig. 16–23. Steps in Oblique Drawing—Skeleton Construction.

depth axis is drawn at any desired angle, usually 30°, 45°, or 60° with horizontal.

II. Set off the width, height, and depth on the axes. In this case, the depth is drawn to full scale.

III. Draw construction box.

IV to VI. Add arcs and circles, with particular attention to points of tangency T.

VII. Darken all required lines.

Note that all shapes lying in the front face of the object, or parallel to it, are shown in true size and shape. Hence, objects with circular shapes can be easily drawn directly with the compass, making oblique drawing much simpler for such shapes than isometric, in which ellipses must be constructed.

If an object is essentially rectangular in shape, it is best drawn by "box construction," Fig. 16–22. Other objects lend themselves to "skeleton construction," Fig. 16–23. Note the points of tangency, III.

16.21 Angle of Depth Axis. As shown in Fig. 16–22 (I), the depth axis may be drawn at any angle. However, the angles usually chosen are 30°, 45°, or 60° with horizontal, since these can be readily drawn with the triangles. The draftsman must decide in each case which angle is most suitable. In Fig. 16–24, the drawings at (a) and (b) are suitable for this particular object, while those at (c) to (e) are not. However, if the object were turned over, as at (f), it is best shown with reversed axes.

16.22 Scale of Depth Axis. The depth axis may be drawn full size or reduced. In Fig. 16–25 (a) is shown an oblique drawing of a cube in which the depth axis is drawn to full scale. The cube appears to be too deep, and the depth lines appear to spread apart as they recede. When the depth axis is drawn to full scale, the oblique drawing is given the special name *cavalier drawing*. Cavalier drawing is perfectly satisfactory for representing many objects, Figs. 16–22 to 16–24, but in others the distortion may be excessive.

If the depth axis is reduced to three-quarter size, (b), or half size, (c), the result is much

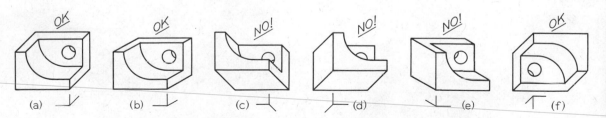

Fig. 16–24. Various Angles of Depth Axis.

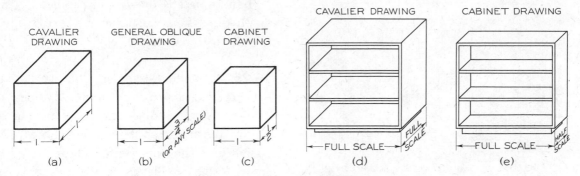

Fig. 16–25. Scale of Depth Axis.

more natural. These reductions can be easily set off with the architects scale. When the depth axis is reduced to half size, the name *cabinet drawing* is given because of its early use in the furniture industry.

A comparison between a cavalier drawing and a cabinet drawing of a bookcase is shown at (d) and (e). Note that if a cabinet drawing is drawn to half scale, the depth axis would be drawn to quarter scale.

16.23 Choice of Position. The chief advantage of oblique drawing is the ease with which circular shapes can be drawn. For example, in Fig. 16–26 (a), all circular shapes of the wheel are faced toward the front, and therefore these shapes are drawn easily with the compass. If the circular shapes are not faced toward the front, (b), the circles become ellipses that are distorted and tedious to draw. If one does not object to drawing ellipses, the wheel can be drawn in isometric, (c), with better results. *Rule: In oblique drawing, always face contours toward the front where they appear in true size and shape.*

The eye is accustomed to seeing parallel lines tend to converge as they recede into the distance. But in oblique drawing, the receding lines are drawn parallel, and the result is sometimes very unnatural. A striking comparison

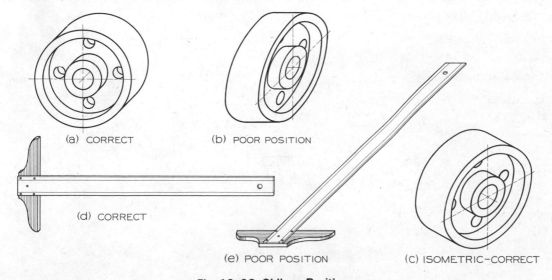

Fig. 16–26. Oblique Positions.

THESE 4 CHAIRS DRAWN IN OBLIQUE – ALL THE SAME SIZE. REST OF SCENE IS IN PERSPECTIVE.

Fig. 16–27. Unnatural Appearance of Oblique Drawing.

between oblique drawing and perspective (the way the eye sees things) is shown in Fig. 16–27. Oblique drawing should not be used in such cases where distortion offends the eye. To minimize this distortion in oblique drawing, observe the following: *Draw long objects with the long dimension perpendicular to the line of sight,* Fig. 16–26 (d).

16.24 Angles in Oblique. In Fig. 16–28 (a) is shown an object having two 30° angles. When drawn "in cavalier," (b), the angle that is faced toward the front is drawn true size. However, the angle that is in a receding plane will not be true size, and must be drawn by constructing the triangle full size, as shown, and transferring distance *x*.

If the object is drawn "in cabinet," every depth dimension, including distance *x*, must be drawn to half scale, (c).

16.25 Arcs and Circles in Oblique. Circles that are faced toward the front will appear as true-size circles; those that are not faced toward the front will appear as ellipses, Fig. 16–29 (a). The four-center ellipse requires an enclosing parallelogram with equal sides; hence it can be used only in cavalier drawing. The four centers are found by simply erecting perpendicular bisectors to the sides of the parallelograms, as shown.

An application of the four-center ellipse in a cavalier drawing is shown at (b). Note that the depth axis is full scale and that the sides of the parallelogram are therefore equal. If the depth axis is drawn to a reduced scale, such as half size (as in cabinet drawing), the four-center ellipse cannot be used. Instead, it is necessary to plot points on the ellipse, as shown at (c), and to draw the ellipse with the irregular curve, Sec. 3.30. If more points on the ellipse are needed, the method of Fig. 16–14 (d) and (e) may be used.

The four-center method can be used to draw arcs in cavalier drawing only, Fig. 16–29 (d). Simply set off given radius R from each

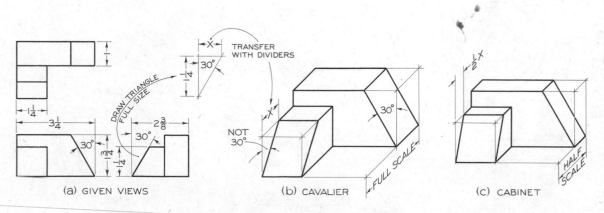

Fig. 16–28. Angles in Oblique Drawing.

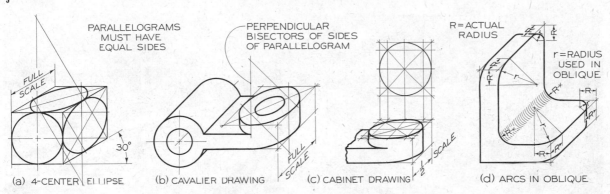

Fig. 16-29. Arcs and Circles in Oblique.

corner and erect perpendiculars to locate centers, as shown.

16.25 Oblique Sections. Where necessary to expose interior shapes in oblique drawings, sections may be drawn, Fig. 16–30. Draw the section lines in opposite directions in a half section, as shown at (a). In general, avoid drawing section lines parallel or perpendicular to the visible lines bounding the sectional areas.

16.27 Oblique Dimensioning. Oblique drawings may be dimensioned if desired, as shown in Figs. 15–30, 15–48, and 15–51. Dimension lines, extension lines, arrowheads, and dimension figures should be drawn to lie in the corresponding planes (extended) of the object. Notes should always be lettered horizontally and "in the plane of the paper."

16.28 Perspective. The eye and the camera both are constructed so that objects appear

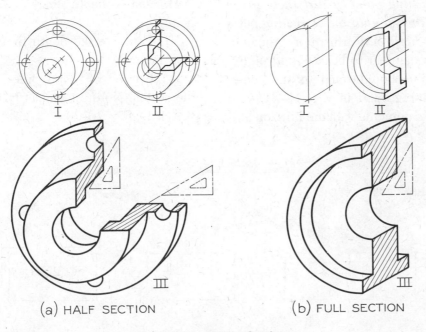

Fig. 16-30. Oblique Sections.

Fig. 16–31. One-Point Perspective (Photograph).

progressively smaller as they are farther away. For example, in Fig. 16–31, the spacing between the rails, the lengths of the ties, and the heights of the telephone poles all appear to diminish with distance. Also, the rails and other lines parallel to them, such as the tops of the telephone poles, all converge at a point on the horizon called the *vanishing point*. Hence, this rule in perspective: *All parallel lines have the same vanishing point, and if they are on or parallel to the ground, the vanishing point will be on the horizon.* Furthermore, the horizon will always appear to be at eye level. For example, if you should stand in the center track in the foreground in Fig. 16–31, your eye level would coincide with the horizon line.

16.29 Perspective Sketching. The object sketched in oblique in Fig. 16–20 may be easily sketched in *one-point perspective* (one vanishing point), as shown in Fig. 16–32.

I. Sketch front face of object true size and shape, and select a vanishing point for the converging depth lines. Before deciding upon the location of the vanishing point, experiment with it in several different places.

II. Sketch depth lines toward vanishing point.

III. Estimate the depth by eye to make it look natural.

IV. Cut off all depth lines and complete the sketch. Note the similarity to the oblique sketch in Fig. 16–20.

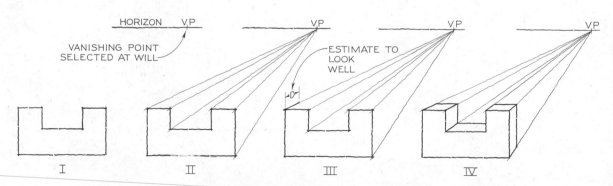

Fig. 16–32. One-Point Perspective Sketch.

The object sketched in isometric in Fig. 16–3 can be easily sketched in two-point perspective (two vanishing points), as shown in Fig. 16–33.

I. Sketch front corner of object true height, and locate two vanishing points by eye where you think they will produce the best picture. Both must be on a horizontal "horizon" line.

II. Estimate width and depth, and sketch enclosing box.

III. Block in the right-angled notch.

IV. Darken all required lines. Note the similarity of this sketch to the isometric sketch in Fig. 16–3.

To gain further understanding, try sketching the same object with the horizon higher or lower than in Fig. 16–33. See what happens if the horizon is placed below the perspective. Also try out the effects of placing vanishing points closer together or farther apart.

16.30 Theory of Two-Point Perspective. In order to draw mechanically a correct perspective, it is first necessary to consider the theoretical method of projection. As shown in Fig. 16–34, a transparent *picture plane* (PP)

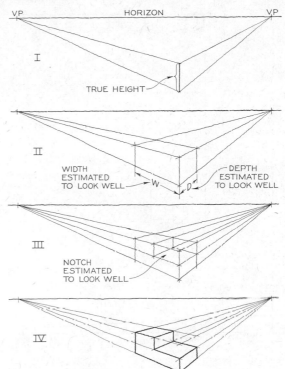

Fig. 16–33. Two-Point Perspective Sketch.

is placed between the observer's eye or *station point* (SP) and the object. *Visual rays* extend from SP to all points on the object. Collectively, the piercing points of the visual rays in

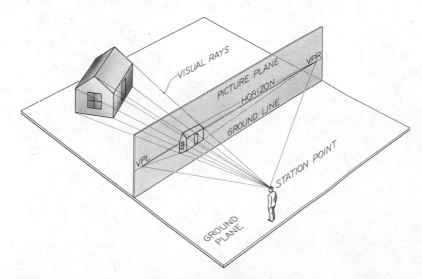

Fig. 16–34. Perspective Projection.

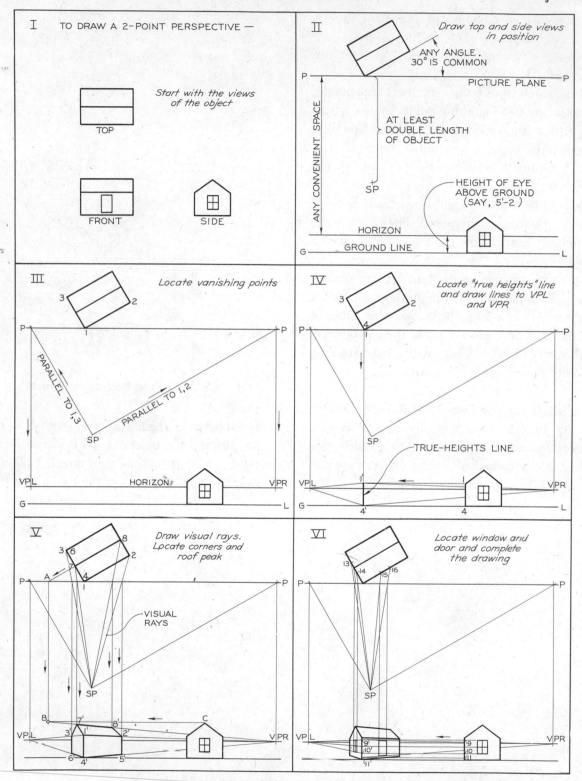

Fig. 16–35. Mechanical Drawing of Two-Point Perspective.

PP form the perspective or picture as seen by the observer. The *horizon line* on PP is drawn at eye level, and the vanishing points VPL and VPR will be on this line.

16.31 To Draw Two-Point Perspective.
The steps in drawing a two-point perspective are illustrated in Fig. 16–35.

I. The views of the object are given.

II. Draw the picture plane PP, horizon, and ground line GL. To simplify the construction, the front corner of the house (top view) is drawn touching PP, and at a convenient angle of 30° with PP. Draw the side view or front view resting on GL and to one side of the drawing. Height dimensions will be projected from this view across to the perspective. Locate the station point SP in front of the house, as shown.

III. Locate vanishing points by drawing lines from SP parallel to lines 1–2 and 1–3, and from their intersections with PP, project down to the horizon to get VPL, the left-hand vanishing point, and VPR, the right-hand vanishing point. Notice that the closer SP is drawn to the top view the closer the vanishing points will be, and vice versa.

IV. Locate true-heights line by projecting down from corner 1–4 in PP to GL. Project across from points 1 and 4 in the side view to the true-heights line to establish 1′–4′, the perspective of the front corner of the house. From 1′ and 4′, draw lines to VPL and VPR, as shown.

V. Draw visual rays from SP to the various points in the top view. From the intersections of these with PP, project down to locate corners 3′–6′ and 2′–5′.

The method of finding the roof peak is a general method which is applied to finding the perspective of any horizontal line: First, extend line 7–8 (top view) until it intersects

PP at A; then project downward from A and across from the peak C to locate B, the piercing point of the line. Then draw line B–VPR. To determine the ends of the peak line 7′–8′, draw visual rays SP–7 and SP–8. Where these intersect PP, project down to locate 7′ and 8′, as shown.

VI. Locate window and door. Project true heights 9, 10, 11 of window and door across from the side view to the true-heights line at 9′, 10′, 11′, and then draw lines toward the two vanishing points, as shown. Draw visual rays from SP to 13, 14, 15, and 16 (sides of window and door in the top view), and project down from the intersections of these lines with PP to establish the widths of the window and door in the perspective.

16.32 To Draw One-Point Perspective.
In Fig. 16–36 (a) is shown a mechanical drawing in one-point perspective of the same object sketched in Fig. 16–32. The front face of the block is placed *in* PP so that it will be drawn in true size and shape. SP is located in front and to one side of the object, at a distance away from it equal to about twice its length. The horizon is placed well above the ground line, and the single vanishing point is on the horizon directly above SP.

To determine the depth, construct the perspective of any convenient point on the back side of the object, as corner 2. Draw visual ray SP–2, intersecting PP at point A. Project down to point 2′, which is the perspective of corner 2.

A one-point perspective of a cylinder with a hole is shown at (b). Note that the radius AC of the back rim of the cylinder is reduced to A′C′ in the perspective. All circles are drawn with the compass, since they are parallel to PP.

In the one-point perspective shown at (c),

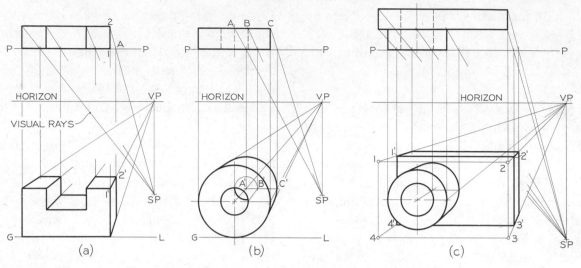

Fig. 16–36. Mechanical Drawing of One-Point Perspective.

it is necessary to construct the base in the picture plane at 1–2–3–4 and then to locate it at the proper depth as shown at 1′–2′–3′–4′ by projecting down from the intersection in PP of the visual rays to the corners.

16.33 Methods of Shading. The purpose of an industrial pictorial drawing is to show clearly the shape of the object and not necessarily to be artistic. Shading should, therefore, be simple and limited to producing a clear picture. Although art training is required to produce professional results, the ordinary draftsman can learn to do all the shading it is necessary for him to do.

Some of the most common types of shading are illustrated in Fig. 16–37. Pencil or ink lines are drawn mechanically at (a) or freehand at (b). Two methods of shading fillets and rounds are shown at (c) and (d). Shading

produced by pen dots is shown at (e), and pencil "tone" shading is shown at (f). Pencil shading is often applied to pictorial drawings on tracing paper. Such drawings can be reproduced with good results by making Ozalid prints or other prints in which the background is white. In blueprinting, the darks and lights are reversed, but even then the results are quite satisfactory.

Examples of line shading on pictorial drawings in industrial sales literature are shown in Fig. 16–38. An "exploded assembly" is shown at (c) in which the several parts are drawn in positions indicating how they are assembled.

16.34 Production Illustration. *Production illustration* is the term applied to a variety of pictorial drawings used in industry, particularly in the aircraft industry. In general, the need for pictorial illustrations arises from the

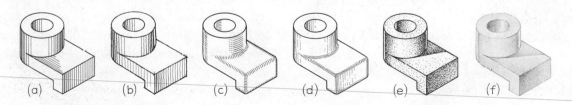

(a)   (b)   (c)   (d)   (e)   (f)

Fig. 16–37. Methods of Shading.

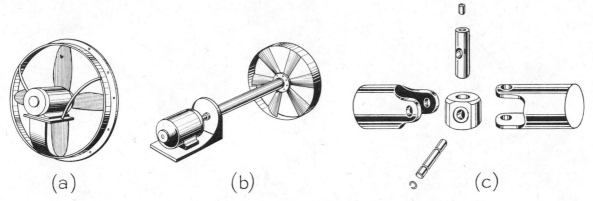

(a)   (b)   (c)

(a) and (b) Courtesy Power Fan Manufacturers Assn., (c) Courtesy Boston Gear Works.

Fig. 16–38. Examples of Mechanical Line Shading.

fact that many manufactured items are becoming so complex—the airplane especially—that it is difficult to follow clearly all details from working drawings alone. This situation is particularly acute when a large number of workers cannot read complicated blueprints.

Production illustrations are used on the production lines, especially in assembling, to show the way parts fit together and the sequence of operations to be performed. In Fig. 16–39 is

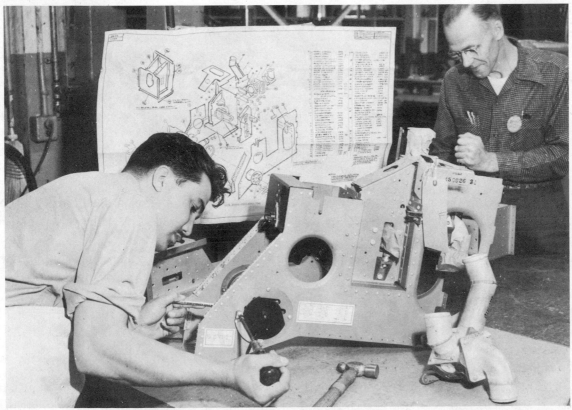

Courtesy Lockheed Aircraft Corp.

Fig. 16–39. Using a Production Illustration to Assemble an F–94C Plane.

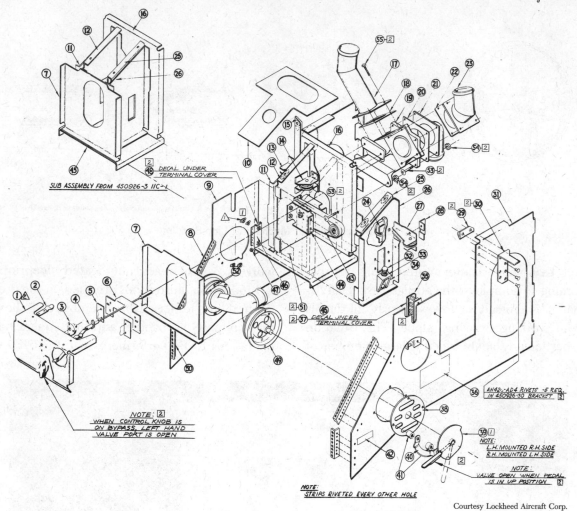

Fig. 16–40. Production Illustration.

shown a worker on the assembly line at Lockheed Aircraft Corporation as he follows a production illustration in assembling a "Control Stand" for an F-94C airplane. This production illustration itself is shown in Fig. 16–40. Note that a small-scale pictorial of assembled parts appears at the upper left, while the main part of the drawing is an "exploded assembly" showing in a clearer manner each part and where it fits.

An exploded assembly of a Boeing Stratofreighter, showing how an airplane is divided into sections so that each section may be constructed independently of the others and the whole assembled later, is shown in Fig. 16–41.

Production illustrations are also used extensively in parts catalogs, maintenance instruction handbooks, and similar publications. A typical parts catalog exploded assembly, showing a drill press, is reproduced in Fig. 16–42.

16.35 Pictorial Drawing Problems. A large number of sketching problems are given in

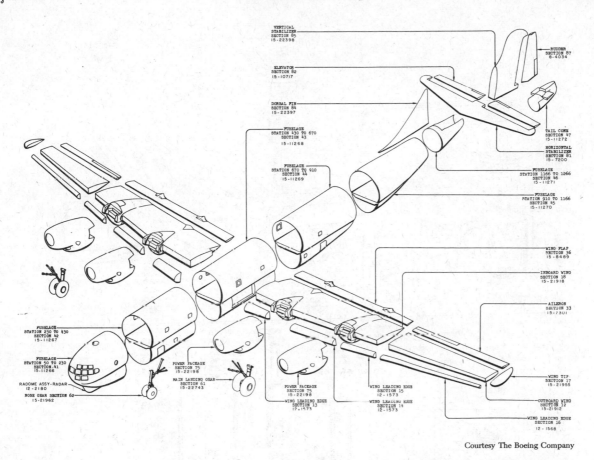

Courtesy The Boeing Company

Fig. 16–41. Structural Breakdown of Boeing Stratofreighter.

Figs. 16–45 and 16–46. Additional problems may be assigned to be sketched in isometric or oblique from Figs. 6–28, 6–29, 7–27 to 7–29, and 9–39 to 9–41. Sketches may be made on cross-section paper or plain paper, as desired by the instructor.

In Figs. 16–48 to 16–53 are given problems to be drawn with instruments in isometric or oblique, as indicated. However, any of the isometric problems may be drawn in oblique, or oblique problems in isometric, and any problem may be assigned to be drawn freehand on cross-section paper or on plain paper. Many additional problems to be drawn mechanically may be assigned from Figs. 7–27 to 7–29, 9–39 to 9–41, and 12–18.

Isometric sectioning problems are given in the lower portion of Fig. 16–50, and oblique sectioning problems in the lower portion of Fig. 16–53. However, isometric sectioning problems may be assigned from the oblique group, or oblique sectioning problems from the isometric group. Many additional isometric or oblique sectioning problems may be assigned from Figs. 11–24 to 11–28.

Perspective problems are given in Figs. 16–54 and 16–55. Additional problems may be selected from those on preceding pages.

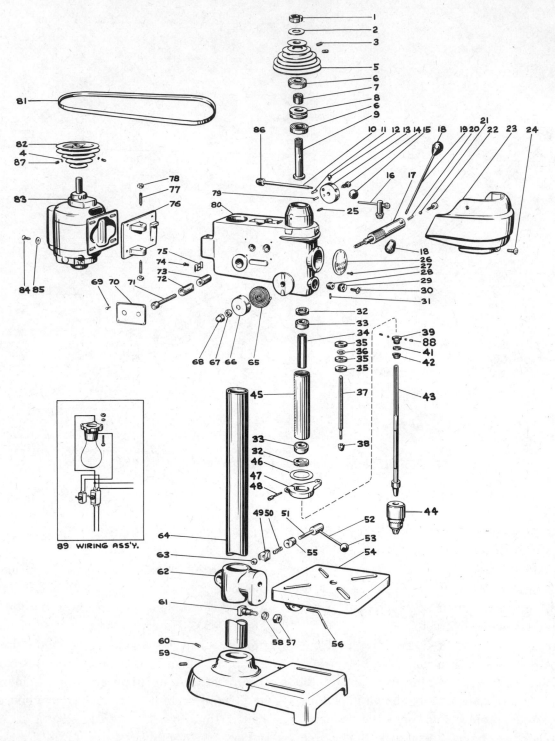

Courtesy South Bend Lathe Works

Fig. 16–42. Parts Catalog Illustration.

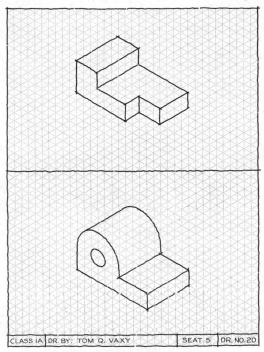

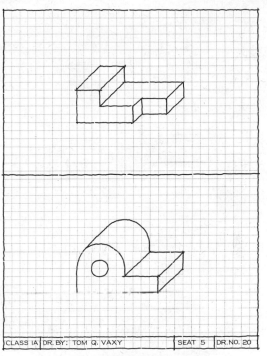

CLASS IA | DR. BY: TOM Q. VAXY | SEAT 5 | DR. NO. 20

Fig. 16–43. Isometric Sketches.

CLASS IA | DR. BY: TOM Q. VAXY | SEAT 5 | DR. NO. 20

Fig. 16–44. Oblique Sketches.

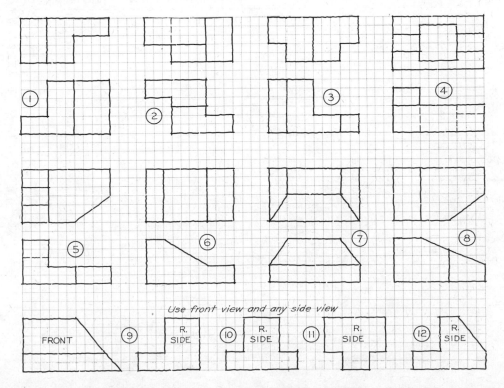

Use front view and any side view

Fig. 16–45. Isometric and Oblique Sketching Problems. Using Layout B (Appendix, page 474), divided into two parts as in Figs. 16–43 and 16–44, sketch problems in isometric or oblique as assigned. Each square = $\frac{1}{4}''$.

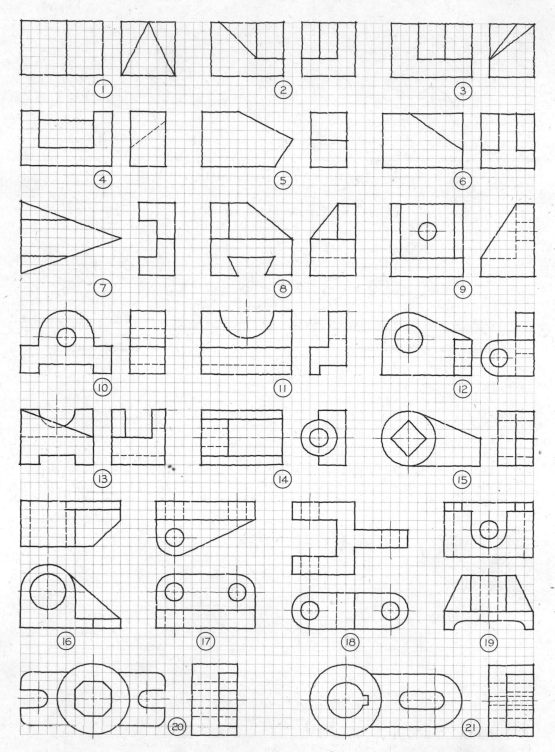

Fig. 16–46. Isometric and Oblique Sketching Problems. Using Layout B (Appendix, page 474), divided into two parts as in Figs. 16–43 and 16–44, sketch problems in isometric or oblique as assigned. Each square = $\frac{1}{4}''$.

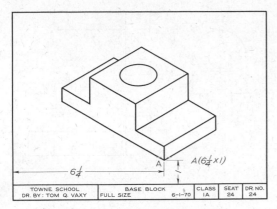

Fig. 16–47. Isometric Problem.

In Figs. 16–48 to 16–50 are given views of objects which are to be drawn in isometric with instruments. Use Layout C (Appendix, page 474). In each problem the location of the starting corner A is given by two dimensions, the first measured from the left border, and the second measured up from the top of the title strip. For example, in Fig. 16–47, point A is $6\frac{1}{4}$″ from the left border and 1″ up from the title strip.

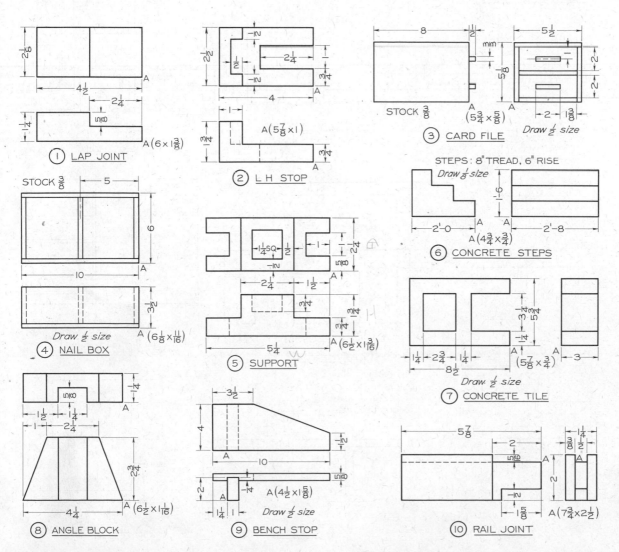

Fig. 16–48. Isometric Problems. Locate starting corners A as explained above. Move titles to title strip and omit dimensions unless assigned.

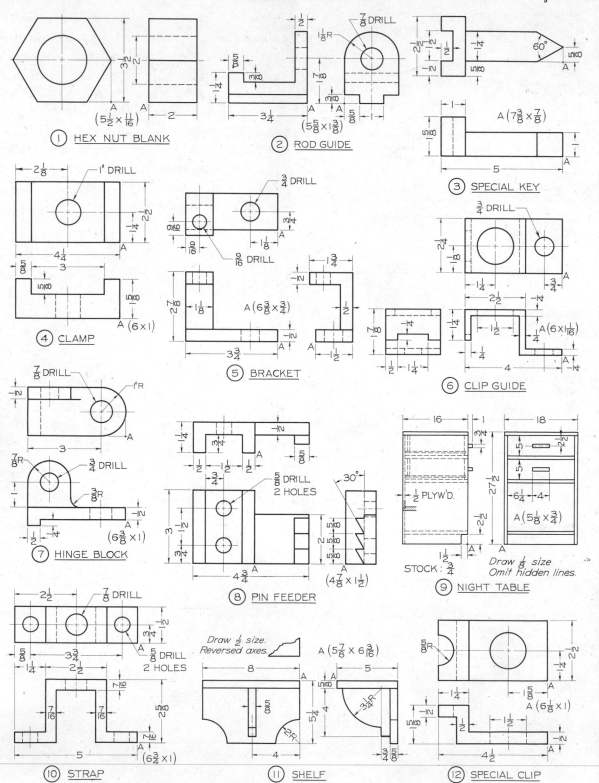

Fig. 16–49. Isometric Problems. Locate starting corners A as explained on page 333. Move titles to title strip and omit dimensions unless assigned.

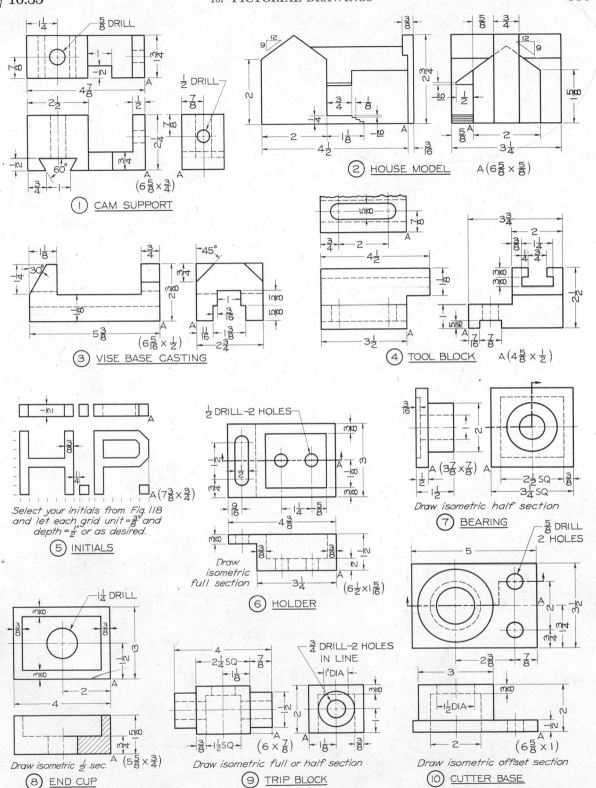

① CAM SUPPORT

② HOUSE MODEL   A(6⅝ × ⅝)

③ VISE BASE CASTING

④ TOOL BLOCK   A(4⅝ × ½)

⑤ INITIALS

Select your initials from Fig. 118 and let each grid unit = ⅜" and depth = ½" or as desired.

⑥ HOLDER

Draw isometric full section

⑦ BEARING

Draw isometric half section

⑧ END CUP

Draw isometric ½ sec.

⑨ TRIP BLOCK

Draw isometric full or half section

⑩ CUTTER BASE

Draw isometric offset section

Fig. 16–50. Isometric Problems. Locate starting corners A as explained on page 333. Move titles to title strip and omit dimensions unless assigned.

In Figs. 16–52 and 16–53 are views of objects to be drawn in oblique with instruments. Use Layout C (Appendix, page 474). In each problem the starting corner A is located by two dimensions, the first measured from the left border, and the second measured up from the top of the title strip. For example, in Fig. 16–51, point A is $4\frac{7}{8}''$ from the left border and $3\frac{1}{8}''$ up from the title strip. In all problems, assume the depth axis at full scale unless otherwise indicated.

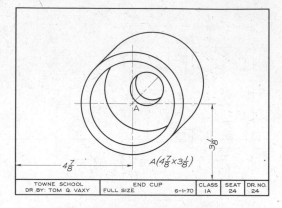

Fig. 16–51. Oblique Problem.

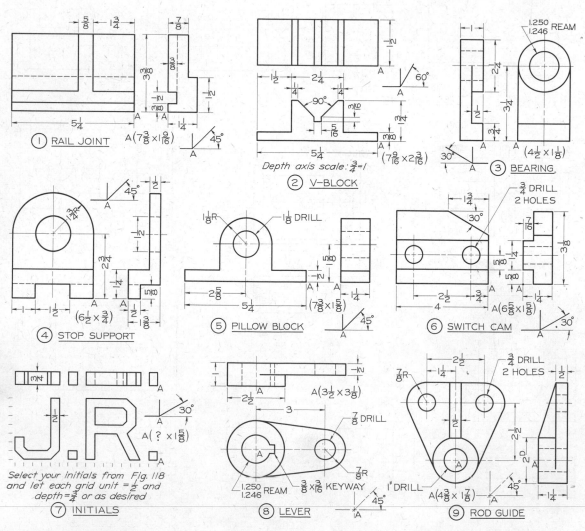

Fig. 16–52. Oblique Problems. Locate starting corners A as explained above. Move titles to title strip and omit dimensions unless assigned.

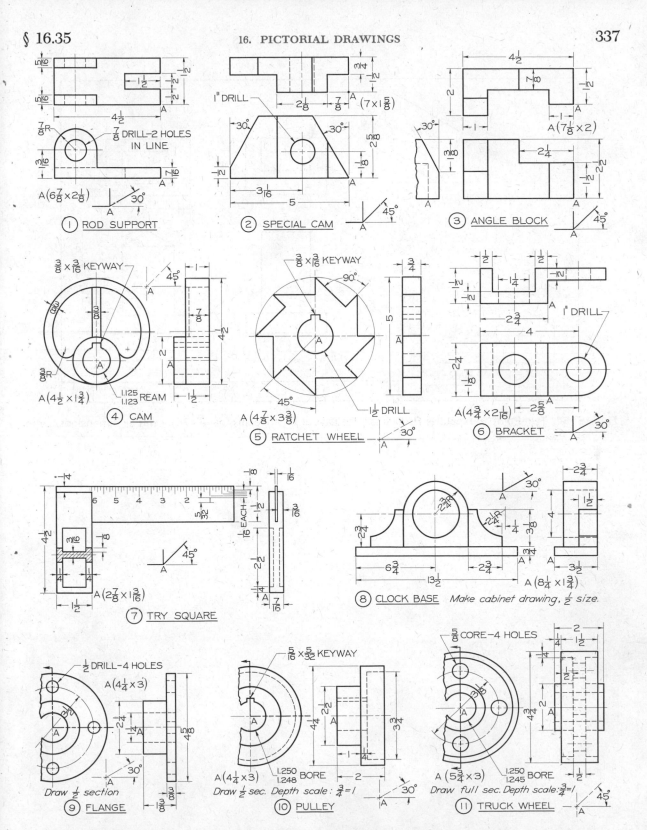

Fig. 16–53. Oblique Problems. Locate starting corners A as explained on page 336. Move titles to title strip and omit dimensions unless assigned.

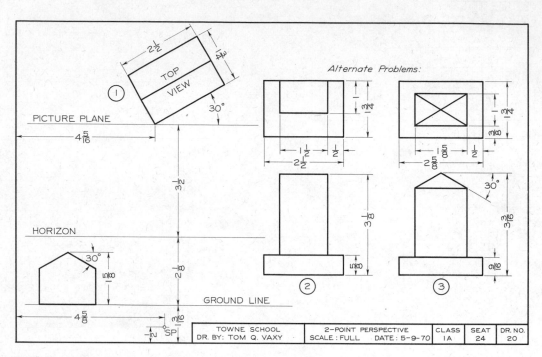

Fig. 16–54. Two-Point Perspective Problems. Use Layout D (Appendix, page 475). Omit all dimensions. Letter VPL, SP, etc.

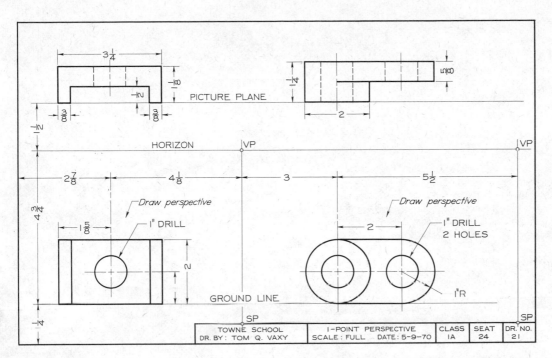

Fig. 16–55. One-Point Perspective Problems. Use Layout D (Appendix, page 475). Omit all dimensions. Letter VPL, SP, etc.

CHAPTER 17

DEVELOPMENTS AND INTERSECTIONS

17.1 Developments. The *development* of an object is the surface of the object laid out on a plane. For example, if an ice cream carton is unfolded and laid out on a table, the result is a development. In the sheet-metal trade, a development is usually referred to as a *pattern* or a *stretchout*. Thousands of different manufactured objects are made by cutting out patterns and then folding them into shape, including pipes, air-conditioning ducts, heating ducts, pans, hoppers, bins, buckets, and even milk cartons and paper cups. A striking example of sheet-metal work in the oil industry is shown in Fig. 17–1.

Fig. 17–1. Catalyst Collector or "Cyclone" for Petroleum Refinery.

341

PARALLEL-LINE DEVELOPMENTS

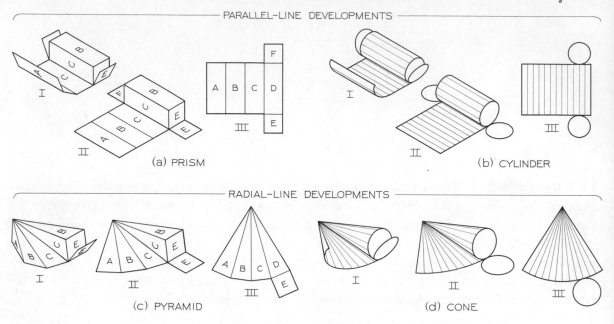

(a) PRISM   (b) CYLINDER

RADIAL-LINE DEVELOPMENTS

(c) PYRAMID   (d) CONE

Fig. 17-2. Developments.

The developments of the four most common solids are shown in Fig. 17–2. Other forms are usually more difficult and more expensive to make and are avoided if possible. The patterns of the prism and pyramid are merely the sides and ends unfolded onto a plane surface, while the patterns for the cylinder and cone are simply the surfaces and ends rolled out or unfolded onto a plane surface. Note that the prism and cylinder roll out into rectangular patterns, which are called *parallel-line developments,* and the pyramid and cone roll out into pie-shaped patterns, which are called

radial-line developments. The various geometrical solids are shown in Fig. 5–1.

17.2 Sheet-Metal Work. Patterns are made of paper, cardboard, plastic, and especially of sheet metal, such as steel, brass, copper, and aluminum. After the metal is cut and folded or rolled into shape, the pieces are fastened together with solder, welds, rivets, or seams of various kinds. Where thickness of metal is a factor, some allowance must be made for stretching or crowding of metal at the bends. Also, extra material must be pro-

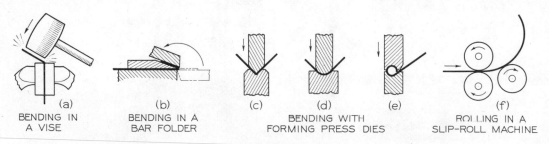

(a)   (b)   (c)   (d)   (e)   (f)
BENDING IN   BENDING IN A   BENDING WITH   ROLLING IN A
A VISE   BAR FOLDER   FORMING PRESS DIES   SLIP-ROLL MACHINE

Fig. 17-3. Bending Metal.

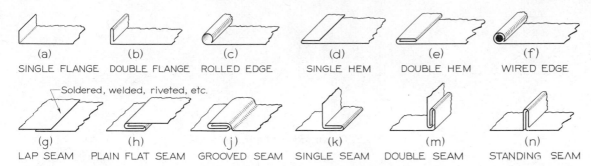

(a) SINGLE FLANGE (b) DOUBLE FLANGE (c) ROLLED EDGE (d) SINGLE HEM (e) DOUBLE HEM (f) WIRED EDGE

—Soldered, welded, riveted, etc.

(g) LAP SEAM (h) PLAIN FLAT SEAM (j) GROOVED SEAM (k) SINGLE SEAM (m) DOUBLE SEAM (n) STANDING SEAM

Fig. 17–4. Sheet Metal Edges and Seams.

vided for laps and other kinds of joints, Fig. 17–4. In the following pages, these allowances for bends and seams will be disregarded.

It is customary to draw patterns so that the *inside* surfaces are up, as shown in Fig. 17–2. Thus, when the object is folded into shape, all fold lines will be on the inside.

After the pattern has been laid out from the drawing onto the metal, the metal is cut by means of hand snips, chisels, circle cutters, ring and circular shears, or other tools or methods.

Sheet metal may be bent, folded, or rolled by hand in a number of ways. These in general consist of hammering the sheets over or around wood blocks or iron *stakes* (anvils of various shapes) with a mallet, Fig. 17–3 (a), or in bending them with a *hand seamer*. A machine used for making narrow bends is the *bar folder*, (b). For folding wide pieces, such as the sides of a box, a large machine called a *brake* is used.

Many types of bends are made on the *press brake* by means of various shapes of *forming-press dies*, some of which are shown at (c) to (e). Conical or cylindrical forms are made by hand-hammering over a rounded stake or by rolling on a *slip-roll forming machine*, as shown at (f).

Certain objects, such as automobile fenders and the warped skin of aircraft, are *non-developable* (cannot be laid out flat on a plane).

They are deformed into shape by pressing flat sheets into dies under heavy pressure.

Exposed edges, as on rims of pans or buckets, are usually flanged, hemmed, rolled, and so forth, as shown in Fig. 17–4 (a) to (f). Various methods of joining metal at seams are shown at (g) to (n). Usually a pattern is fastened along the shortest edge to save labor and materials. However, the seams may be made along any convenient edge, and often the arrangement is simply the one that can be cut out most economically.

17.3 Model Construction. Paper or cardboard models should be made of at least a few of the earlier projects assigned. The final appearance and the fit of the pattern when folded and fastened together will depend upon the accuracy of your drawing.

For paper models, use any stiff paper, such as drawing paper. Draw tabs $\frac{1}{4}''$ wide along edges to be fastened, clipping the corners at 45°, as shown in Fig. 17–5 (a). For curved edges, notch the tabs as shown at (d) or (e).

To cut out the pattern, use a razor blade or a sharp knife, but be sure to cut over a piece of heavy cardboard so as not to damage your drawing board or table top. *Never cut along a triangle or T-square*, as a single nick will ruin them. To make the corners and the tabs fold smoothly, score along the fold lines by draw-

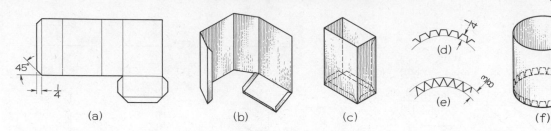

Fig. 17–5. Construction of a Paper Model.

ing the divider point along the lines, using the triangle as a guide. Do not press hard, and keep the leg of the divider almost flat on the paper.

Fasten the seams of the model together with paste, glue, rubber cement, or cellulose tape. The model can be made more attractive by painting or spraying with colored lacquer or with airplane dope.

PARALLEL-LINE DEVELOPMENTS

17.4 Pattern of Truncated Prism—Fig. 17–6. Prisms have plane faces that intersect to form edges that are parallel, Fig. 5–1. The

development of a square prism is shown in Fig. 17–2 (a), the side surfaces folding out into a simple rectangle.

If a prism is cut off at an angle, Fig. 17–6 (a) and (b), it is said to be *truncated*. The top and front views of the truncated prism are shown at (c), together with an auxiliary view of the inclined surface. The lower end of the prism will develop into a straight line, 1–1, called the *stretchout line*, (d) and (e). The stretchout line is the perimeter of the base, or the total distance around the base, laid out in a straight line. The upper end of the prism will develop irregularly, as shown in the figure.

On the stretchout line, set off distances 1–2,

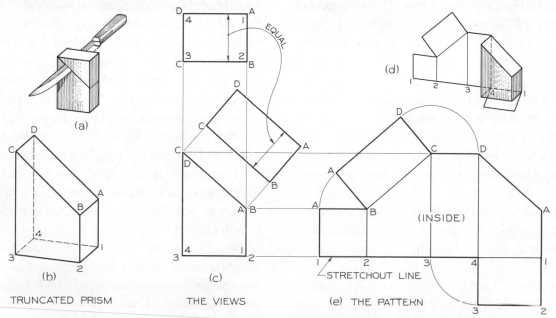

TRUNCATED PRISM   THE VIEWS   (e) THE PATTERN

Fig. 17–6. Pattern of Truncated Prism.

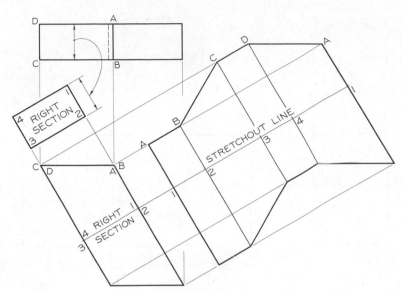

Fig. 17–7. Pattern of Oblique Prism.

2–3, 3–4, and 4–1, taken from the top view. Through these points draw the edges, or fold lines, perpendicular to the stretchout line. These are parallel; hence the term "parallel-line development." The upper ends A, B, C, etc., of these lines are found by projecting across from the front views of the corresponding points.

The true size of the bottom is shown in the top view, and the true size of the inclined surface is shown in the auxiliary view. These are transferred, if desired, to the pattern and attached along any convenient joining edges.

17.5 Pattern of Oblique Prism—Fig. 17–7. If both ends of a prism are cut off at an angle other than 90°, Fig. 17–7, neither end will roll out into a straight line. However, if an imaginary sectioning plane is passed through the prism at right angles to the edges, a right section 1–2–3–4 is produced, as shown, and this right section will roll out into a straight line. The true size of the right section is shown in the auxiliary view.

To draw the pattern, extend the stretchout line, as shown. Set off on the stretchout line distances 1–2, 2–3, 3–4, and 4–1, taken from the auxiliary view where they are shown true length. Through these points draw the edges perpendicular to the stretchout line. Establish end-points A, B, C, etc., by projecting from corresponding points in the front view.

17.6 Cylinders. A cylinder may be a *right cylinder* or an *oblique cylinder,* Fig. 5–1. Cylinders usually are circular, but may be elliptical or otherwise. A right cylinder, whose bases are perpendicular to its center line, will develop into a simple rectangle. A good way to demonstrate this is to take a paint roller and apply one revolution of the roller on a wall. The painted area will be a rectangle, Fig. 17–8 (a).

17.7 Circumference. As shown in Fig. 5–1, the circumference of a circle is the distance around the circle. The circumference of a right circular cylinder is the distance around the

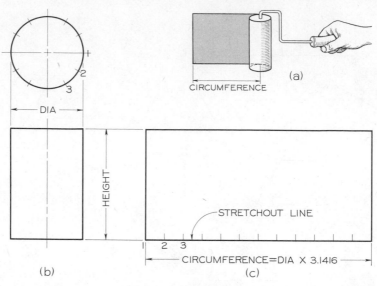

Fig. 17-8. Circumference.

base, as shown in the top view in Fig. 17–8 (b). If the cylinder is rolled out on a plane, the length of the pattern will be the circumference of the cylinder, and the height of the pattern will be the height of the cylinder, Fig. 17–8 (c). The circumference may be laid out approximately by setting off a number of equal divisions 1–2, 2–3, etc., on the circle in the top view and then stepping off with the bow dividers the same number on the stretchout line on the pattern, as shown in the figure. However, the distances set off would be chords of the arcs, not the actual lengths of the arcs; the slight errors for each distance would add up to a sizable error on the total length.

The circumference of any circle divided by its diameter is 3.1416, or about $3\frac{1}{7}$. This number, 3.1416, is known to mathematicians as π. It is the Greek letter *pi* (pronounced *pie*). Thus, if you know the diameter of a circle, you can always get the circumference by multiplying the diameter by 3.1416. For example, if the diameter of the base of the cylinder in Fig. 17–8 is 2″, the circumference of the base and the length of the pattern will be 2″ × 3.1416

= 6.2832″, or almost exactly $6\frac{9}{32}$″. For most practical purposes, you can multiply by 3.14 instead of 3.1416, or even multiply by $3\frac{1}{7}$. Thus, if the diameter is 2″, the circumference is 2″ × $3\frac{1}{7}$, or 2″ × $\frac{22}{7}$, or $\frac{44}{7}$, or 6.28″, which again is $6\frac{9}{32}$″ (to the nearest $\frac{1}{64}$″).

17.8 Elements. An *element* of a cylinder is an imaginary straight line on the surface parallel to the axis. A cylinder may be thought of as a prism with an infinite number of edges. Even if the prism has as few as 12 sides, Fig. 17–9 (a), the result is close to an actual cylinder. If lines are marked on the cylinder in the corresponding places, as shown at (b), the lines are *elements*, and are useful as will be shown in the following pages. Elements should be drawn as construction lines on the views and on the pattern. A sharp hard pencil should be used.

Practical methods used in dividing a circle into a number of equal parts are shown in Fig. 17–9 (c) to (f). At (c) the compass is used, with centers at points 1, 4, 7, and 10, and with a radius equal to the radius of the circle. At (d)

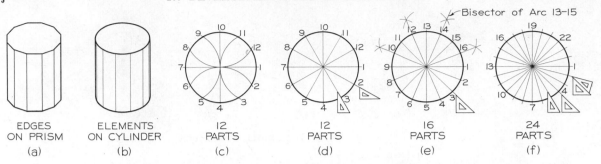

Fig. 17–9. Elements and Divisions of a Circle.

the 30° × 60° triangle is used as in Fig. 3–14 (d). At (e) the 45° triangle is used as in Fig. 3–14 (b) to get 8 divisions, and then each of these is bisected with the compass, as in Fig. 5–6. At (f) the two triangles are used in combination, as seen in Fig. 3–18.

Elements of a cone are illustrated in Fig. 17–17.

17.9 Pattern of Truncated Cylinder—Fig. 17–10. If a cylinder is truncated, or cut off

at an angle, the angled end will develop into a curved line, as shown in Fig. 17–10 (a). The lower end will develop into a straight line 1–1, which will be the stretchout line.

To develop the cylinder, divide the top view, (b), into any convenient number of equal parts by one of the methods shown in Fig. 17–9, and project down to draw the elements in the front view. As shown at (c), draw the stretchout line 1–1, and set off the true circumference on it. Divide it into the same num-

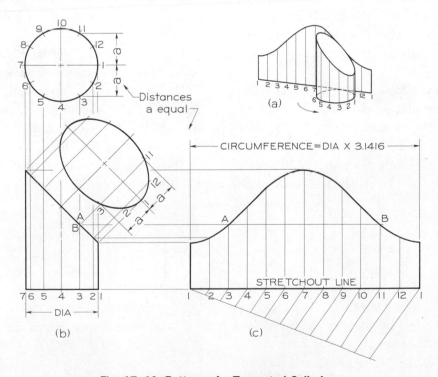

Fig. 17–10. Pattern of a Truncated Cylinder.

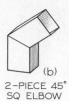

(a)	(b)	(c)	(d)	(e)	(f)	(g)
2-PIECE 90° SQ ELBOW	2-PIECE 45° SQ ELBOW	3-PIECE 90° SQ ELBOW	2-PIECE 90° ELBOW	2-PIECE 45° ELBOW	3-PIECE 90° ELBOW	4-PIECE 90° ELBOW

Fig. 17–11. Elbows.

ber of equal parts as in the top view, using the parallel-line method, Fig. 5–2 or 5–3. At the division points, draw elements perpendicular to the stretchout line, and locate the top ends of the elements by projecting across from the top ends of the corresponding elements in the front view. Note that each time you project across, you can locate two points, as A and B, in the pattern. Sketch a light smooth curve through the points, and heavy-in the final curve with the aid of the irregular curve, Sec. 3.30.

If the bases are needed in the pattern, they can be cut out separately. The true size of the bottom is shown in the top view, while the true size of the inclined surface is an ellipse and is shown true size in the auxiliary view.

17.10 Elbows. *Elbows,* Fig. 17–11, are common in sheet-metal work. Some are made up of prismatic shapes and others of cylindrical shapes; they may be composed of two, three, or more pieces. Each piece in the elbows at (a), (b), and (c) can be developed in a manner

similar to that seen in Figs. 17–6 and 17–7. Each end piece in the elbows at (d) to (g) can be developed as shown in Fig. 17–10. The method of developing the center pieces at (f) and (g) is shown in Fig. 17–12.

A number of elbows in practical work are illustrated in Fig. 17–1.

17.11 Pattern for Three-Piece Elbow—Fig. 17–12. It is necessary first to draw the views of the elbow. The two end pieces are the same shape, and their patterns will be identical. The middle piece is double the size of an end piece, or the same as the two end pieces put together.

I. Draw *heel* and *throat radii,* and square off ends with light construction lines.

II. Draw vertical, horizontal, and 45° construction lines tangent to arcs as shown; then through the intersections, draw the *mitre lines* to the center of the elbow.

III. Draw a semicircular half view adjacent to either end piece. A full circle is unnecessary,

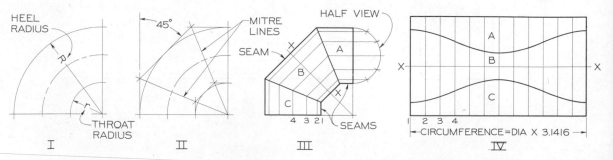

Fig. 17–12. Pattern of a Three-Piece Elbow.

and no additional views of the elbow are needed. Divide the semicircle into equal parts, say 6, as shown in Fig. 17–9 (d). From these points, draw elements on all three sections, as shown. Draw line X–X, the right section of the central piece.

IV. Draw the pattern. Pieces A and C are drawn in the way shown in Fig. 17–10. All elements are shown true length in the front view; those in piece C are projected directly across from the front view, while those in pieces A and B are transferred with dividers. All three pieces can be cut out of a rectangular piece, as shown. If desired, piece B could be developed in a manner shown for the oblique prism in Fig. 17–7.

17.12 Pattern of a Gutter—Fig. 17–13.

Gutters are good examples of the extensive use of sheet metal work in building construction. Gutters are composed of various combinations of prisms and cylinders; the patterns are therefore parallel-line developments.

In Fig. 17–13 (a) is shown a pictorial view

of an *ovolo* gutter, sectioned at A–A. The sectional view, (b), shows the true right section of piece B. In the top view, the right section appears as a line A–A, which will roll out into a straight stretchout line. Piece B is imagined to be rolled to the right so that the inside of the pattern will be up. The plane surfaces are developed as for the truncated prism in Fig. 17–6, and the quarter-cylinder as for the truncated cylinder in Fig. 17–10. The stretchout for the quarter-cylinder will be one-fourth the circumference of the complete cylinder:

$$\frac{\text{Dia.} \times 3.1416}{4} \quad \text{or} \quad \frac{R \times 3.1416}{2}$$

The stretchout of the quarter-cylinder may be closely approximated by stepping off on the stretchout line the chord distances 4–5, 5–6, 6–7, and 7–8, taken from the sectional view at (b).

Other styles of gutters are shown in Fig. 17–13 (d).

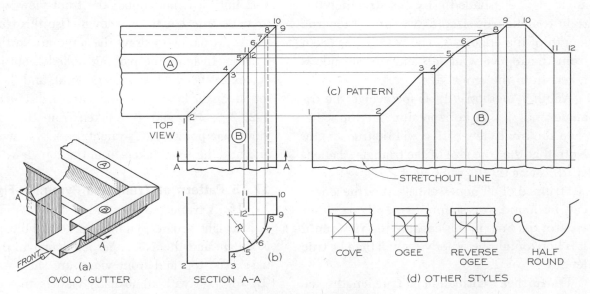

Fig. 17–13. Pattern of an Ovolo Gutter.

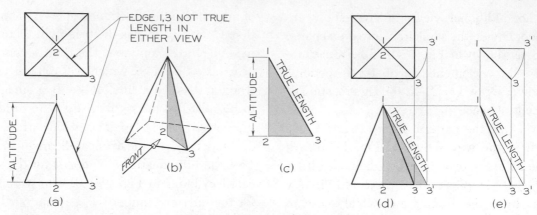

Fig. 17–14. True Length of Line.

RADIAL-LINE DEVELOPMENTS

17.13 True Length of Line. In radial-line developments, the edges or elements "radiate" like spokes in a wheel from a point, instead of being parallel, Fig. 17-2 (c) and (d). These lines usually do not show true length in the regular views; hence the true lengths to be used in the patterns must be found. For example, in Fig. 17–14 (a) the edge 1–3 does not appear true length in either view. As shown at (b), edge 1–3 is the hypotenuse of right triangle 1–2–3 (shaded). By constructing this right triangle true size, (c), we can get the true length of the edge 1–3. The base 2–3 is taken from the top view at (a), and the altitude is taken from the front view.

Another method, (d), is to revolve the triangle until 1–3 is horizontal in the top view; then the front view of 1–3 will be true length. Note that the true size of the triangle (shaded) at (d) is exactly the same as at (c).

Stripped of all nonessentials, the true length can be found, (e), simply by revolving either view of the line (in this case the top view) until it is horizontal; the other view will then be true length.

For further information on true lengths, see Figs. 12–14 and 13–7.

17.14 Pattern of a Pyramid—Fig. 17–15. *Pyramids* have flat triangular faces that intersect at a common point called the *vertex*. See Fig. 5–1. Bases of pyramids are polygons of three or more sides. In Fig. 17–15 (a) is shown a *right rectangular pyramid*—"right" because its axis is perpendicular to the base, and "rectangular" because the base is a rectangle.

The top and front views of the pyramid are shown at (b). All inclined edges are the same length, but none is shown true length. To get the true length, revolve the top view of edge V–2 until it is horizontal; the front view will then be true length, as shown. Use this true length as radius to draw the large arc in the pattern. Imagine the pyramid rolled about its vertex to the right as shown at (c), and then set off, on the large arc in the pattern, distances 1–2, 2–3, 3–4, and 4–1, taken from the top view. Join points with straight lines, as shown, and add the base, taken from the top view.

17.15 Pattern of Truncated Pyramid—Fig. 17–16. A truncated pyramid is shown at (a). It is a right square pyramid because the base is square and the axis is perpendicular to the base. The top and front views are shown at (b), together with an auxiliary view of the cut surface.

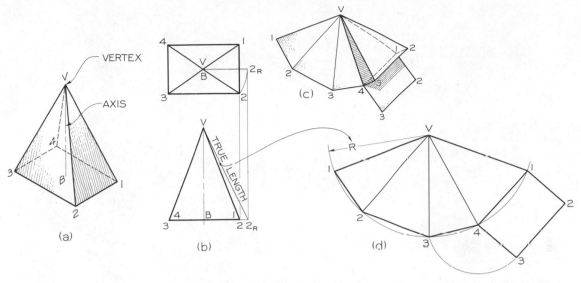

Fig. 17–15. Pattern of a Pyramid.

To draw the pattern, (d), imagine the pyramid to be rolled to the right, as shown at (c). In the pattern, radius R is the true length of one of the inclined edges of the pyramid, as shown in the front view. Along the large arc in the pattern, set off distances 1–2, 2–3, 3–4, and 4–1, taken from the top view; then join the points to each other and to the vertex with straight lines. In the front view, (b), the true lengths from vertex V down to points B and C on the cut surface are shown. Transfer the true lengths VB′ and VC′ to the pattern, and complete the pattern by adding the true size of the base, taken from the top view, and the true size of the cut surface, taken from the auxiliary view.

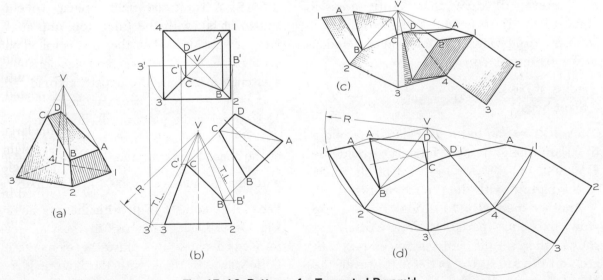

Fig. 17–16. Pattern of a Truncated Pyramid.

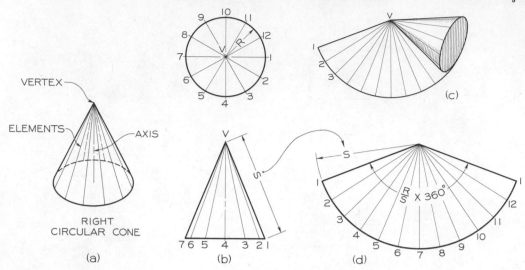

Fig. 17–17. Pattern of a Cone.

To transfer the auxiliary view ABCD to the pattern, draw diagonal CA so as to form two triangles CAD and CAB. Transfer the triangles by the method shown in Fig. 17–19.

17.16 Pattern of a Cone—Fig. 17–17. A *right circular cone* is shown at (a), with the top and front views at (b). When the cone is rolled out on a plane, the pattern will be a sector of a circle, or pie-shaped, (c). To draw the pattern, (d), draw an arc with radius S equal to the slant height of the cone, taken from the front view. The total angle included in the pattern is equal to

$$\frac{\text{Radius of base}}{\text{Slant height}} \times 360° \qquad \text{or} \qquad \frac{R}{S} \times 360°$$

Thus, if the cone has a 2″ radius base and a 5″ slant height, the formula would be $\frac{2}{5} \times 360°$, or $\frac{720°}{5}$, or 144°. Set off the angle in the pattern with the protractor, Fig. 3–19.

Another method is to divide the base (top view) into equal parts and draw elements as shown; then set off chord distances 1–2, 2–3, etc., on the arc in the pattern. However, the chord of an arc is slightly shorter than the arc, and when a number of chords are set off, there

may be considerable cumulative error. If the bow dividers are used and set *very slightly* larger than the chord distance, the resulting error will be small and the method will be satisfactory in most cases.

The true size of the base is shown in the top view and may be added to the pattern or cut out separately.

17.17 Pattern of a Truncated Cone—Fig. 17–18. A truncated right circular cone is shown at (a), with the front, top, and auxiliary views at (b). When the cone is rolled out on a plane, (c), the lower end develops into a circular arc, while the upper, or truncated, end develops into an irregular curve.

To develop the pattern, (d), draw the large arc with the radius S equal to the slant height of the cone. Then compute the included angle with the formula $\frac{R}{S} \times 360°$, as described in Sec. 17.16, and set it off with the protractor, Fig. 3–19. Divide the base in the top view into equal parts, and draw the elements in both views. By trial, with the bow dividers, space off the same number of equal parts on the base arc in the pattern, and draw the ele-

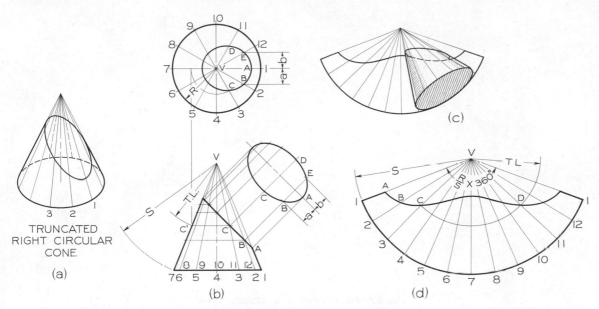

Fig. 17–18. Pattern of a Truncated Cone.

ments. If care is exercised, you may use the spacing between divisions in the top view to determine the complete angle of the pattern instead of having to compute the angle. Then set off from V on the elements in the pattern the true lengths VA, VB, VC, etc., down to the inclined cut, taken from the front view. For example, in the front view, the true length of VC is VC'. This is transferred to the pattern to give points C and D, and each true length in the front view will give two points in the pattern in a similar manner. Note, at (b), that you obtain each true length merely by drawing a horizontal line from the point to an outside element of the cone. This is equivalent to revolving the element until it appears true length in the front view. When all points in the curve have been found, trace a smooth curve through them, using the irregular curve, Sec. 3.30.

The intersection of a plane and a cone, in this case, is a true ellipse, and will appear as a true ellipse in the top and auxiliary views. The true size of the ellipse is shown in the auxiliary view. In the auxiliary view, equal

distances *a* and *b*, and others on each side of the center line, are transferred from corresponding points in the top view. The inclined face and the base may be cut out separately if needed in the pattern.

17.18 Triangulation. *Triangulation* is the process of dividing a surface into a number of triangles and then transferring each of them in turn to the pattern. To transfer a triangle, say ABC in Fig. 17–19 (a), draw side AB in the desired new location at (b). With the ends A and B as centers, and the lengths of the other sides of the given triangle as radii, strike two arcs to intersect at C. Then, as shown at (c), join point C to points A and B.

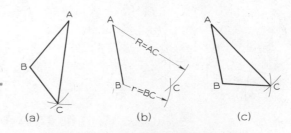

Fig. 17–19. Transferring a Triangle.

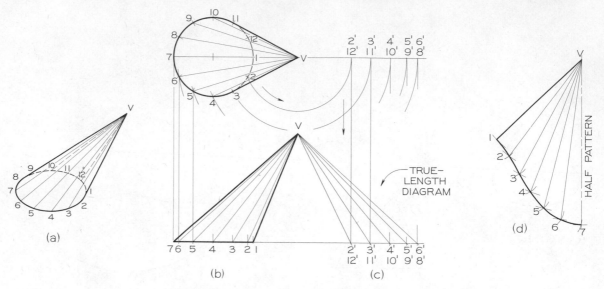

Fig. 17–20. Pattern of an Oblique Cone.

17.19 Pattern of Oblique Cone—Fig. 17–20.

An *oblique cone* is shown at (a), with the top and front views at (b). Divide the base (top view) into equal parts, and draw elements to the vertex as shown. Only elements V–1 and V–7 are true length, as shown in the front view. The surface of the cone is thus divided into triangles by the elements, and the pattern will be composed of these triangles laid out next to each other on a plane.

The simplest way to get true lengths is to construct a true-length diagram, as shown at (c). Revolve the top view of each element until it is horizontal. Then project down to the base line to get points 2′, 3′, etc., and con-

nect with lines to the front view of the vertex V. These are true lengths to be used in the pattern. It is not necessary to find true lengths of elements V–1 and V–7, since they are already shown true length in the front view.

If the pattern is divided on element V–1, the pattern is symmetrical, and only one half need be drawn. From V in the pattern draw V–1 equal to V–1 in the front view. Then from V in the pattern, strike arc V–2 taken from the true-length diagram, and arc 1–2 taken from 1–2 in the top view of the base of the cone. This triangle V–1–2 was transferred in the same manner as was the triangle in Fig. 17–19. Complete the half-development by transfer-

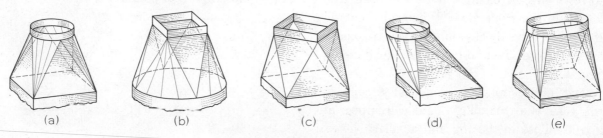

Fig. 17–21. Transition Pieces.

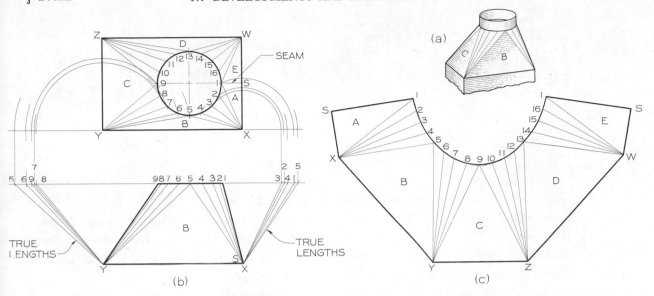

Fig. 17–22. Pattern of a Transition Piece.

ring the remaining triangles in the same way. Connect the points with a light freehand curve, and then heavy-in the curve with the aid of the irregular curve, Sec. 3.30.

17.20 Transition Pieces—Fig. 17–21.
A *transition piece* is one that connects two different-shaped or skewed-position openings—as, for example, a round opening to a square opening. Transition pieces are widely used in air-conditioning, ventilating, heating, and similar installations. In most cases, transition pieces are composed of a combination of plane surfaces and conical surfaces. Therefore, the methods given above for pyramids and cones can be applied.

17.21 Pattern of Transition Piece—Fig. 17–22.
A transition piece having a round opening at the top and a rectangular opening at the bottom is shown at (a), with the top and front views at (b). The surface is composed of four triangular plane surfaces and four conical surfaces. The conical surfaces are

divided into narrow triangles so that they can be transferred to the pattern.

Assume the seam at 1–S (see top view). Triangle 1–S–X is a right triangle, which can be easily drawn in the pattern, with the true length of 1–S taken from the front view, and the true length of SX taken from the top view. Then triangle 1–X–2 and all others can be transferred by taking the small bases from the circle in the top view and the long sides from the true length diagrams, transferring in the manner of Figs. 17–19 and 17–20.

INTERSECTIONS

17.22 Intersections.
A line intersects a surface in a point. Two surfaces intersect in a *line of intersection*. The complete intersection between two solids is called a *figure of intersection*. Such intersections are common in building construction, sheet-metal work, machine construction, and like projects, as illustrated in Fig. 17–23, and the draftsman or designer must know how to construct them.

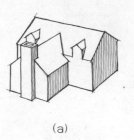 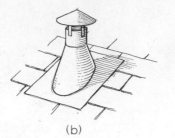  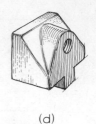

(a)   (b)   (c)   (d)

Fig. 17–23. Intersections.

Intersections are found by using one method over and over: finding the point where a line pierces a surface. In Fig. 17–24 (a) a nail is shown penetrating a prism, with two piercing points A and B. At (b) the top view shows clearly where the nail intersects the surfaces, because the surfaces appear edgewise. Project down to the front view of the nail to get piercing points A and B, as shown.

The same procedure applied to a cylinder is shown at (c) and (d).

17.23 Intersection of Two Square Prisms— Fig. 17–25.

Two intersecting square prisms are shown pictorially at (a), with the three views at (b). The points in which edges W, X, Y, and Z of the horizontal prism pierce the vertical prism are points on the figure of intersection. Points 1 and 3 of the intersection are already evident in the front view. In the top

view, the edges X and Z of the small prism intersect the large prism at points 2 and 4. Project down to get the front view of points 2 and 4. Join the points of the intersection with straight lines, as shown.

To draw the pattern of the small prism, (c), set off on the stretchout line W–W the widths of the faces WX, XY, etc., taken from the side view, and draw the edges through these points as shown. Set off from the stretchout line the lengths of the edges W–1, X–2, etc., taken from either the front or the top view, and join the points 1, 2, 3, etc., with straight lines.

To develop the pattern of the large prism, (d), set off on the stretchout line A–A the widths of the faces AB, BC, etc., taken from the top view, and draw the edges through these points as shown. Set off on the stretchout line distances BE and DF, taken from the top view. To locate points 1, 2, 3, and 4

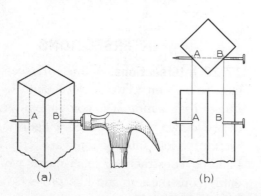 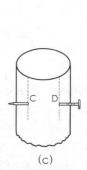 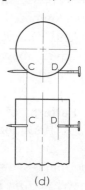

(a)   (h)   (c)   (d)

Fig. 17–24. Intersection of a Line and a Solid.

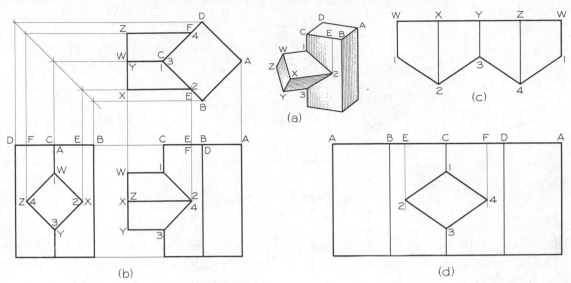

Fig. 17-25. Intersecting Prisms.

in the pattern, project down from points E, C, and F, and across from points 1, 2, 3, and 4 in the front view. Join the points with straight lines.

17.24 Intersection of Cylinders—Fig. 17-26.
To obtain the intersection, divide the circle of the top view, (b), into a number of equal parts, and draw the elements in both the front and side views. Their points of intersection with the surface of the half-cylinder are shown in the side view at A, B, C, etc., and are located in the front view by projecting across to the corresponding elements in the front view. The accuracy of the curve depends, of course, on the number of points found. Connect the points smoothly with the aid of the irregular curve, Sec. 3.30.

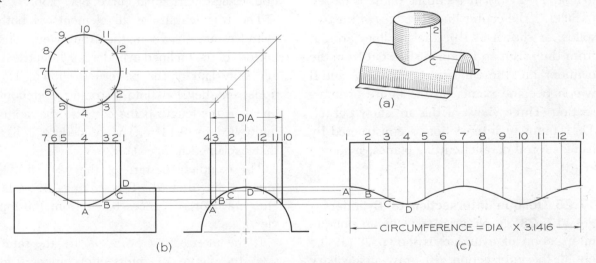

Fig. 17-26. Intersection of Cylinders.

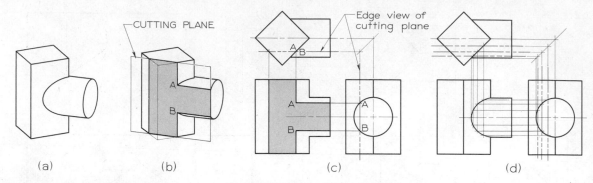

Fig. 17-27. Cutting Planes.

The pattern of the vertical cylinder is shown at (c). The method used is the same as described in Sec. 17.9.

For other intersections of cylinders, see Fig. 7-22. Note especially at (d) that if the cylinders are the same size, the figure of intersection appears as straight lines.

17.25 Cutting Planes. Imaginary *cutting planes,* similar to those used in sectioning, Sec. 11.1, can be used to great advantage in drawing intersections. For example, consider the intersection between a prism and a cylinder in Fig. 17-27 (a). If a cutting plane is passed parallel to the center lines or edges of the two solids, as shown at (b), straight lines are cut from the prism and elements are cut from the cylinder, and these intersect at points A and B which become points on the figure of intersection. Three views of this are shown at (c). The complete intersection is constructed by using several parallel cutting planes, as shown at (d).

17.26 Oblique Intersection of Cylinders— Fig. 17-28. A pictorial view of an oblique intersection of cylinders is shown at (a). To obtain the intersection, (b), draw an auxiliary view of the inclined cylinder, and divide it

into equal parts as shown. Draw cutting-plane lines through the divisions, and then draw the corresponding cutting-plane lines in the top view. Spacings between cutting-plane lines in the top and auxiliary views must be equal, as shown for distances *a* and *b*. Draw elements cut by the planes, and locate points where corresponding elements intersect. For example, element G of the vertical cylinder intersects elements 5 and 3 of the inclined cylinder at points X and Y. This is shown pictorially at (c). Trace a smooth curve through the points to establish the figure of intersection, using the irregular curve, Sec. 3.30.

The true lengths of all elements of both cylinders are shown in the front view. The pattern of the inclined cylinder is symmetrical, and only half of the pattern is shown. The upper end develops into a straight stretchout line 1-7, the length being equal to the radius multiplied by 3.1416. The procedure is like that explained in Sec. 17.7.

The complete pattern of the vertical cylinder is shown at (d). The spacings BC, CD, etc., are chord-distances taken from the top view.

If the intersecting cylinders are the same size, the figure of intersection appears as straight lines, Fig. 7-22 (d).

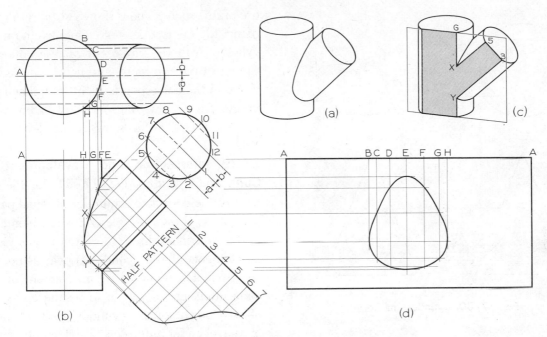

Fig. 17-28. Oblique Intersection of Cylinders.

17.27 Intersection of Prism and Pyramid—Fig. 17-29.

The surfaces of the pyramid do not appear edgewise in any view, but the surfaces of the prism appear edgewise in the side view. Draw cutting planes *containing* these surfaces and cutting straight lines on the pyramid. These lines are lines of intersection between the surfaces of the pyramid and the surfaces of the prism. For example, as shown at (a), a cutting plane containing a vertical surface of the prism cuts lines 1–2 and 2–3 on the pyramid, and these intersect edges of the prism at points A and F. Edge VX of the pyramid is seen in the front view to intersect the top and bottom surfaces of the prism at points B and E. Locate the points of intersection in all views and connect them with straight lines to complete the intersection as shown.

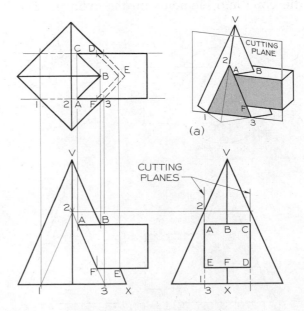

Fig. 17-29. Prism and Pyramid.

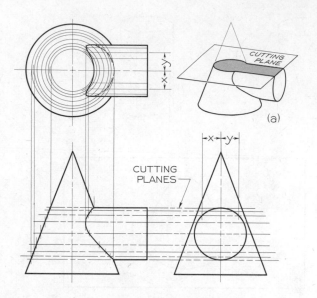

Fig. 17–30. Cylinder and Cone.

17.28 Intersection of Cone and Cylinder— Fig. 17–30. A convenient way to find the intersection of a cone and a cylinder is to use a series of cutting planes that cut circles on the cone and elements on the cylinder. The use of one such plane is shown at (a). For each plane, the intersections of elements cut on the cylinder with the circle cut on the cone are shown in the top view, and can be projected down to the corresponding cutting-plane lines in the front view. This method is not convenient if the surfaces are to be developed.

17.29 Intersection and Development of Cone and Cylinder—Fig. 17–31. Divide the base of the cone into a number of equal parts, say 16, as shown at (a). Draw the elements of the cone in all three views. In the side view, all elements of the cylinder appear as points. Imagine each point where an element of the cone intersects the circle to be the end view of an element of the cylinder. Each element of the cylinder intersects an element of the cone, giving a point on the figure of intersection. Join all points with a smooth curve in the top and front views to complete the intersection.

Another method is to use cutting planes. Draw elements of the cone as before. In the

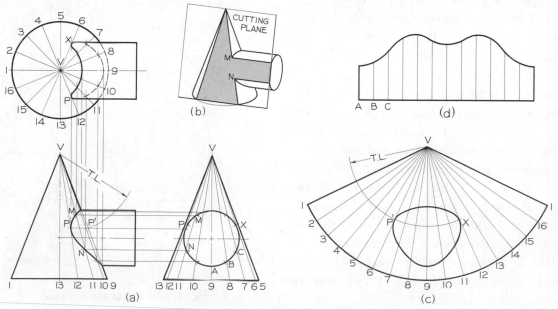

Fig. 17–31. Cone and Cylinder.

side view, think of these elements as also the edge views of a series of cutting planes. Each plane will cut from each solid two elements, which intersect to give points on the figure of intersection. One such cutting plane, producing two points M and N, is illustrated at (b).

To draw the pattern of the cone, (c), draw a large arc with radius equal to the slant height of the cone. Set off on this arc the chord-distances 1–2, 2–3, 3–4, etc., taken from the top view, or compute the angle of the pattern as explained in Sec. 17.16. To locate points on the opening in the pattern, take true lengths in the front view from the vertex down to points on the figure of intersection. The true length (TL) of VP is shown at VP′ in the front view and transferred to the pattern to get points P′ and X.

To draw the pattern of the cylinder, divide the circle in the side view into an equal number of parts, and draw the elements. To simplify the illustration, these elements are omitted in the figure. Draw the stretchout line equal to the circumference of the cylinder (dia. × 3.1416), and divide it into the same number of equal parts. The true lengths of the elements are shown in the front view, and can be transferred directly to the pattern.

17.30 Development and Intersection Problems. A very wide range of problems on developments and intersections is provided in Figs. 17–32 to 17–37. The first four groups, Figs. 17–32 to 17–35, consist of layouts (Layout D, Appendix, page 475) in which there are many alternate problems to provide different assignments for students. Following these are a number of problems applying the principles of this chapter. These also are to be drawn on Layout D or 11″ × 17″ sheets.

In the illustrations of this chapter, cylinders and cones are usually divided into only 12 elements to keep the presentation as simple as possible. In the problems, the instructor may wish the student to use 16 or 24 divisions for greater accuracy.

Methods of constructing paper models are explained in Sec. 17.3. It is suggested that at least a few of the early problems in developments and later in intersections be actually cut out and formed into models.

In the problems that follow, disregard allowances of extra material for seams, rolled edges, and thickness of materials. Several of these problems are included by permission of Mr. Philip Burness, to whom acknowledgment is gratefully given.

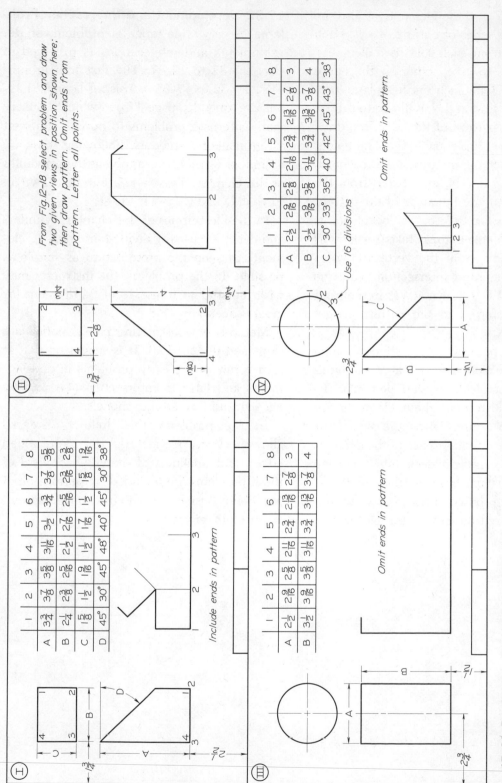

Fig. 17–32. Parallel-Line Development Problems. Using Layout D (Appendix, page 475), draw given views and pattern of problem assigned by instructor. Omit table and all spacing dimensions and instructional notes.

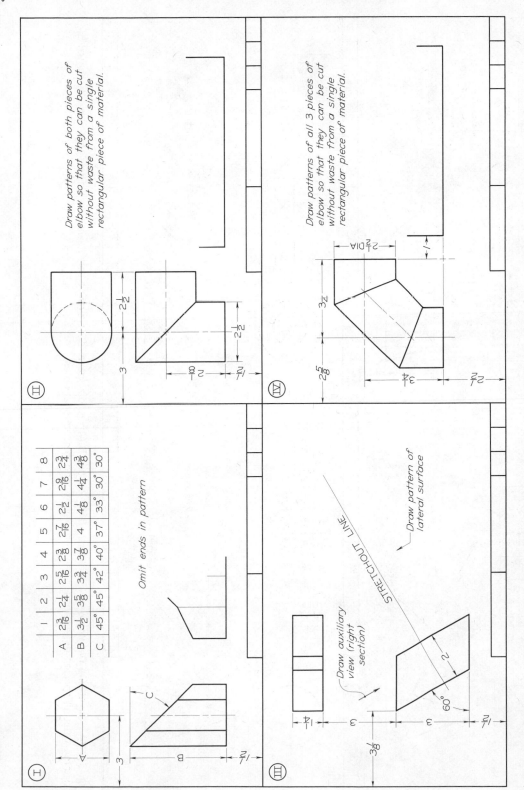

Fig. 17-33. Parallel-Line Development Problems. Using Layout D (Appendix, page 475), draw given views and pattern of problem assigned by instructor. Omit table and all spacing and instructional notes.

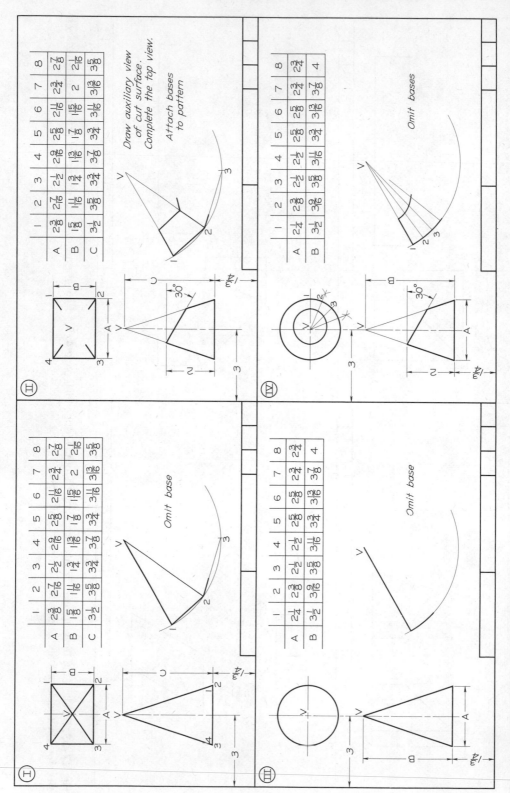

Fig. 17-34. Radial-Line Development Problems. Using Layout D (Appendix, page 475), draw given views and pattern of problem assigned by instructor. Omit all spacing dimensions, table dimensions, and instructional notes.

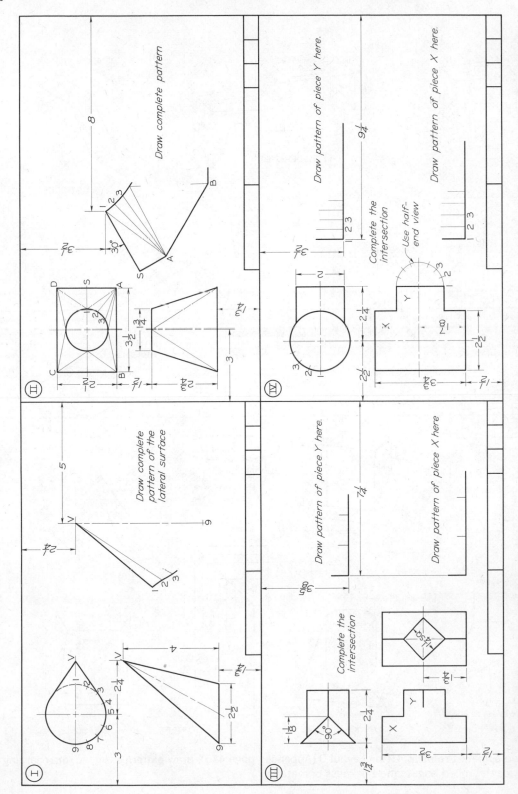

Fig. 17-35. Development and Intersection Problems. Using Layout D (Appendix, page 475). draw views and pattern(s) of problem assigned by instructor. Omit all spacing and instructional notes.

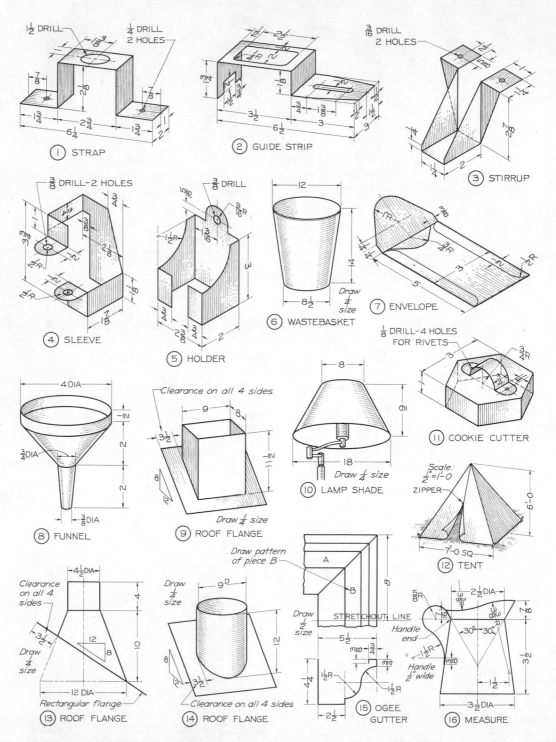

Fig. 17–36. Development Problems. Using Layout D (Appendix, page 475), draw pattern assigned, disregarding allowances for seams, rolled edges, and thickness of material.

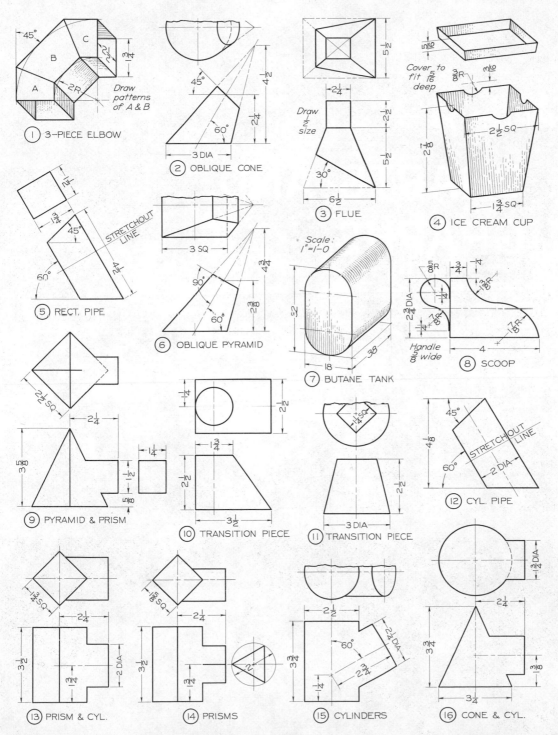

Fig. 17–37. Development and Intersection Problems. Using Layout D (Appendix, page 475), draw pattern assigned. Disregard allowances for seams, rolled edges, and thickness of material.

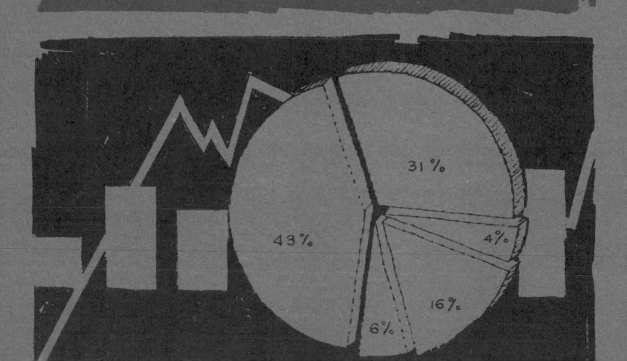

CHAPTER 18

CHARTS AND GRAPHS

18.1 Graphical Presentation. We are constantly concerned with numbers, quantities, and comparisons of amounts. Statistics—facts expressed by numbers—are often regarded as "cold" or uninteresting because their meaning is not immediately apparent. The financier watches the Dow Jones stock averages, the sales manager keeps his eye on dollar volume of sales, and the housewife certainly is interested in the ups and downs of the cost of living.

If numerical facts are presented graphically—that is, by means of drawing—the information catches the eye immediately, and means more than a simple tabulation of numbers. Such drawings are variously called *charts*, *graphs*, or *diagrams*. There are many more kinds than can be discussed here, but the most common are *bar charts*, *line charts*, and *pie charts*.

18.2 Bar Charts. The bar chart is an effective device for showing a comparison between amounts. It is easily understood by everyone and is used extensively in newspapers, magazines, and books.

The simplest form of bar chart is the *100 percent bar*, Fig. 18–1, which shows the per-

centage ratio of various parts to a given whole. The total length of the bar represents 100 percent, and this distance is divided into segments that are proportional parts of the whole. The bars may be drawn horizontally or vertically, and appropriate shading or crosshatching is used to distinguish between the segments.

The bar chart in Fig. 18–2 shows very clearly the relative populations in our four largest cities. Bar charts are useful in comparing amounts either from large to small or the reverse. As a rule, the bars should be arranged in order of height and not at random.

A tabulation of points scored by three football teams in a given year is as follows:

School	Total Points, 1966
State College	78
Western University	246
Smith Institute	325

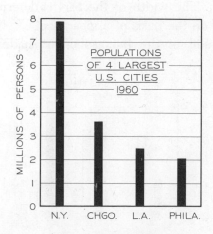

Fig. 18–2. Bar Chart.

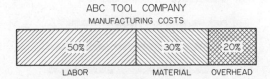

Fig. 18–1. 100 Percent Bar Graph.

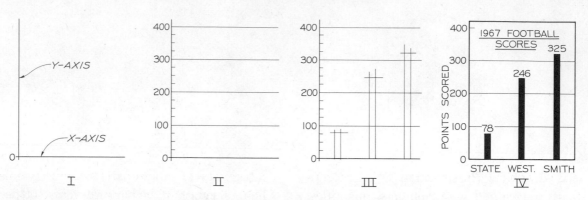

Fig. 18–3. Steps in Drawing a Bar Chart.

The steps in drawing a bar chart that gives this information in graphic form are shown in Fig. 18–3.

I. Draw a vertical line called the Y-axis or *ordinate* and a horizontal line called the X-axis or *abscissa*. The point of intersection will be zero on the chart.

II. Select a suitable scale for the points along the Y-axis. For example, using the architects scale, let $\frac{1}{8}'' = 10$ points; or using the engineers scale, let $1'' = 100$ points ($\frac{1}{10}'' = 10$ points). Letter the numbers at convenient intervals, say 100, 200, and 300, as shown. Draw horizontal lines across the chart at these locations.

III. Block in bars, spacing them far enough apart to permit lettering the names at the bottom. The width of the bars is determined simply on the basis of appearance.

IV. Fill in the bars solid, or shade with colored pencil, or use section-lining. Add necessary lettering, including a title.

Many variations of the bar chart are used. For example, suppose we want to show the scoring record of the Smith Institute in comparison with the scores made against Smith by all of its opponents. Such a chart is shown in Fig. 18–4. The bars are arranged in pairs, with one set filled in solid and the other section-lined to distinguish them from each other.

Bar charts may be drawn with the bars in a vertical position or in a horizontal position. A horizontal bar chart is shown in Fig. 18–5. In this case the quantities (billions) are expressed by means of bars and also in figures at the ends of the bars.

Bar charts may be drawn in a variety of ways. For example, the bars can be drawn

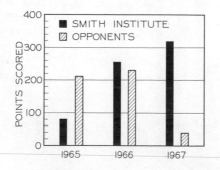

Fig. 18–4. Bar Chart—Scoring Records.

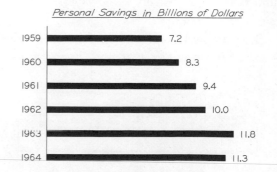

Fig. 18–5. Bar Chart—Personal Savings per Year.

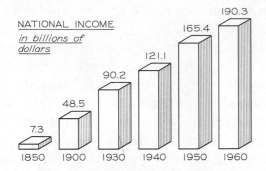

Fig. 18–6. Bar Chart—National Income.

pictorially in oblique, Fig. 18–6, or in isometric, or they may be emphasized by means
of shading or by colors.

18.3 Cross-Section Paper.
Much time can
be saved if the charts or graphs are constructed
upon cross-section paper such as we have already used for freehand sketching. Paper
having $\frac{1}{8}''$ or $\frac{1}{4}''$ squares may be used, but generally it is desirable to use paper with $\frac{1}{10}''$ or
$\frac{1}{20}''$ squares.

18.4 Line Charts.
Line charts are used espe-
cially to show *trends*—as, for example, the ups
and downs of the stock market.

The following figures indicate the expenditures, in billions of dollars, for space activities
by the U.S. government:

Year	Billion $
1958	0.249
1959	0.521
1960	0.960
1961	1.468
1962	2.390
1963	4.077
1964	6.176

A line chart showing the trend of increasing
expenditures for these activities is shown in
Fig. 18–7. To make such a chart, draw the
X- and Y-axes as in Fig. 18–3 for bar charts.

Select suitable scales to be used along both
axes, and draw horizontal and vertical grid
lines. Then plot the numbers on the charts.
For example, for 1962 the point for 2.390 is
found on the vertical line of '62 and above the
horizontal marked 2. Actually, the amount
above 2 is approximately $\frac{4}{10}$ of the distance
between 2 and 3. Connect all plotted points
with straight lines.

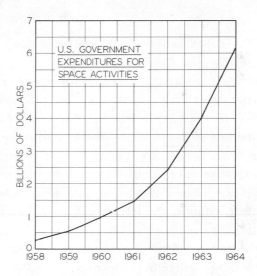

Fig. 18–7. Line Chart.

In the chart in Fig. 18–7, regular $\frac{1}{4}''$ cross-
section paper was used. Along the vertical
scale each two squares were taken to equal
1 billion dollars. Along the horizontal scale the
years were spaced two squares apart.

Line charts are used in an endless variety of
ways. For example, if we want to compare the
records of the Chicago Cubs and the Chicago
White Sox, Fig. 18–8, a solid line can be used
to represent one team and a broken line the
other team. This chart not only shows the trend
for each team, but compares the records at
all points.

18.5 Pie Charts.
The idea for pie charts undoubtedly came from the custom of cutting
pies into portions. For showing how the

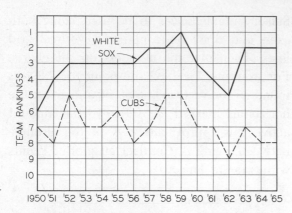

Fig. 18–8. Line Chart Comparing Two Teams.

whole is split up into several unequal portions, the pie chart is excellent, Fig. 18–9. It is desirable to place the lettering horizontally in the sectors of the chart, as shown in the figure. Therefore, the pie chart is most suitable when there are no more than five or six divisions. If there are more divisions, so that each is too small to contain the lettering, leaders can be used as in (b).

To draw a pie chart as shown at (a), draw a circle large enough to contain the lettering without undue crowding. Then determine the angles of the sectors by using the protractor, Fig. 3–19. To convert percentage into degrees, multiply the percentage by 360°.

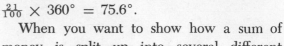

For example, for the 21 percent cross traffic, $\frac{21}{100} \times 360° = 75.6°$.

When you want to show how a sum of money is split up into several different amounts, you often think of the whole as a dollar and the parts as cents—which is another way of expressing percent or units of 100. In these cases, the "dollar" may be drawn in oblique, as shown at (b), to look like an actual coin.

18.6 Pictographs. A *pictograph* is a graph or chart in which pictures are used as symbols to represent units or quantities. For example, in Fig. 18–10 each symbol of a person represents 10 million people. These symbols are then repeated to show the various totals. Actually, the result is a form of bar chart. Another variation is to show one symbol opposite each year instead of several, but to vary the size of the symbol from small to large according to the number represented.

18.7 Problems. It is suggested that before drawing any charts, each student bring to class at least one example of each type of chart discussed here. These can be found in newspapers, magazines, and in schoolbooks. The

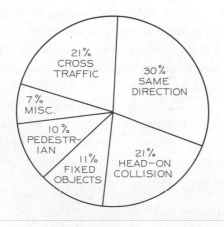

(a) CAUSES OF AUTO ACCIDENTS   (b) EST. FEDERAL BUDGET DOLLAR, 1966

Fig. 18–9. Pie Charts.

POPULATION

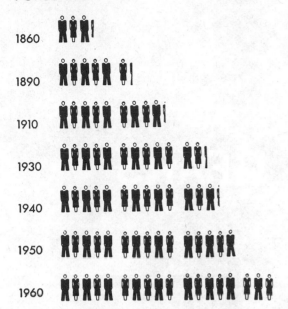

Each symbol represents ten million people

Adapted from graph by Twentieth Century Fund

Fig. 18–10. Pictograph—Growth of U.S. Population.

various charts can then be passed around and discussed at length by the class.

BAR CHARTS

PROB. 18/1. A typical family budget dollar is spent as follows: Food—28.2¢, Housing—21.7¢, Transportation—11.8¢, Household Operations—10.9¢, Clothing—8.8¢, Recreation—6.0¢, Medical Care—5.3¢, Personal Care—2.8¢, and Miscellaneous—4.5¢. Make a vertical bar chart showing this information.

PROB. 18/2. Draw a bar chart showing, in terms of percentages, the final standings last year of baseball teams in either the National or the American League.

PROB. 18/3. Draw a bar chart showing the numbers of freshmen, sophomores, juniors, and seniors in your school.

PROB. 18/4. Same as Prob. 18/3, except that chart is to show a pair of bars for each grade, one bar for girls and one for boys.

LINE CHARTS

PROB. 18/5. Make a line chart showing how your grades have changed from month to month in a given subject throughout the school year.

PROB. 18/6. List the total points scored by a football team last year in each game. Make a line chart showing the progress of the team through the season.

PROB. 18/7. U.S. foreign aid for the years indicated was as follows (in billions of dollars): 1957—5.1, 1958—4.9, 1959—5.3, 1960—4.6, 1961—4.3, 1962—4.6, 1963—5.1, 1964—4.7. Draw a line chart showing these facts. Let each inch on the vertical scale equal 1 billion dollars.

PIE CHARTS

PROB. 18/8. It is estimated that the 1966 federal "budget dollar" was spent as follows: National Defense—40¢, Space—4¢, Fixed Interest Charges—7¢, Veterans—4¢, Social Security and Other Trust Funds—26¢, and Other Costs—19¢. Make a pie chart showing this distribution.

PROB. 18/9. In 1964 employed people in the United States were distributed as follows: Manufacturing—33%, Agriculture—8%, and All Others—59%. Make a pie chart showing this information.

PROB. 18/10. A student has an allowance of $10 per month, and in addition he earns $18 per month working for a grocery. He budgets his expenditures as follows: School lunches—$8.00, Supplies—$2.00, Amusement—$6.00, Clothing—$7.00, Savings—$5.00, Total—$28.00. Make a pie chart showing this distribution. To convert into percentages, divide the part by the whole, and multiply by 100. For example, for Clothing—$7.00, figure as follows:

$$\tfrac{7}{28} \times 100 = \tfrac{700}{28} = 25\%$$

CHAPTER 19
CAMS AND GEARS

19.1 Cams. A *cam* is a machine element used to obtain an irregular or unusual motion. Cams are manufactured in an endless variety of forms, some of which are shown in Fig. 19–1. *Plate cams,* or *disk cams,* are essentially flat with irregular edges, Fig. 19–2 (a) to (c). A *cylindrical cam,* (d), is a cylinder with an irregular groove cut around it. The basic principle of a cam is illustrated at (a). A uniformly rotating *camshaft* has mounted upon it an irregularly shaped disk or plate, which is the cam. As the cam rotates counterclockwise, as shown by the arrow, the *follower* moves up gradually, then down more rapidly, and finally remains "at rest" until the starting point is reached again.

The *roller* on the follower, which provides smooth operation, is held in contact with the cam by gravity or a spring. The draftsman or designer must design a cam that will produce a desired motion of the follower.

19.2 Cam Followers. The type or shape of cam follower selected is determined by the requirements of the mechanism or the operation that is to be performed. The three most

Fig. 19–1. A Variety of Cams.

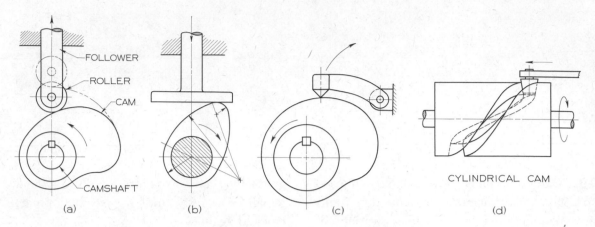

Fig. 19–2. Cam Types and Terminology.

common types of cam followers are shown in Fig. 19–2 (a) to (c). A roller follower is shown at (a), a *flat-faced* follower at (b), and a *pointed* follower at (c). The axis of the follower may be located on the vertical center line of the cam, as shown in (a) and (b), or may be offset as shown in (c). Many other special cam followers can also be designed or modified from the basic types to simplify a complicated cam profile or mechanism.

19.3 Cam Motion. The desired motion of the follower of a cam can be shown in a *displacement diagram*, Fig. 19–3, and may be *uniform* (*constant velocity*), *harmonic*, or *uniformly*

accelerated and *retarded* (*parabolic*). These motions may be used separately or in combination with one another. The horizontal base line (travel) on the diagram represents one revolution (360°) of the cam, and any convenient length can be used to represent this distance. The vertical distances (rise or fall) on the diagram are drawn to scale and show the actual follower displacement.

The dashed line AB in Fig. 19–3 represents uniform motion in which both the rise and travel are divided into the same number of equal distances. This is the simplest of all motions, but since it makes the follower start and stop suddenly it is also the least practical.

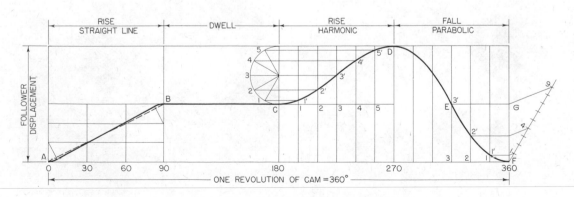

Fig. 19–3. Typical Cam Displacement Diagram.

This motion can be modified by an arc at each end of the motion, as shown by the heavy line.

Line BC represents a period of *dwell*, during which time the follower does not rise or fall.

To construct the curve CD, which gives harmonic motion to the follower, draw a semicircle whose diameter is equal to the desired rise. Divide the semicircle and the travel into the same number of equal parts. Points on the curve are located by drawing horizontal lines from the points on the semicircle to intersect the vertical lines drawn from the corresponding points on the travel, as shown. Sketch a smooth curve through the points, and heavy-in the curve with the aid of the irregular curve, Sec. 3.30.

The parabolic curve DEF gives the follower uniformly accelerated and retarded motion, and consists of two halves, with the half of the curve from D to E being exactly the reverse of the curve from E to F. To construct the curve EF, divide the vertical height from G to F into parts proportional to 1, 3, 5, etc. The travel is divided into the same number of equal parts. Points on the curve are located by drawing horizontal lines from the points on line GF to intersect the vertical lines drawn from the corresponding points on the travel, as shown. Sketch a smooth curve through the points and heavy-in the curve with the aid of the irregular curve, Sec. 3.30. The displacement diagram is usually constructed before the actual cam profile is drawn.

19.4 Layout of a Plate Cam Profile. The method of laying out a typical plate cam profile is illustrated in Fig. 19–4 (a). In this case it is desired to move the follower a certain distance with uniform motion and then to return the follower to the starting point with uniform motion. This cam is to rotate counterclockwise, as shown by the arrow.

The speed of rotation is constant; therefore, *equal angles of rotation will be equivalent to equal units of time.* Thus, the follower will move the same distance for each angle of rotation of the cam.

The displacement diagram, (b), indicates the desired motion of the follower. The horizontal base line represents one revolution of the cam, and is divided into 12 equal parts corresponding to the 12 equal angular divisions of the base circle. The vertical distances show the actual rise and fall of the follower. Note that the diagram has been modified by including arcs at each end of the follower motion, as shown in Fig. 19–3.

The steps in laying out the cam are:

1. Draw vertical and horizontal center lines, and then draw the *base circle*. The radius of the base circle is the distance from the center of the cam to the center of the roller when the two parts are closest together.

2. Divide the base circle into 12 equal angles of 30° each, as shown in Fig. 17–9 (d). For greater accuracy, draw 24 divisions of 15° each, as shown at (f).

3. On the center line of the follower, from zero (center of roller), set off the desired "rise" of the follower, and divide this into 6 equal parts, as shown.

4. Imagine the cam to be stationary, and the roller moving *clockwise* around the cam. From cam center C, strike construction arcs C–1, C–2, C–3, etc., to intersect 30° radial lines at 1′, 2′, 3′, etc. At each point, draw a construction arc representing the roller at that position.

5. Sketch a smooth curve tangent to the small arcs, and heavy-in the curve with the aid of the irregular curve, Sec. 3.30.

NOTE: For a more detailed discussion of cams, see a book on mechanism or any standard text on the subject.

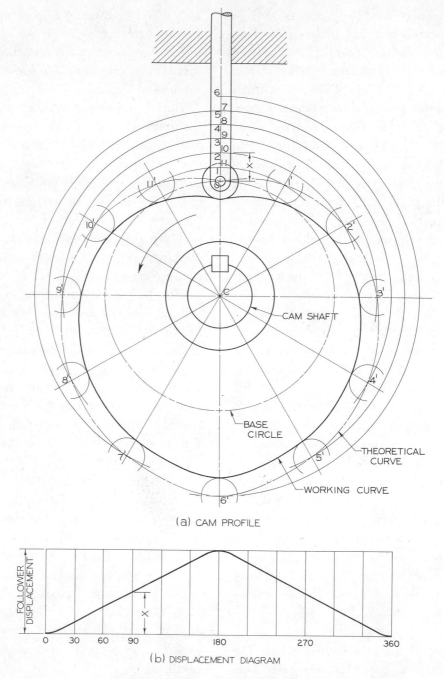

(a) CAM PROFILE

(b) DISPLACEMENT DIAGRAM

Fig. 19–4. Layout of a Plate Cam Profile.

19.5 Gears. If two round cylinders or *friction gears,* Fig. 19–5 (a), are placed in contact, one, if rotated, will transmit motion to the other. Note that the two will turn in opposite direc- tions. If they are the same size and one is turned around once, the other will revolve once. If one is twice the diameter of the other, it will turn halfway around for one revolution

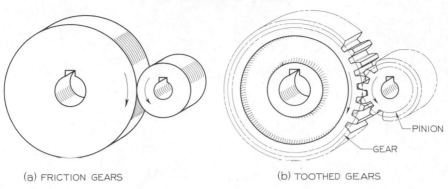

(a) FRICTION GEARS   (b) TOOTHED GEARS

Fig. 19–5. Friction Gears and Toothed Gears.

of the other. This points up one of the chief reasons for using gears; namely, to reduce or increase speeds.

However, friction gears are subject to slipping, and excessive pressure between them is required to obtain the necessary "traction." If teeth are provided, the resulting gears, Fig. 19–5 (b), will transmit motion without slipping.

There are a great many kinds of gears, some of which are illustrated in Fig. 19–6. *Spur gears* connect parallel shafts, *bevel gears* connect shafts whose axes intersect, and *worm gears* connect shafts whose axes ordinarily are at right angles to each other but are nonintersecting. The gears in Fig. 19–5 (b) are spur gears, the smaller one being commonly called a *pinion* and the larger a *gear*. If teeth are cut

Fig. 19–6. A Variety of Gears.

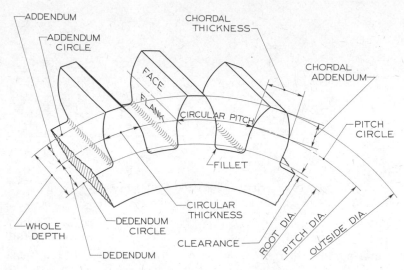

Fig. 19–7. Gear Tooth Terms.

on a flat piece, the part is called a *rack*, as shown at the lower right corner of Fig. 19–6.

Spur gear proportions, and the shape of the teeth, are well standardized, and the terms used, Fig. 19–7, are common to practically all spur gears. The dimensions relating to tooth height are for "full-depth $14\frac{1}{2}°$ involute teeth," which is the most common type.

The *outside diameter* (OD) is the overall diameter of the gear. The *pitch diameter* (PD) is the diameter that corresponds to the friction gear without teeth. The *root diameter* (RD) is the diameter at the bottom of the tooth.

The *circular pitch* (CP) is the distance measured along the pitch circle from a point on one tooth to the corresponding point on the next tooth. The *diametral pitch* (DP), or simply *pitch*, is the number of teeth on the gear per inch of pitch diameter. The *addendum* (A) is the height of the tooth above the pitch circle, and the *dedendum* (D) is the depth of the tooth below the pitch circle. The following formulas can be used to obtain various dimensions, where N equals the number of teeth:

$$\text{Diametral pitch DP} = \frac{N}{PD}$$

$$\text{Circular pitch CP} = \frac{3.1416}{DP}$$

$$\text{Pitch diameter PD} = \frac{N}{DP}$$

$$\text{Outside diameter OD} = \frac{N+2}{DP}$$

$$\text{Root diameter RD} = PD - 2\,D$$

$$\text{Addendum A} = \frac{1}{DP}$$

$$\text{Dedendum D} = \frac{1.157}{DP}$$

$$\text{Clearance C} = \frac{0.157}{DP}$$

$$\text{Whole depth WD} = A + D$$

$$\text{Circular thickness CT} = \frac{CP}{2}$$

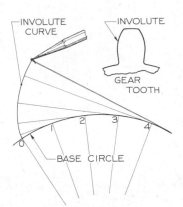

Fig. 19–8. Involute.

19.6 The Shape of the Teeth. The so-called *involute system* is the most widely used gear tooth form in use today. An *involute* may be described as the path traced by the end of a cord as it is unwound, while kept taut, from a circle called the *base circle*, Fig. 19–8.

The exact form of a standard involute tooth, however, is seldom drawn. For display drawings, or whenever it is desired to show the teeth but not necessary to represent them exactly, the involute curves may be approximated by circular arcs as in Fig. 19–9. Draw the addendum circle, pitch circle, and root circle, as shown. At any point P on the pitch circle, draw a tangent line; then draw a second line through P at $14\frac{1}{2}°$ with the tangent

line. This may be drawn 15° to simplify the construction. This is the so-called "pressure angle." Now draw the base circle tangent to the $14\frac{1}{2}°$ line. Along the pitch circle, set off by trial, with the bow dividers, the spacing of the teeth. With radius one-eighth the pitch diameter, and with centers on the base circle, draw circular arcs as shown. The lower portions of the teeth are radial lines ending in small fillets.

The pressure angle of $14\frac{1}{2}°$ is most common. However, if the angle is increased to 20° and the height of the teeth reduced, the resulting teeth are *stub teeth*.

19.7 Working Drawings of Spur Gears. A typical working drawing of a spur gear is shown in Fig. 19–10. A sectional view is usually used, and in many cases no circular view is needed. The gear teeth are not drawn, except as they appear (not section-lined) in the sectional view. In the circular view, if drawn, the addendum circle and the root circle are phantom lines, and the pitch circle is a center line.

The dimensions are given for the "gear blank" in the views and the gear tooth information is given in a note or in a table. The gear-tooth data shown in the table in Fig.

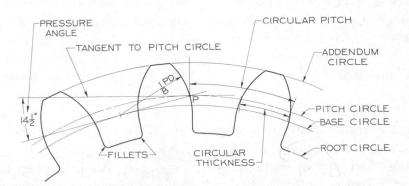

Fig. 19–9. Approximate Representation of Involute Spur Gear Teeth.

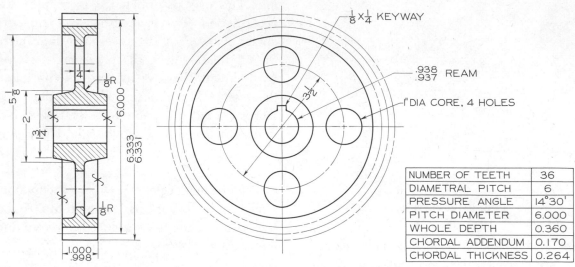

Fig. 19–10. Working Drawing of a Spur Gear.

NUMBER OF TEETH	36
DIAMETRAL PITCH	6
PRESSURE ANGLE	14°30'
PITCH DIAMETER	6.000
WHOLE DEPTH	0.360
CHORDAL ADDENDUM	0.170
CHORDAL THICKNESS	0.264

19–10 are minimum, with details left to the gear manufacturer. If more complete gear tooth information on the drawing is desired, the circular thickness, addendum, dedendum, tooth thickness, and other data must be provided.

19.8 Bevel Gears. *Bevel gears,* Fig. 19–6 (second from upper left), transmit power between two shafts whose axes intersect. If the axes intersect at right angles, as is usually the case, the gears are often called *mitre gears.* The comparable friction wheels would be a pair of cones rolling against each other, with their vertexes at a common point. A technical treatment of bevel gears is outside the scope of this text. For complete information, see a standard text on mechanism.

19.9 Working Drawings of Bevel Gears. As in the case of spur gears, a working drawing of a bevel gear gives only the dimensions of the gear blank, while the data for cutting the teeth are given in a note or table, Fig. 19–11.

Usually only a single sectional view is drawn, and if a second view is needed, only the gear blank or blanks are drawn, with the teeth omitted. It is common to draw a pair of mating gears together, as shown, although they are often drawn separately. The gear teeth are not section-lined. As in the case of spur gears, the larger is called the *gear,* and the smaller the *pinion.*

19.10 Problems

CAMS

PROB. 19/1. Using Layout D (Appendix, page 475), draw a displacement diagram, similar to that shown in Fig. 19–3, with the following follower motions: rise $1\frac{1}{4}''$ with modified uniform motion during 90°, dwell during 90°, rise $1\frac{1}{4}''$ with harmonic motion during 90°, and fall $2\frac{1}{2}''$ during the remaining 90°. Make the horizontal base line 10″ long and the follower displacement $2\frac{1}{2}''$ high.

PROB. 19/2. Using Layout B (Appendix, page 474), design a plate cam with roller follower, similar to that shown in Fig. 19–4,

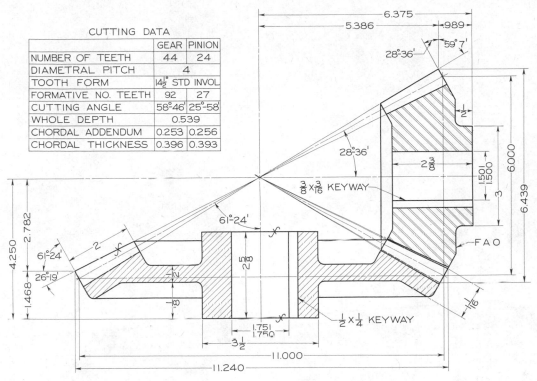

CUTTING DATA	GEAR	PINION
NUMBER OF TEETH	44	24
DIAMETRAL PITCH	4	
TOOTH FORM	$14\frac{1}{2}°$ STD INVOL	
FORMATIVE NO. TEETH	92	27
CUTTING ANGLE	58°46'	25°58'
WHOLE DEPTH	0.539	
CHORDAL ADDENDUM	0.253	0.256
CHORDAL THICKNESS	0.396	0.393

Fig. 19–11. Working Drawing of Bevel Gears.

with the following motions: rise $1\frac{1}{2}''$ with modified uniform motion during 180°, and fall $1\frac{1}{2}''$ with modified uniform motion during the remaining 180°. The diameter of the base circle is 3", and the diameter of the roller is $\frac{1}{2}''$. Draw a displacement diagram for this cam.

PROB. 19/3. Same as Prob. 19/2, except the follower has the following motions: rise $\frac{3}{4}''$ with modified uniform motion during 90°, dwell during 90°, rise $\frac{3}{4}''$ with modified uniform motion during 90°, and fall $1\frac{1}{2}''$ with modified uniform motion during the remaining 90°.

GEARS

PROB. 19/4. Using Layout D (Appendix, page 475), make a full-size drawing of a spur gear, similar to that shown in Fig. 19–16, with 48 teeth and a diametral pitch of 8. Completely dimension the drawing and compute new values for the gear tooth data table appearing on the drawing.

PROB. 19/5. Same as Prob. 19/4, except that the gear has 60 teeth and a diametral pitch of 10.

PROB. 19/6. Same as Prob. 19/4, except that the gear has 72 teeth and a diametral pitch of 12.

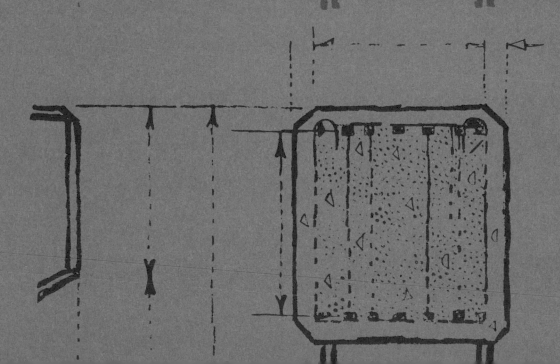

CHAPTER 20

STRUCTURAL DRAWING

20.1 Introduction to Structures. A spider web is one of nature's fine examples of a structure. The web is made of many parts connected to form a unit strong enough to support the spider. Man-made structures, Fig. 20–1, also have many parts connected together to form a framework strong enough to support loads. Figure 20–2 (a) shows a simple bridge structure. The portion of the structure lettered ABCDE is a *truss*. This truss has seven parts connected to each other at the *joints* A, B, C, D, and E. The parts are called *members,* and all of the truss members are in the same vertical plane. AB is an *upper chord member;* CD and DE are *lower chord members;* AE, AD, BD, and BC are *inclined members.* The bridge is made up of the truss ABCDE, the similar truss on the far side, the cross members con-

Fig. 20–1. Rib-type Dome of the Auditorium and Ice Rink at Brown University.

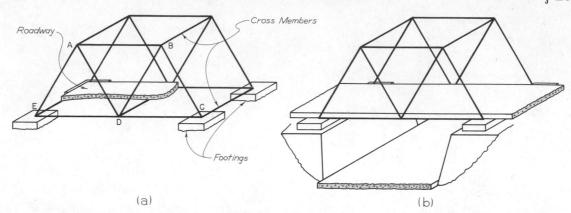

Fig. 20–2. A Bridge Structure.

necting the two trusses at corresponding joints, and the roadway. The structure is shown being used as a highway bridge over a highway in Fig. 20–2 (b). The width of the highway is the *span* of the bridge.

If the bridge were loaded with a truck, the weight of the truck would be transmitted by the bridge to the supports at the ends of the bridge. These supports are called *footings.* The truck would be supported by the roadway, the roadway by the lower cross members, the lower cross members by the trusses, and the trusses by the footings.

Each structural member performs a different function. The lower cross members have *bending loads* and are called *beams.* The roadway load acts perpendicular to the axes of these members and tends to bend them. The truss members have *axial loads,* the load on each truss member being in the direction of that member's axis. They will be either *tension members* or *compression members* depending upon whether they have pulling or pushing loads.

A *loading diagram* for the lower cross member at the center of the bridge is shown in Fig. 20–3 (a). The roadway load is called a *distributed load;* it is distributed over the length of the member. The forces designated

R_1 and R_2 are called *reactions.* The cross-member reactions are the supporting forces supplied by the trusses. A loading diagram for one of the trusses is shown in Fig. 20–3 (b). The truss loads designated P_1, P_2, and P_3 are called *concentrated loads;* they are concentrated at the lower joints. Note that R_1 in (a) would be equal to P_2 in (b); they both represent the force between the center cross member and the truss. The truss reactions are the supporting forces supplied by the footings.

A multi-story building is another example of a structure. Its members are *beams, girders,* and *columns.* The girders are large beams supporting other beams. The structural loads are transmitted through the members to the foundation of the building. The weight of a safe on one of the upper floors would be transmitted from the floor to the floor beams, from the floor beams to the girders, from the girders to the columns, and then from the columns to the foundation.

Drawings are necessary at all stages of structural work. A complete set of drawings for a large structure might include *layout drawings* showing the overall dimensions of the structure and its location; *foundation plans* showing the footings or the location of piles to be driven; *design drawings* indicating the design loads,

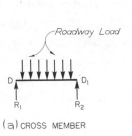

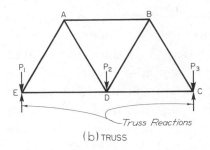

Fig. 20–3. Loading Diagrams.

showing the general arrangement of the structure, and giving specifications of the different members; *detail drawings* giving sufficient information about members for them to be made; and *erection diagrams* showing piece markings and indicating the sequence to be followed in the final assembly of the structure.

A structural designer must know the strength and other properties of the materials in his structure, and he must know the different types of possible structural failure. He must consider *buckling, twisting, shear, vibrations,* and various combinations of loading, in addition to tension, compression, and bending.

20.2 Wood Construction. Wood was undoubtedly the first material used for structures. It will always have a place in the structural field. It is in common use for sills, studs, rafters, and roof trusses. It is often used in the construction of railway and highway trestles. In remote areas, wood may be the only available material for structures. The principal woods used in construction are white pine, yellow pine, fir, cypress, and oak.

Wood construction makes a satisfactory substitute for steel and reinforced concrete construction during shortages of steel. Wood is also used for formwork and supports in reinforced concrete construction.

20.3 Steel Construction. The high strength and ductility of steel make it an ideal structural material. It has equal strength in either tension or compression, and it can be rolled into shapes specifically designed for structural members. Its ductility facilitates fabrication and erection, and enables a completed steel structure to withstand limited overloading without complete structural failure.

Cross-sections of the principal rolled structural shapes are shown in Fig. 20–4. The symbols beneath each section are used to specify

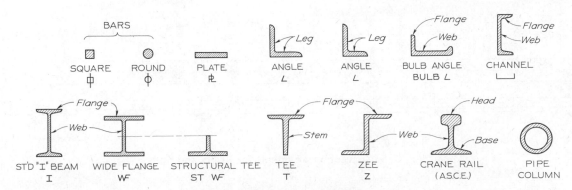

Fig. 20–4. Structural Steel Shapes.

particular shapes on drawings and in technical literature. Each shape is designed for some structural function. The standard "I"-beam and the wide flange beam are designed to resist bending with minimum cross-sectional area and minimum weight of the member. All of the shapes are available in different sizes, and in lengths suitable for structural members. Steel *fabrication* is the process of *riveting, welding,* or *bolting* structural steel shapes into structures.

A complete listing of standard sizes for each structural shape is given in *Manual of Steel Construction* (1965) published by the American Institute of Steel Construction (AISC), 101 Park Ave., New York, N.Y. 10017. This is the standard handbook in the field. It includes specifications and standards.

20.4 Structural Steel Drawings. Detail (working) drawings for steel structures show the various parts as they are shipped. The shipping units for large steel structures are *sub-assemblies* of members with connecting parts attached. It is economical to do as much of the connecting as possible in the fabricating

shop because of the favorable working conditions there. The drawings must specify the connections in detail and indicate which connections are to be made in the shop and which are to be made in the field during erection.

Connections are either *riveted, bolted,* or *welded.* Figure 20–5 shows the AISC conventional symbols used on drawings to indicate the different kinds of structural rivets and bolts. *Shop rivets* are those applied to the structure in the fabricating shop before delivery to the site. *Field rivets* are those applied at the site. The center line of a line of rivets is referred to as a *gage line.* The uniform distance between centers of rivets is called the *pitch* of the rivets. These terms also apply to bolts.

Standard methods of making beam connections are shown in Fig. 20–6. The connections are between the I-beam shown and a column or girder perpendicular to the beam. In the framed-beam connection the short structural angles (called *clip angles*) have been shop-riveted to the web of the beam and are shipped attached to the beam. In the seated-beam connection the clip angles have been shop-riveted

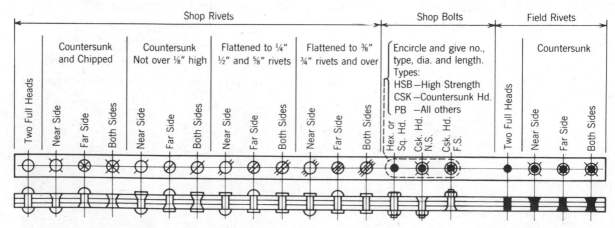

From AISC Manual of Steel Construction

Fig. 20–5. Conventional Symbols for Rivets and Bolts.

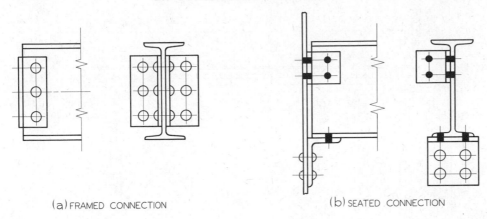

(a) FRAMED CONNECTION   (b) SEATED CONNECTION

Fig. 20–6. Beam Connections.

to the web of the column or girder; the lower clip angle provides a seat for the beam during erection.

Figure 20–7 shows a detail drawing of a shipping unit for a large structure. The unit consists of an I-beam with several short structural angles shop-welded to its web. The structural angles at the left end are for connection to a column during erection. The intermediate angles are for seated connections to other beams framing into the beam shown. The right end of the beam frames into a girder. Note that the lower view is a section viewed from the top. This is customary procedure in structural drawing. In a machine drawing this view would be placed above the front view.

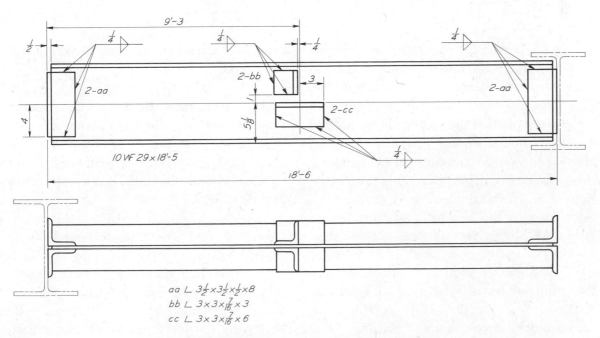

aa L $3\frac{1}{2} \times 3\frac{1}{2} \times \frac{1}{2} \times 8$
bb L $3 \times 3 \times \frac{7}{16} \times 3$
cc L $3 \times 3 \times \frac{7}{16} \times 6$

Fig. 20–7. Detail Drawing of a Welded Shipping Unit.

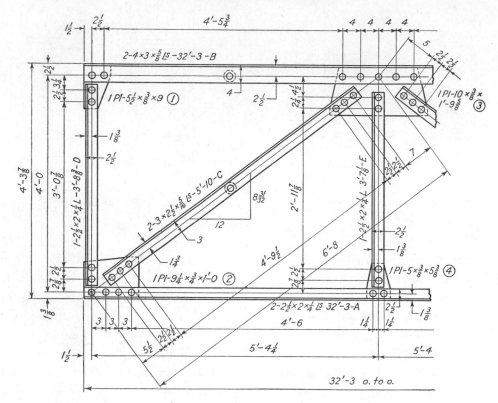

Fig. 20–8. A Riveted Structure.

If a steel structure is small enough to be shipped and erected as a unit, both design and details may be shown on the same drawing. Figure 20–8 shows a partial drawing of such a structure. Note that the inch symbol is omitted and that the dimension figures are lettered above continuous dimension lines instead of being placed in a gap. Observe that the inclination of a structural member is shown by a right triangle, the lengths of whose sides are given. All of the rivet symbols indicate shop rivets. The plates designated 1, 2, 3, and 4 are called *gusset plates*. The members are structural angles and have the designations A, B, C, and D. The AISC publishes the authoritative, two-volume *Structural Shop Drafting*, which gives further details of the standard methods of making structural steel drawings.

20.5 Reinforced Concrete Construction.

The method of reinforcing concrete with metal had its origin in France in about 1850 when Lambot built a boat of reinforced mortar. Today we find reinforced concrete structures everywhere.

The design of reinforced concrete structures is complicated. The designer must know the entire construction process to be able to design a structure that may be built economically. He should know how the reinforcing bars are bent into shape, how they are tied together, how forms will be erected and supported, how the reinforcement is placed in

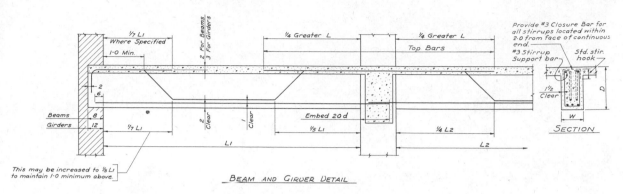

Courtesy American Concrete Institute

Fig. 20–9. Beam and Girder Detail.

the forms, how the concrete is to be poured, and how the forms are stripped. He should also know the methods of testing, curing, and inspection.

20.6 Reinforced Concrete Drawings.

In order to assure uniformity in reinforced concrete drawing, the American Concrete Institute (ACI), P. O. Box 4754, Redford Station, Detroit, Mich. 48219, publishes the approved *Manual of Standard Practice for Detailing Reinforced Concrete Structures.* This manual shows typical drawings, Fig. 20–9, for reinforced concrete buildings, bridges, piers, retaining walls, culverts, and arches.

A typical reinforced concrete structural drawing is shown in Fig. 20–10. Note that the right-side view is in full section. The solid dots in the sectioned areas are sections through the rods of reinforcing steel. In the front view, the left half is shown in section, but the concrete is not indicated because to show it would take time and would not improve the clearness of the drawing. The long lines within the section represent the rods of reinforcing steel. Note that Sections A–A and B–B are both views "looking down" and would be placed above the front view in a machine drawing.

Concrete is often used in connection with structural steel, as seen in Fig. 20–11. In such buildings, the floor loads are carried by the steel skeleton, while the concrete is used for floors and to cover structural members to obtain fireproofing and better appearance.

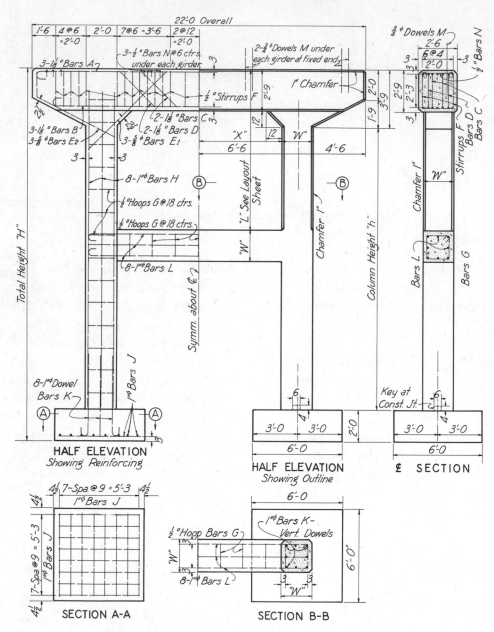

Fig. 20-10. An Interior Bent of a Highway Bridge.

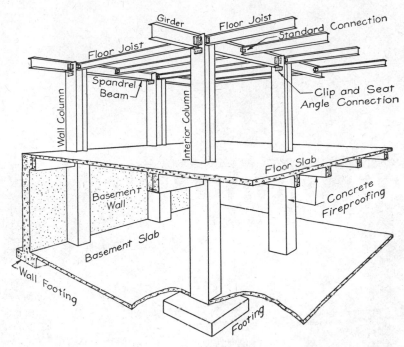

From *Theory of Modern Steel Structures* by L. E. Grinter (Macmillan)

Fig. 20–11. Fireproofed Structure of a Steel-Frame Building.

20.7 Structural Drafting Problems

PROB. 20/1. Make a detail drawing of diagonal member C in Fig. 20–8 with gusset plates 2 and 3 attached. (Assume that the structure is to be shipped "knocked down" instead of assembled.)

PROB. 20/2. Same as Prob. 20/1, except make a detail drawing of the vertical member D in Fig. 20–8 with gusset plate 1 attached.

PROB. 20/3. Same as Prob. 20/1, except make a detail drawing of member E in Fig. 20–8 with gusset plate 4 attached.

PROB. 20/4. Assume that the structure shown in Fig. 20–8 is 4 ft. deep and that there are six panels similar to the one shown. Make an oblique drawing of the complete structure, showing very little detail. Make it similar to Fig. 20–2, with dimensions added.

PROB. 20/5. Make a bill of material for the complete structure in Fig. 20–8.

CHAPTER 21
MAP DRAFTING

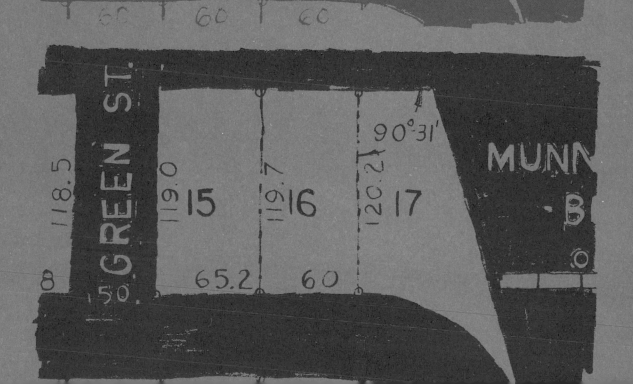

21.1 Maps. A *map* is a drawing of the earth's surface or a part of it. When maps are bound together in book form, the result is an *atlas*. A map prepared primarily for navigation is generally known as a *chart*. A map of a small "parcel" of land is called a *plat* or a *plot*. A typical plot of farm land is shown in Fig. 21–1.

The *lines* of a plot are established by a surveyor with a device called a *transit*. Iron stakes are driven at corner points if no other fixed points are available. Actual distances are measured with a tape. The *bearings* of the lines, or the angles they make with due north or south, are also established by the surveyor. All of this information is taken down in the form of *field notes*, and the plot is drawn from these notes. Observe, in Fig. 21–1, that the bearing and the length are lettered along each line. For example, the line at the upper left near "T. C. Martin" is 44° 52′ east of due north, and is 3791.4 ft. long. Note the arrow indicating true north. As a rule, a map is drawn with "north" toward the top.

Another common type of map is a *city map*, Fig. 21–2, showing streets and lots. Note the method of indicating the scale of the drawing.

A *topographic map* shows the physical features, such as lakes, streams, forests, mountains, and man-made structures like bridges, dams, and buildings. Generally, the elevations of the various portions are represented by *contour lines*. Figure 21–3 is a topographic map of a country estate.

Keep in mind that the purpose for which a map is intended determines what features should be shown on it and how much detail should be included. Maps of large areas, such as towns, counties, or states, will not include all the details that may appear on a map of a small area. The scale to which a map is drawn depends upon the size of the area to be represented and the amount of detail necessary to be shown. The scale for the map of a farm in Fig. 21–1 is 1″ = 300′, whereas the scale of a map by the United States Geological Survey, covering a large area, would often be 1:62,500. In inches, this would be approximately 1″ to a mile.

We represent topographic features with *symbols*, as shown in Fig. 21–3. For example, trees in general are always sketched in a certain way, grasslands are indicated by groups of short marks that suggest grass, and streams and bodies of water are symbolized by free-hand parallel lines along the water's edge. See any standard text on topographic drawing for other symbols.

A *contour* is an imaginary line on the earth's surface, passing through points of equal elevation, or height above sea level. If a series of equally spaced planes are passed through a hill, Fig. 21–4, each plane will produce a contour line which is the intersection between the plane and the earth's surface. The distance between the planes is called the *contour interval*, in this case 10 ft. For a large rugged

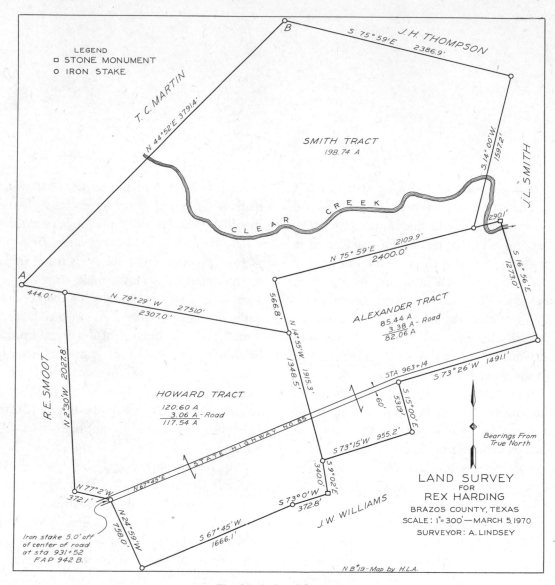

Fig. 21–1. Land Survey.

area, the contour interval may be as much as 100 ft. For a small area, such as a park, a garden, or a building site, where the surface is relatively flat, the contour interval may be less than a foot. The *elevations* are lettered on each contour line, either along each line, as in Fig. 21–3, or in gaps left in each line, as in Fig. 21–4.

A *profile* is a sectional view through the earth produced by a vertical sectioning plane. In Fig. 21–4 the edge view of the imaginary plane is indicated in the top view by the line X–X. Note how each intersection of this plane with a contour produces a point on the profile when projected down to the corresponding elevation. When a profile is drawn, a larger

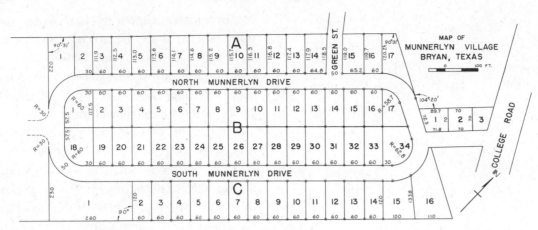

Fig. 21–2. A Partial City Map.

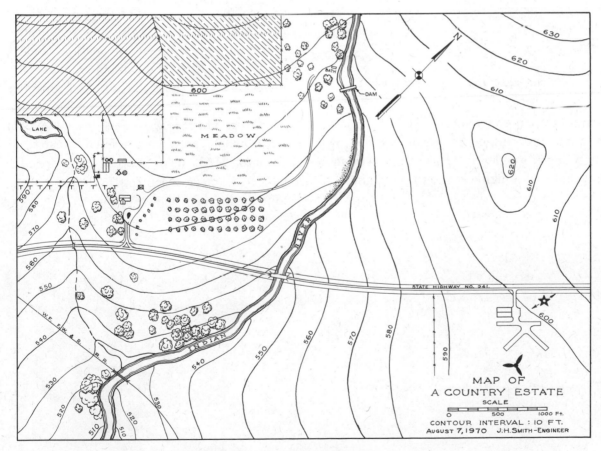

Fig. 21–3. A Topographic Map.

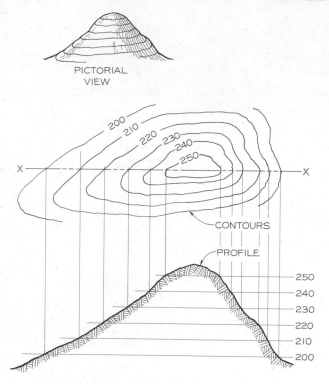

Fig. 21–4. Contours and Profile.

scale is often used for vertical distances in order to exaggerate surface irregularities so they will show up more clearly. This is quite common in profiles of a highway showing the grades of the roadway.

Contours are drawn freehand with an ordinary pen, or with the aid of a *contour pen*. Note, in Fig. 21–3, that contour lines close together indicate a steep slope, whereas widely separated lines show a gentle slope. Also note in the same figure that contour lines tend to run upstream. Thus, even if no water is shown, a depression that at times may carry water is shown by the contour lines—as, for example, at the left of the figure. Anyone experienced in making or reading maps can easily visualize the shape of the earth's surface from a glance at the contour lines.

21.2 Map Drafting Problems

PROB. 21/1. Using Layout E (Appendix, page 475), draw a plat of the survey shown in Fig. 21–1 to as large a scale as practicable. Use a protractor to set off bearings and an engineers scale to set off distances.

PROB. 21/2. Using Layout D (Appendix, page 475), draw a city map, similar to that shown in Fig. 21–2, of your own town or neighborhood area. Select a suitable map scale.

PROB. 21/3. Using Layout D (Appendix, page 475), make a double-size drawing of the topographic map shown in Fig. 21–3.

CHAPTER 22

ELECTRICAL DRAFTING

22.1 Electrical Drafting. *Electrical drafting* may be defined as the drawing of *electrical wiring diagrams, interconnection diagrams, block diagrams, layout diagrams,* and *schematics.* A draftsman, with the aid of drawing instruments, will draw these electrical diagrams, using standard graphical symbols to represent the various electrical devices that are to be used. The completed electrical drawings provide the necessary information for the connection, installation, control, and maintenance of electrical devices.

The electrician who is responsible for the installation of the electrical work in a new home, or public building, must have electrical drawings or wiring plans, Fig. 23–19, that show the location of lights, switches, receptacles, and the power source. The electronics technician, when assembling or repairing a television set, requires a wiring diagram or schematic diagram to assist and direct him through the electronic circuits. The electrician and technician will study and follow the lines and symbols of the drawing prepared by the draftsman, in analyzing the various circuits. The ease with which these drawings can be interpreted depends on how well the wiring or schematic diagram is laid out and drawn by the draftsman. The electrical draftsman and the drawings that he produces are a very important link between the electrical or electronic designer and engineer and the actual construction of their ideas and designs. The

primary purpose of the draftsman is to relay this information from the designers and engineers to the electricians, technicians, and manufacturers in the electrical and electronic industries.

An electrical draftsman must not only be well versed in the principles and techniques of technical drawing, but must also possess a working knowledge of electricity and electrical and electronic circuitry. A student will find that courses in physics and electric shop are of considerable help in understanding and drawing electrical diagrams.

22.2 Graphical Symbols and Techniques. The lines appearing on an electrical wiring diagram indicate the connection of many different types of electrical devices. Since it would be impractical and time-consuming to draw the many electrical devices as they actually appear, *graphic symbols* are used to simplify representation. These symbols have been standardized by the United States of America Standards Institute* and are referred to by the electrical draftsman before drawing any electrical diagrams.

The symbols must be placed on the electrical diagram in the same place where the electrical device is to be indicated, and the sizes of the symbols must be proportionate throughout the

---

* USAS Y32.2—1967 and USAS Y32.9—1962 (United States of America Standards Institute, 10 East 40th Street, New York, N.Y. 10016).

401

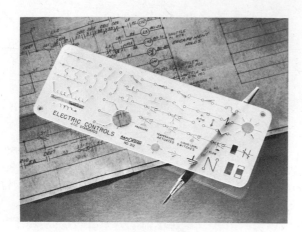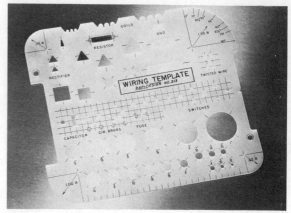

(Courtesy RapiDesign, Inc.)

Fig. 22–1. Template Guides for Electrical Symbols.

diagram. Symbols are not to be crowded, but should be drawn clearly and distinctly. If a symbol is identified with a letter and number prefix, no other symbol should be identified with the same letter and number. For example, a relay symbol may be drawn and identified as ──(ICR)──, and a second relay would be drawn and identified as ──(2CR)──, etc. Some of the more important symbols used on electrical diagrams are shown in Tables 16 and 17 (Appendix, pages 467–69).

Plastic *templates*, Fig. 22–1, are available for drawing electrical symbols, and are definite time-saving devices. However, their use may be limited if symbols smaller or larger than those appearing on the template are required.

In general, electrical diagrams need not be drawn to any specific scale, although the smallest size standard layout compatible with the nature of the diagram should be selected. Line thicknesses and letter sizes should be selected on the basis of producing a drawing, conforming to USA Standards, that is pleasing and legible. A line thickness of *medium* weight is recommended for general use on electrical diagrams, with *thin* lines being used

to represent brackets, leader lines, etc. A heavy line may occasionally be used to emphasize a particular portion of a circuit.

22.3 Electric Circuits. An *electric circuit* may be defined as the path through which a current flows. *Current* is the time rate of flow of electricity, and is measured in terms of *amperes*. However, we cannot have a current flowing in a circuit unless there is a closed circuit with a *voltage* and a *resistance*. The voltage, sometimes called the *electromotive force* (emf) or *potential difference* (pd), is the difference in electrical pressure which causes the flow of electricity, and is measured in *volts*. The *resistance* is the opposition to the flow of electricity, and is measured in *ohms*.

The relationship among current, voltage, and resistance is known as *Ohm's law* and may be expressed by the following formula:

$$\text{Current (amperes)} = \frac{\text{Potential difference (volts)}}{\text{Resistance (ohms)}}$$

Expressed in symbols, it is

$$I = \frac{E}{R}$$

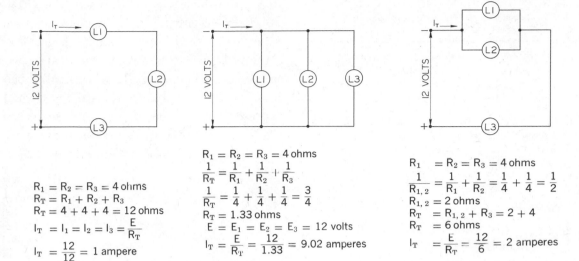

$R_1 = R_2 = R_3 = 4$ ohms
$R_T = R_1 + R_2 + R_3$
$R_T = 4 + 4 + 4 = 12$ ohms
$I_T = I_1 = I_2 = I_3 = \dfrac{E}{R_T}$
$I_T = \dfrac{12}{12} = 1$ ampere

(a) SERIES CIRCUIT

$R_1 = R_2 = R_3 = 4$ ohms
$\dfrac{1}{R_T} = \dfrac{1}{R_1} + \dfrac{1}{R_2} + \dfrac{1}{R_3}$
$\dfrac{1}{R_T} = \dfrac{1}{4} + \dfrac{1}{4} + \dfrac{1}{4} = \dfrac{3}{4}$
$R_T = 1.33$ ohms
$E = E_1 = E_2 = E_3 = 12$ volts
$I_T = \dfrac{E}{R_T} = \dfrac{12}{1.33} = 9.02$ amperes

(b) PARALLEL CIRCUIT

$R_1 = R_2 = R_3 = 4$ ohms
$\dfrac{1}{R_{1,2}} = \dfrac{1}{R_1} + \dfrac{1}{R_2} = \dfrac{1}{4} + \dfrac{1}{4} = \dfrac{1}{2}$
$R_{1,2} = 2$ ohms
$R_T = R_{1,2} + R_3 = 2 + 4$
$R_T = 6$ ohms
$I_T = \dfrac{E}{R_T} = \dfrac{12}{6} = 2$ amperes

(c) SERIES–PARALLEL CIRCUIT

Fig. 22–2. Series and Parallel Electric Circuits.

For example, suppose we want to find the amount of current passing through a resistance of 10 ohms connected to a 110-volt source.

Solution: $I = \dfrac{E}{R} = \dfrac{110}{10} = 11$ amperes

Ohm's law can be applied not only to an entire circuit, but also to any part of a circuit, and it is one of the most important principles in all electrical work. While it applies only to direct current (dc) circuits, a special form of this law is used also in alternating current (ac) work.

Electric circuits may be connected in *series, parallel,* or a combination of both, Fig. 22–2. In a series circuit, (a), the various electrical devices are connected one after the other in a single line or path. In a series circuit the current is the same in every part of the circuit, the resistance is the sum of the separate resistances, and the voltage is equal to the sum of the voltages across the separate parts.

In a parallel circuit, (b), the current may flow through two or more paths or branches. In a parallel circuit the voltage is the same across each branch of the circuit, the total current is the sum of the currents through the branches, and the total resistance is equal to the voltage divided by the total current. In a parallel circuit the total resistance (R_T) may be determined by the following formula:

$$\frac{1}{R_T} = \frac{1}{R_1} + \frac{1}{R_2} + \frac{1}{R_3} + \text{etc.}$$

A simple explanation of series and parallel circuits might best be illustrated by comparing two Christmas-tree light sets, one wired in series and the other in parallel. In the series set, if one light burns out, all the remaining lights will also be out, while in the parallel set, if one light burns out, the remaining lights will continue to operate.

An example of a combination series-parallel circuit is shown at (c).

22.4 Electrical Diagrams. Several different types of electrical diagrams are USA Standard.*

* USAS Y14.15—1966.

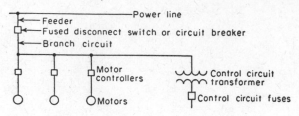

Fig. 22–3. Single-Line Diagram of a Typical Branch Circuit Protective Arrangement for a Multi-motor Machine.*

The *single-line* or *one-line diagram,* Fig. 22–3, presents the various circuit information simply by means of single lines and standard graphical symbols.

The *schematic* or *elementary diagram,* Figs. 22–4 and 22–5, shows the electrical connections and functions of a circuit, by means of standard graphical symbols, without any regard to the actual physical size, shape, or location of the electrical devices or parts which make up the circuit.

The *connection* or *wiring diagram,* Fig. 22–6, shows the connections of an installation, or the electrical devices or parts which comprise the circuit. It includes such detail as is necessary to make or trace internal and/or external connections that are involved and usually shows the general physical arrangement of the component devices or parts.

The *interconnection diagram* shows only the external connections between unit assemblies or equipment. In this type of connection, or wiring, diagram the internal connections are usually omitted.

22.5 Schematic Diagrams. A schematic diagram is one of the simplest forms of electrical

---

* Reproduced with permission from JIC Electrical Standards for Industrial Equipment, 1957 Joint Industrial Conference, National Machine Tool Builders Association, 2139 Wisconsin Avenue, Washington, D.C. 20007.

diagrams, and is frequently used to illustrate an electrical circuit. In a schematic diagram, sometimes called an elementary diagram, no attempt is made to show the electrical devices in their actual positions. The electrical devices may be shown between vertical lines which represent the source of power. A schematic diagram of a three-way and four-way switch controlling two lamps is shown in Fig. 22–4.

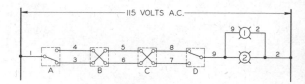

Fig. 22–4. Schematic Diagram of a Three-Way and Four-Way Switch Controlling Two Lamps.

It is appropriate at this point to state several important principles of electricity. First, electricity flows from one terminal (negative) of the power source, through the electrical devices, and around to the other side (positive) of the source. Next, a path must be provided for the flow of electricity, and this generally is a copper wire called a *conductor.* The nonmetallic parts of an electrical circuit are called *insulators.* Finally, a voltage is necessary to produce a flow of electricity through the electrical circuit.

22.6 Drawing the Schematic Diagram. To draw a schematic diagram, the draftsman will first make a freehand sketch, Fig. 22–5 (a). Then, using the sketch as a guide, the draftsman will locate the graphical symbols of the component devices and parts and determine the size of symbols to be used throughout the schematic, as shown at (b). The ability to do this with ease and speed can be acquired only by experience.

The size of the freehand sketch will also be used as a basis for determining the size of drawing paper required for the mechanical drawing. If larger than standard size, the rough sketch will have to be divided into sections. The symbols on the schematic diagram can be positioned in approximately the same location as determined on the sketch. There must be an orderly layout of evenly spaced horizontal lines for each symbol from the top of the drawing to the bottom. The symbols representing main components, such as lights, coils, bells, motors, etc., should be at the right of the drawing between the two vertical lines representing the power source. The symbols representing switches, or controlling elements, are drawn to the left in the same horizontal line as the associated main component located at the right of the diagram. The completed schematic diagram is shown in Fig. 22–5 (b).

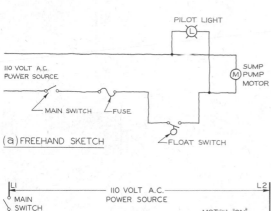

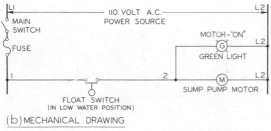

(a) FREEHAND SKETCH

(b) MECHANICAL DRAWING

Fig. 22–5. Drawing the Schematic Diagram of a Sump Pump Motor Circuit.

22.7 Wiring Diagrams. A wiring diagram, Fig. 22–6, is a type of electrical diagram in which the graphical electrical symbols are arranged in the same physical relationship as the actual equipment. The lines drawn connecting the symbols represent the actual wires connecting the electrical devices. These wires may be in conduits or inside the enclosure of the electrical device. The connection lines show every individual connection made between components in a precise manner and indicate the shortest path.

A wiring diagram can be drawn by a draftsman after he has arranged all the symbols, representing the electrical devices, on the drawing. The relative location of these symbols should approximate the actual physical and/or mechanical arrangement of the electrical devices. The schematic is the guide that the draftsman will follow to draw the wiring diagram and show the connection of electrical devices.

The development of a schematic into a wiring diagram, Fig. 22–7, is usually done by a senior or advanced electrical draftsman. The conversion of schematic to wiring diagrams is the basis of all electrical drafting and the beginning draftsman should make every effort, through study and practice, to learn how to make these drawings skillfully, correctly, and rapidly.

22.8 Industrial Electrical Diagrams. The electrical control circuit of a motor that operates an industrial machine is a very important and integral part of electrical diagrams. The method of controlling the electrical power to a motor can be indicated on a schematic diagram, Fig. 22–8 (a). The actual wiring of the industrial machine is shown on a wiring diagram, (b).

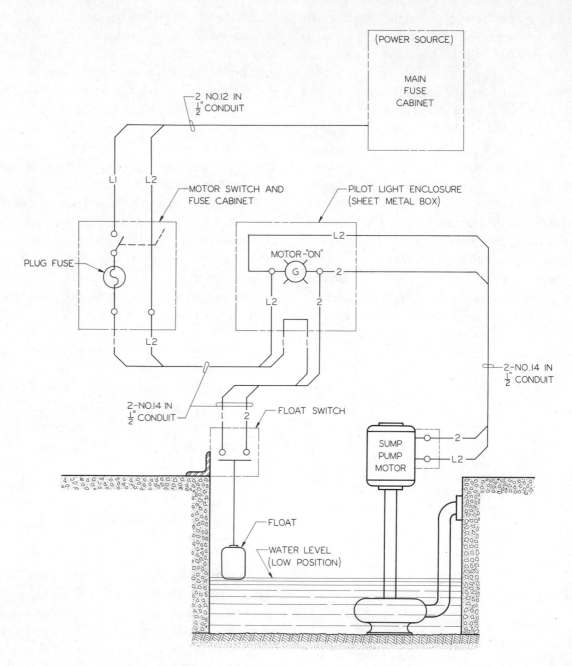

Fig. 22–6. Wiring Diagram of a Sump Pump Circuit.

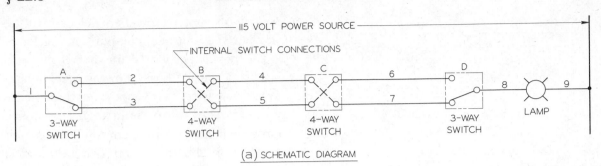

(a) SCHEMATIC DIAGRAM

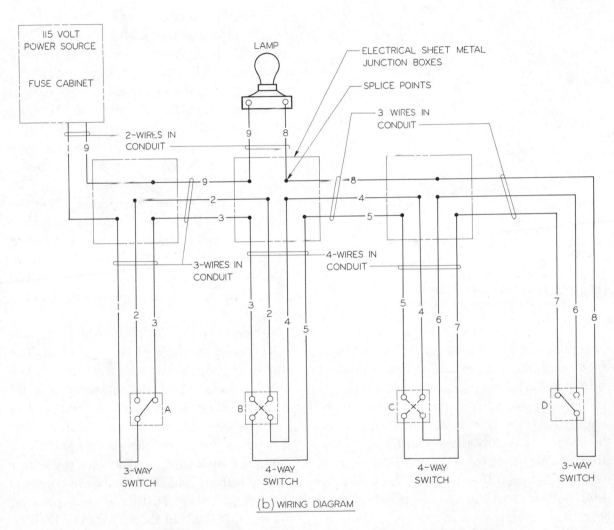

(b) WIRING DIAGRAM

Fig. 22-7. Schematic and Wiring Diagrams of Four Switches Controlling One Lamp.

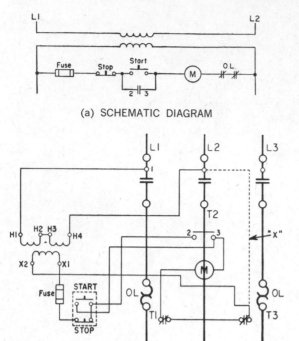

(a) SCHEMATIC DIAGRAM

(b) WIRING DIAGRAM

Courtesy Allen-Bradley Company

Fig. 22–8. Schematic and Wiring Diagrams of a Step-Down Transformer in a Control Circuit.

Industrial electrical diagrams, as shown in *block diagram* form, Fig. 22–9, can be divided into two parts, one portion being the *power* circuit and the other portion the *control* circuit. The power portion is drawn with heavy lines and the control portion with a relatively thinner line in order to provide the desired contrast between the two portions. Generally both portions are shown on one drawing; however, if necessary, each portion may be shown on a separate drawing. The method of applying or regulating power to a motor is shown in the power portion. The devices in the power portion are controlled by a circuit that is shown in the control portion, and any required safety

controls or motor-protective devices will be shown in the control-circuit portion, Fig. 22–8.

The power portion is drawn first and will consist of heavy lines showing single-phase power (two wires to the motor), or three-phase power (three wires to the motor). The control portion will consist of the control devices such as relays, pushbuttons, and switches, and is shown with thinner lines.

The electrical circuit devices are shown by using standard graphical symbols and abbreviations of their common names, and should conform to NEMA* and USA† standards.

A system of identifying the control devices is necessary to show the difference between similar control devices in a circuit wiring diagram or schematic diagram. One widely used system combines a number with the abbreviation of the common name of the control device. The number may precede or follow the abbreviation. For example, if two relays are used in a circuit, they may be identified as 1CR and 2CR or as CR1 and CR2. The contacts and coil of each relay must be identified by the same combination of numbers and letters.

A *motor starter*, Fig. 22–10, is a device that controls and regulates power to a motor. The starter includes parts that will carry heavy currents and also parts that will carry a smaller control current in the same assembly. The motor starter contains contacts that connect the power source to the motor and a coil energized by a small amount of electric current to activate and close the power contacts. A comparison of the picture and drawing shown in Fig. 22–10 should help you become familiar with the motor starter as it is represented in wiring-diagram form. The principal

* NEMA Control Standards—"Device Designations and Symbols," NEMA Standards Publication 1C–3.01 to 1C–3.04.
† USAS Y32.3—1967.

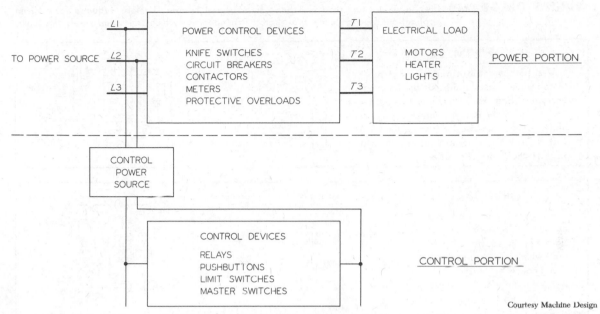

Courtesy Machine Design

Fig. 22-9. Block Diagram of Industrial Electrical Circuit.

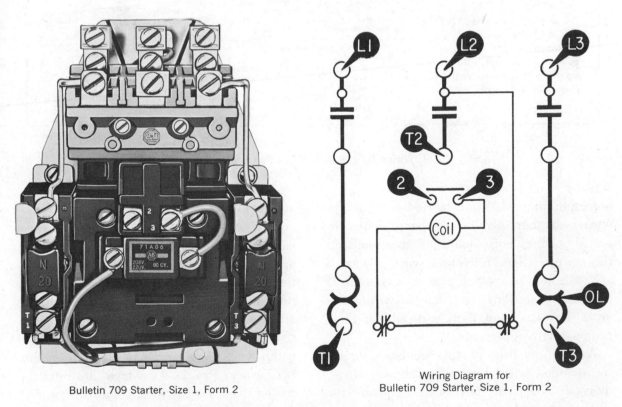

Bulletin 709 Starter, Size 1, Form 2

Wiring Diagram for
Bulletin 709 Starter, Size 1, Form 2

Courtesy Allen-Bradley Company

Fig. 22-10. Wiring Diagram of a Motor Starter.

WIRING DIAGRAMS The Timoflex is furnished in 3 basic forms shown below. The form required for a given application depends upon the pilot switch operation which starts and resets the timer.

FORM 32

Use with a maintained contact pilot switch which closes and holds the timer energized during its timing interval. At the end of the time delay, contact 3-2 opens and contact 3-1 closes. Opening the pilot switch resets the timer and returns the contacts to their normal position.

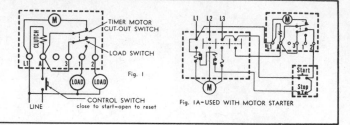

FORM 33

Use with a momentary contact pilot switch. This form has an additional contact which closes when the timer coil is energized. This contact maintains itself when the pilot switch opens.

Fig. 3 shows a special application of Form 33 to provide a short closure at the end of the time period. The closed time for contact No. 1 depends upon the time scale as follows:

TIME SCALE	CONTACT No. 1 CLOSED
55 seconds	¼ seconds approx.
5½ minutes	1½ seconds approx.
27½ minutes	7½ seconds approx.
55 minutes	55 seconds approx.

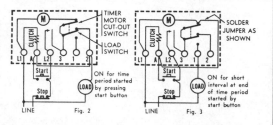

FORM 42

Use this form where (a) voltage failure must not reset timer and (b) closing of control switch resets timer — opening of control switch starts timing period.

This form has the coil action reversed. Closing the pilot switch to energize the timer coil resets the timer. It remains in the reset position (contact 3-2 closed) until the pilot switch is opened. Opening the pilot switch starts the timing period. Contact 3-2 opens and contact 3-1 closes at end of timing period. The timer normally stays in this timed out position until the pilot switch recloses again.

Voltage failure stops timing which restarts from point at which it stopped. Closing the pilot switch during the timing period resets timer to its initial position.

A momentary closure (¼ second approximately) of the pilot switch resets the timer although the pilot switch may stay closed indefinitely. Timing starts when the switch opens.

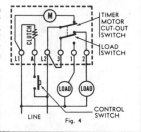

Courtesy Eagle Signal, Division E. W. Bliss Company

Fig. 22-11. Wiring Diagrams of Three Forms of a Reset Timer.

corresponding parts are labeled so that the wiring diagram can be compared with the actual starter. This should aid in visualizing the starter when studying a wiring diagram and will help in making connections when it is actually wired up. Note that the wiring diagram shows as many parts as possible in their proper relative positions.

A *list of materials* or parts tabulation should be included with every schematic or wiring diagram. The materials list is necessary for the maintenance of the electrical equipment and should completely describe the components and their functions. These lists are a valuable aid in understanding the schematic or wiring diagram, since they include information that the component symbol alone cannot provide. Manufacturers' catalogs and bulletins are also useful in providing detailed specifications and wiring diagrams of the individual components. An example of such information is shown in Fig. 22-11, which illustrates the wiring diagrams for several forms of a *reset timer*.

22.9 Electrical Drafting
Problems

PROB. 22/1. An electric circuit has three lamps connected in series having the following resistances: 12, 6, and 4 ohms.

(a) Draw the schematic diagram showing the circuit. Use Layout A (Appendix, page 474), and make the diagram approximately three times as large as the one shown in Fig. 22–5 (a).

(b) What is the total resistance of the circuit?

(c) What amount of current will flow if the entire circuit is connected to a 24-volt source?

PROB. 22/2. The resistances in Prob. 22/1 are all connected in parallel.

(a) Draw the schematic diagram showing the circuit. Use Layout A (Appendix, page 474), and make the diagram approximately three times as large as the one shown in Fig. 22–5 (b).

(b) What is the total resistance of the circuit?

(c) What amount of current will flow through each lamp if the voltage of the source is 9 volts?

PROB. 22/3. Using Layout B (Appendix, page 474), make a double-size drawing of the wiring and schematic diagrams of the step-down transformer in the control circuit shown in Fig. 22–10.

PROB. 22/4. Using Layout A (Appendix, page 474) and a suitable scale, draw an electrical layout wiring diagram of a 9′ × 12′ room containing two wall receptacles and a center ceiling light controlled by a switch at the room entrance.

PROB. 22/5. Using Layout C (Appendix, page 474) and a suitable scale, draw an electrical layout wiring diagram of your home or apartment similar to that shown in Fig. 23–19.

PROB. 22/6. Using Layout A (Appendix, page 474), draw a schematic diagram of two three-way switches controlling one lamp.

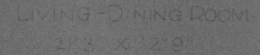

CHAPTER 23

ARCHITECTURAL
DRAFTING

23.1 Architectural Drafting. The story of civilization in its early times is to a large extent the story of man's effort to construct permanent "shelters" or buildings for himself and his activities. The buildings, or *architecture,* of any period of history provide an excellent record of man's way of life and of his ability to solve the problems that are created by a large number of people settled in one place. When men of the Stone Age started their first settlements, they felt the need of practical, useful, and safe shelters. And it soon became necessary to learn new tasks and new skills in order to construct relatively large buildings. When the building was no longer a product of a single man, and other men had to contribute their efforts and their skills, a new way of communication other than language had to be developed. What one could call early *architectural drawings* were not only instructions to workers describing the structure as to its shape and size, but also a permanent record of achievement. This dual function of architectural drawing has remained unchanged through the ages up to the present day.

23.2 Organization of Architectural Drawings. Modern architectural drawings, sometimes called *architectural plans,* inform contractors and workers graphically about the essential parts of the structure and their relationships. They also provide the basis for coordination of the activities of the various trades, and furnish the necessary information for engineering installations such as mechanical and electrical equipment.

As a permanent record they permit the checking of the contractor's performance as to completeness and accuracy, as to the quality and quantity of the materials used in construction, and as to the quality of workmanship. However, a complete and precise description of materials is seldom possible by means of lines and symbols only. For example, a symbol and note on an elevation view may indicate face brick. But nothing is indicated about the texture of the surface or the color. These very special properties of the materials are described in detail in a set of *specifications.**

The architect, as the creator of the drawings, must be guided by all these requirements and proceed to describe as completely as possible the site, the shell or structural skeleton, the outside, and the inside of the building. A complete set of such drawings makes up the heart of the collection of all documents con-

---

* Specifications are written documents accompanying the architectural drawings. They describe in detail the mutual relationships of the owner, the architect, and the builder. They specify the quality and other properties of building materials, and establish the performance standards and quality of workmanship. In general, they describe in words such features on the drawings that cannot be adequately described graphically.

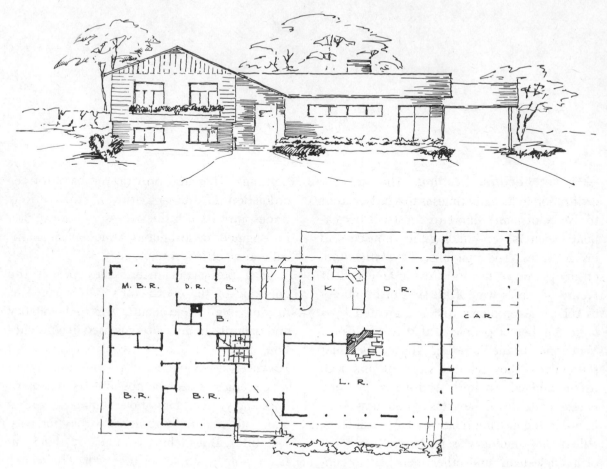

Fig. 23–1. Preliminary Sketch.

nected with the construction. The other parts are the *specifications*, and legal forms such as *contracts, performance bonds, liens, titles*, etc. All together they set forth in a legally binding way the individual responsibilities of the architect, the contractor, and the owner.

The organization of architectural drawings is quite different from that of standard machine drawings, although the objective of both is the same: to describe the object completely as to shape and size. However, the architectural drawing describes an object of comparatively large size—many times larger than the drawing itself—and therefore must be drawn at a considerably reduced scale. For example, a house is usually drawn to the scale $\frac{1}{4}'' = 1'-0''$. At this scale many important details are too small on the drawing and must be shown by some kind of symbol, or a special kind of "shorthand." Many different views of the structure are needed for a complete description, and it is customary to group them into three major divisions: *plans, elevations* and *sections*, and *details*. Each of these divisions may include one or more drawings. Details are customarily drawn to a larger scale than the plans and elevations in order to clarify the elements. In the process of development of the drawings for a house the architect will at first produce a set of *preliminary drawings,* or *preliminary sketches*. These may be just a collection of freehand sketches of room arrangements and perhaps a few sketches of outside elevations, usually not drawn to any exact scale, Fig. 23–1. For larger buildings, however, preliminary drawings are always drawn to scale, Fig. 23–2. In special cases, for promotional purposes, the architect will pro-duce a *rendering* of the building, usually in perspective and sometimes in color, Fig. 23–2. From the preliminary drawings a set of working drawings is made, representing the last stage of the project. Working drawings are very carefully drawn and repeatedly checked because, in case of a disagreement, they become the legal basis for a settlement. In this respect they must be very carefully coordinated with the specifications, so that all problems arising from the construction are properly covered and described.

PERSPECTIVE

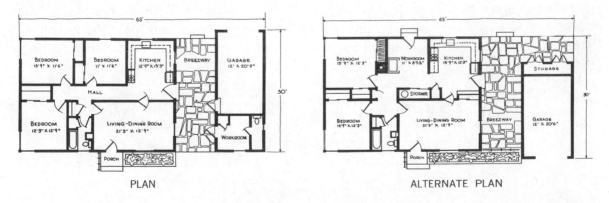

PLAN   ALTERNATE PLAN

Courtesy U.S. Gypsum Co.

Fig. 23–2. Preliminary Drawings.

ABCDEFGHIJKLMNOPQRSTUVWXYZ
ABCDEFGHIJKLMNOPQRSTUVWXYZ¢ 1234567890
ABCDEFGHIJKLMNOPQRSTUVWXYZ¢ 1234567890 ½ ¼ ⅛ ⅜ ¾ ⅝ ⅞

ABCDEFGHIJKLMNOPQRSTUVWXYZ
ABCDEFGHIJKLMNOPQRSTUVWXYZ¢ 123456789
ABCDEFGHIJKLMNOPQRSTUVWXYZ¢ 1234567890 ½ ¼ ⅛ ⅜ ¾ ⅝ ⅞

Fig. 23–3. Architectural Lettering.

23.3 Notes on Architectural Drawings. The most carefully drawn and dimensioned set of architectural drawings still would not describe the structure adequately. For example, the sequence of building operations can never be shown on the drawing itself; the drawing shows only the completed construction. Thus, many explanatory *notes* are found on architectural drawings. These are neatly lettered in vertical or inclined letters of the type and size described in Sec. 23.4. Because of the extensive use of notes on architectural drawings, architectural lettering is a very essential part of the art of architectural drafting. See Figs. 23–3 to 23–6.

23.4 Architectural Lettering and Symbols. Lettering of notes on the drawing, in the title block, and in the legend must be precise in meaning, brief, legible, and neat in appear-

ance. Architectural practice has developed standard wording, standard symbols, a standard alphabet, and many abbreviations. In architectural practice, both vertical and inclined capital letters are used, and good architectural draftsmen will be proficient in both. However, both styles should not be used on the same drawing or set of drawings. Each individual will in time develop a system of strokes and letter forms best suited to his own purposes. No lowercase letters are used on architectural drawings.

The lettering in notes on architectural drawings may be any size from $\frac{1}{16}''$ to $\frac{1}{8}''$ in height. The spacing of horizontal guidelines and the spacing between lines will be done best by use of the Ames Lettering Instrument or Braddock Lettering Triangle, Fig. 4–13. Experienced draftsmen can space the guidelines by eye. The general rules for good engineering let-

ABCDEFGHIJKLMN
OPQRSTUVWXYZ&
1234567890

Fig. 23–4. Architectural Title Letters (Serif Type).

ABCDEFGHIJKLMNOPQRST
UVWXYZ&... 1234567890
ARCHITECTURAL LETTERING

Fig. 23–5. Architectural Title Letters (Gothic Type).

tering are equally applicable to architectural practice. Lettering forms an integral part of architectural drawing, and its proper execution adds not only to legibility but also to professional appearance. Even an excellent drawing can be ruined by carelessly lettered notes and poorly formed numerals for dimensions.

Title blocks and titles will sometimes be lettered in serif forms either in solid or outline type, Fig. 23–4. Gothic-type letters are shown in Fig. 23–5. Title blocks, Fig. 23–6, are somewhat different from title blocks used on machine drawings. Compare this title block with some machine drawing title blocks, for example, that in Fig. 15–12.

Besides notes, a large number of standard abbreviations and graphic symbols are found throughout any architectural drawing. These abbreviations and symbols save a great deal of time, and because of their extensive use are just as readable and understandable as full names or precisely drawn views of objects. See Tables 17 and 19 (Appendix, pages 469 and 472).

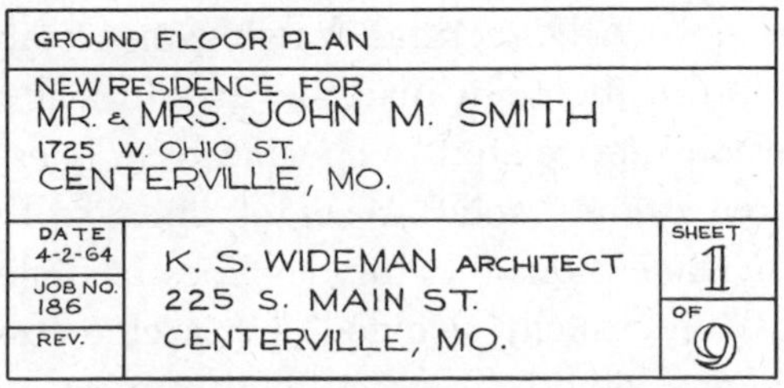

23.5 Elements of a House. A building is composed of many parts or elements. Thus, terms such as *foundations, columns, beams, doorways, ceilings,* etc., have acquired with time precisely defined meanings, some of which are different from the meaning of the word as used in everyday conversation. For example, when we say "Open the window," we mean in architectural terms "Open the sash." Then, again, some meanings of architectural terms are rather obscure and generally unknown. For example, the word *astragal* (a small convex molding) is meaningless to most people, and some terms, such as *meeting rail* (middle horizontal member on a double-hung window) do not describe the part understandably to the average person. For these reasons it will be essential for anyone interested in architectural drawing to become acquainted with the various architectural terms. An immediate identification of symbols on a drawing and the association of such a part with a proper name are expected from any architectural draftsman.

The major elements of a house are: *founda-*

GROUND FLOOR PLAN

NEW RESIDENCE FOR
MR. & MRS. JOHN M. SMITH
1725 W. OHIO ST.
CENTERVILLE, MO.

DATE 4-2-64
JOB NO. 186
REV.

K. S. WIDEMAN ARCHITECT
225 S. MAIN ST.
CENTERVILLE, MO.

SHEET 1
OF 9

Fig. 23–6. Architectural Title Block.

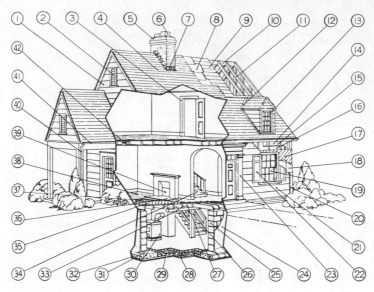

1. Gable end	21. Siding
2. Louver	22. Shutters
3. Interior trim	23. Exterior trim
4. Shingles	24. Waterproofing
5. Chimney cap	25. Foundation wall
6. Flue linings	26. Column
7. Flashing	27. Joists
8. Roofing felt	28. Basement floor
9. Roof sheathing	29. Gravel fill
10. Ridge board	30. Heating plant
11. Rafters	31. Footing
12. Roof valley	32. Drain tile
13. Dormer window	33. Girder
14. Interior wall finish	34. Stairway
15. Studs	35. Subfloor
16. Insulation	36. Hearth
17. Diagonal sheathing	37. Building paper
18. Sheathing paper	38. Finish floor
19. Window frame and sash	39. Fireplace
20. Corner board	40. Downspout
	41. Gutter
	42. Bridging

Courtesy National Bureau of Standards

Fig. 23-7. Elements of a House.

tions (including *footings*), *walls*, *partitions*, *floors*, *roof*, *doorways*, *windows*, *stairs*, *fireplace*, and *chimneys*. The most important structural elements of a house are: *columns*, *beams*, *lintels*, *girders*, and *trusses*. The most common parts of a house are identified by name in Fig. 23-7.

23.6 Plans. The *plans* form the most important part of the architectural drawings because they contain most of the information and show more of the dimensions than do the elevations. What we call a *plan* is in reality a horizontal section through the house taken above the window-sill height. Thus, all walls appear sectioned, and all openings, such as doors and windows, appear as open breaks in walls and partitions, Fig. 23-19. In general, each floor of the house will be shown on a separate floor plan. Such a plan will contain all information about the sizes, as well as refer-

ences to the details, cross-sections, and interior views shown elsewhere. It will also contain designations of finished floor materials and the separation of these materials by edging or moldings. Normally a plan will show the location of lighting fixtures, switches, outlets, radiators or heating registers, and utilities, such as heaters, furnaces, sump pumps, etc.

The dimensions are the most important feature of the plan. There are some significant differences between architectural dimensioning and the dimensioning used in machine drafting. In machine drafting we try to avoid dimensions being placed inside of the outline of a view. Architectural drawings necessarily abound in these. In machine drawing, all dimensions are related to specific base lines or reference lines, such as center lines, so that chain dimensions (continuous lines of dimensions) are usually avoided. On architectural drawings, almost all dimensions are parts of a chain. These differences are clearly a result of the much greater tolerances that building practices allow. See Sec. 9.26. A typical example of a complete plan is shown in Fig. 23–19. The placing of notes in the open areas of the plan and the arrangement of dimensions on the outside and inside should be especially noted. Besides the general plan, a more complex building will require special plans for floor and roof framing, Fig. 23–23, and sometimes for electrical wiring and heating or air-conditioning systems. These are drawn by draftsmen specializing in mechanical equipment drafting or electrical drafting.

Most architectural drawings will contain a small-scale plan, called a *plot plan,* that shows the location of the building with respect to the property lines, sometimes showing the location of existing trees and structures located on the property. The building is shown on a plot plan only by its outline or section-lined areas, as in Fig. 23–8.

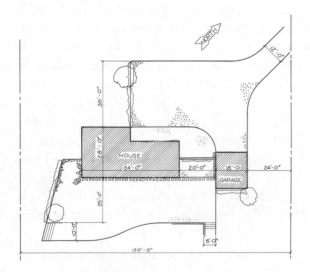

Fig. 23–8. Plot Plan.

23.7 Elevations and Sections. As the plan shows the horizontal shape and dimensions of the building, the vertical shapes and sizes are defined by *elevations* and *sections.* Elevations could be called outside views of the building, and usually all such views will be shown. For a rectangular building four regular elevations will suffice, whereas a U-shaped building may require as many as six. The general rule is that every outside surface of the building containing any openings or special features must be shown on some elevation.

Very often the elevation views cannot be placed on the same sheet as the plan. In order to identify properly each elevation so that it can be coordinated with the respective side of the plan, the architect identifies the sides of the building by the points of a compass, such as NORTH ELEVATION, Fig. 23–20, EAST ELEVATION, Fig. 23–21, etc. The north direction symbol is usually given on the plot plan or the GROUND FLOOR PLAN. Elevations and sections contain only vertical dimensions arranged in chains. No horizontal dimensions should be placed on elevation views. Sections are usually taken through a window, as shown on Fig. 23–22 (a).

Elevation views serve also to identify different materials, or elements, used on the outside of the building, such as *siding, stonework, brick, roof shingles, metal flashing, glass* and *glass blocks, cement* and *concrete, downspouts* and *gutters,* etc. In every case where a symbol would not be sufficient to identify the material or element, an abbreviation or word description should be used. Figures 23–20 and 23–21 show many instances of this type, such as *flash* and *counterflash, T.C. lining (terracotta flue lining).* Heavy long-and-short-dash lines, indicating floor levels, should also be noted.

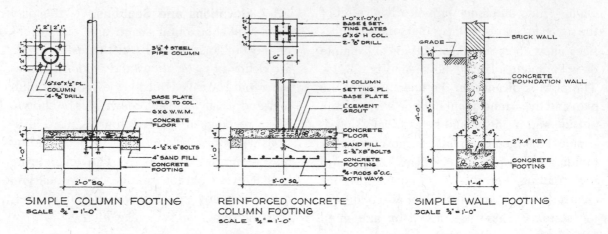

Fig. 23–9. Footings.

23.8 Architectural Details. Next to the plan, the most important drawings in the architectural set are the *details*. Since the plans and elevations are drawn at a scale too small to distinguish the small parts and features, many of these must be shown on the details. The scale must be sufficiently large to identify all component parts. This scale may be any size from $1'' = 1'-0''$ to $3'' = 1'-0''$ to full size.

Detail drawings may be placed on sheets containing plans, elevations, and sections, but in large sets of drawings they are placed on separate sheets. As shown on Figs. 23–19 and 23–22, each detail is identified by the same title on the small-scale drawing and under the detail drawing itself. This cross-identification of small-scale drawings and details is essential in order that the checker, the specification writer, the estimator, and the contractor will fully understand the construction, the appearance of the structure, and the materials of which it is constructed.

The details drawn by the architect serve as a basis for the shop drawings (showing manufacturing details and fabrication and installa-

tion procedures) that are prepared by the contractor and his subcontractors. In this way the architect is able to control not only the exact execution of the design but also the quality of the construction of specific parts of the structure. Architectural practice follows the conventional or standard manufacturing detailing practices for many parts of the house, and such standard details can be readily found in any appropriate handbook, such as *Architectural Graphic Standards* by Ramsey and Sleeper,* *Don Graf's Data Sheets,*† and *Time-Saver Standards.*‡ Yet, because each house or structure is based on an individual design, many special conditions must be covered by details adapted for a particular structure or condition. However, in general it will be found that these special details follow the arrangements and practices shown on standard details. For this reason a careful study of detailing

* John Wiley & Sons, 605 Third Avenue, New York, N.Y. 10016.
† Reinhold Book Division, 430 Park Avenue, New York, N.Y. 10022.
‡ F. W. Dodge Corp., New York, N.Y. 10036.

methods and standard detailing practices will be very helpful.

23.9 Principal Details. Included in this category are usually all details that repeatedly form a component part of every house or structure. The most important ones are: *footings* and *foundations, wall sections, floor* and *ceiling framing, doors, chimneys* and *heating tracts, stairs, roof framing* and *roofs;* miscellaneous structural details, such as *lintels, beams, girders,* and *columns;* and *details of built-in installations.*

23.10 Footings and Foundations. The stability of the structure depends on the strength, size, and quality of the footings. Climate variations dictate that the footings be placed at a certain depth below the grade to insure their placement on a frost-free base. The size of the footings depends on the loads that they are to carry. These loads are calculated to include the complete weight of the structure, including the equipment, and the loads that will be placed on the floors after the completion of the building. In cases of flat roofs the possible *snow load,* and of steep roofs the possible *wind load,* must be added. Municipal and county building codes usually require footings of a sufficient size for residential buildings to insure the safety of the structure under normal loading conditions. Examples of footings of a residential type are shown in Fig. 23–9. Footings support the foundation walls, which in turn support the ground floor framing and the structure above. These walls are usually poured concrete and are waterproofed by different methods from the outside or inside, or both. The continuity of a foundation wall is essential to prevent seepage into the spaces that are placed below the grade. A typical detail of a foundation wall is shown in Fig. 23–10.

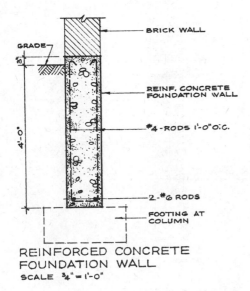

REINFORCED CONCRETE
FOUNDATION WALL
SCALE ¾" = 1'-0"

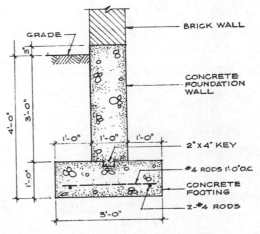

CONCRETE FOUNDATION WALL
ON REINFORCED CONCRETE FOOTING
SCALE ¾" = 1'-0"

Fig. 23–10. Foundation Walls.

23.11 Wall Sections. The most common types of residential construction are the *frame wall*, Fig. 23–11, *brick veneer wall, brick wall, brick and concrete block wall*, Fig. 23–12, and *concrete block wall*. The frame wall shown in Fig.

23–11, which is constructed entirely of wood, is no longer allowed in most localities because of its low fire resistance. Up to the time of World War I this was the most common type of residential construction. Some wood fram-

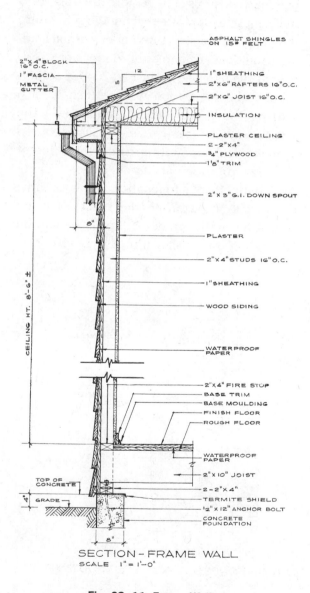

Fig. 23–11. Frame Wall.

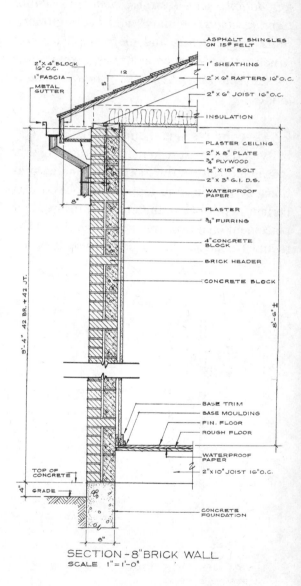

Fig. 23–12. Brick and Concrete Block Wall.

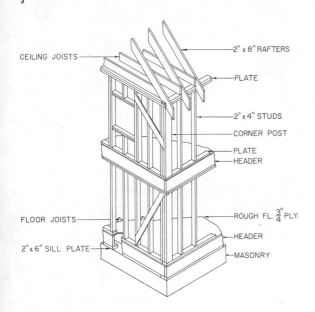

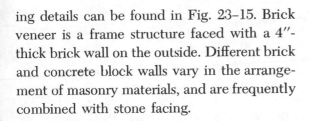

Fig. 23-13. Western Framing.

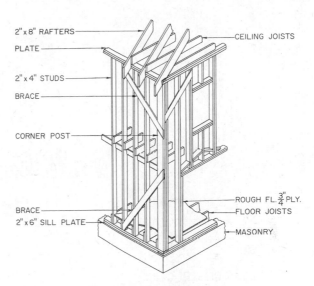

Fig. 23-14. Balloon Framing.

ing details can be found in Fig. 23–15. Brick veneer is a frame structure faced with a 4″-thick brick wall on the outside. Different brick and concrete block walls vary in the arrangement of masonry materials, and are frequently combined with stone facing.

23.12 Floor and Ceiling Framing. Different materials and their combinations create various types of floor and ceiling framing. The sizes of joists and structural members are usually specified in the building codes for normal residential conditions. Figures 23–13 and 23–14 show the most common types of such framing.

23.13 Doors and Windows. If made of wood, doors and windows are classified as *mill work* and are usually factory-made and delivered to the building site as complete units. Their

details will vary from one manufacturer to another and can be found readily in manufacturers' catalogs. Standard terms have been established by usage, such as *double-hung window, casement sash, solid wood door, hollow core door, solid core door,* and many others. An architectural draftsman must be acquainted with these terms and also with the specific manner of their construction. All such details can be found in architectural handbooks and manufacturers' literature. Figures 23–16 and 23–22 show examples of this type of detailing.

23.14 Stairs. When the structure is composed of two or more stories, a means of vertical communication must be used, usually a stairway. Interior and exterior stairways are made of wood, steel, concrete, or combinations of these materials. Wood stair details are shown in Fig. 23–17.

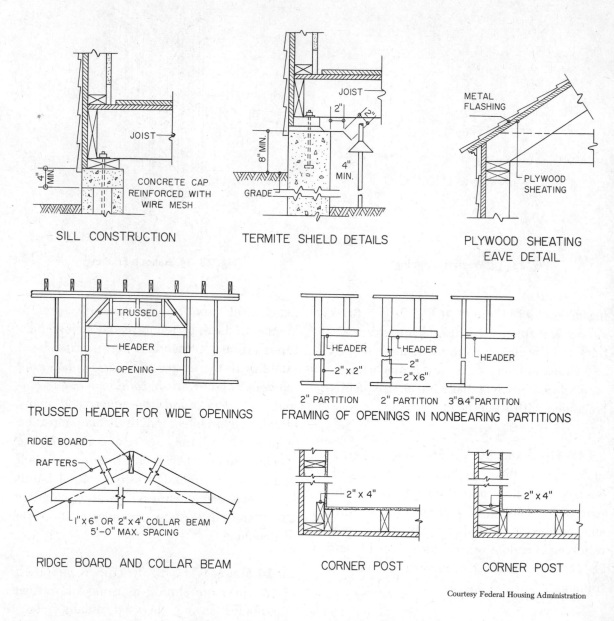

SILL CONSTRUCTION   TERMITE SHIELD DETAILS   PLYWOOD SHEATING EAVE DETAIL

TRUSSED HEADER FOR WIDE OPENINGS   FRAMING OF OPENINGS IN NONBEARING PARTITIONS

2" PARTITION   2" PARTITION   3"&4" PARTITION

RIDGE BOARD AND COLLAR BEAM   CORNER POST   CORNER POST

Fig. 23–15. Miscellaneous Framing Details.

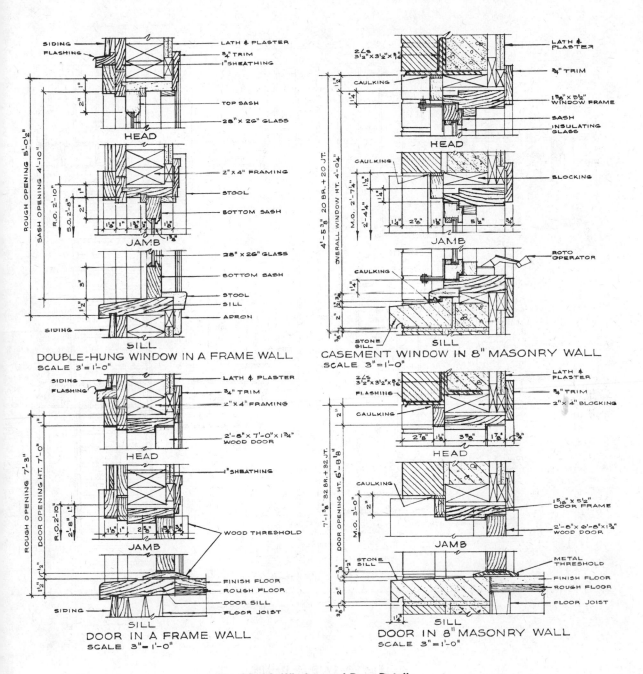

DOUBLE-HUNG WINDOW IN A FRAME WALL
SCALE 3"=1'-0"

CASEMENT WINDOW IN 8" MASONRY WALL
SCALE 3"=1'-0"

DOOR IN A FRAME WALL
SCALE 3" = 1'-0"

DOOR IN 8" MASONRY WALL
SCALE 3" = 1'-0"

Fig. 23-16. Window and Door Details.

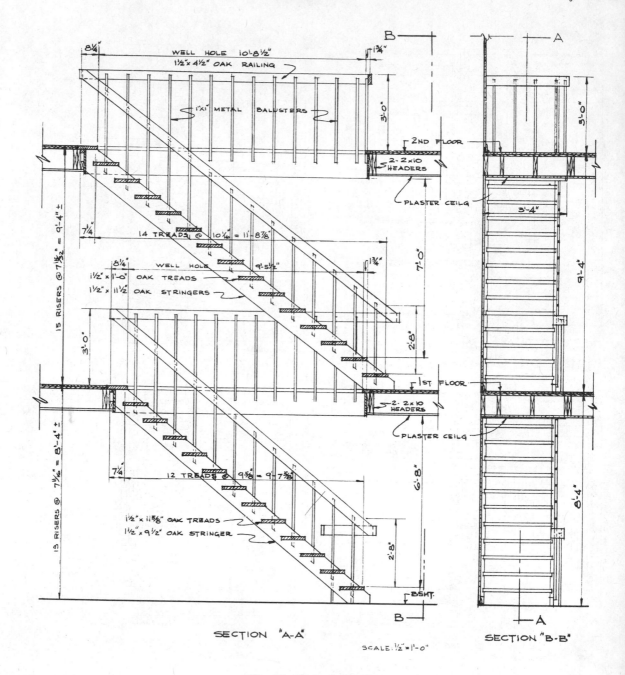

Fig. 23–17. Stair Details.

23.15 Miscellaneous Structural Details. A modern residential structure, with its large windows and open plan, requires somewhat more structural detailing than the conventional old-type brick residential building. In many cases it requires a steel skeleton and steel roof framing. Other structural details are details of lintels, girders, and beams. See Fig. 23–18.

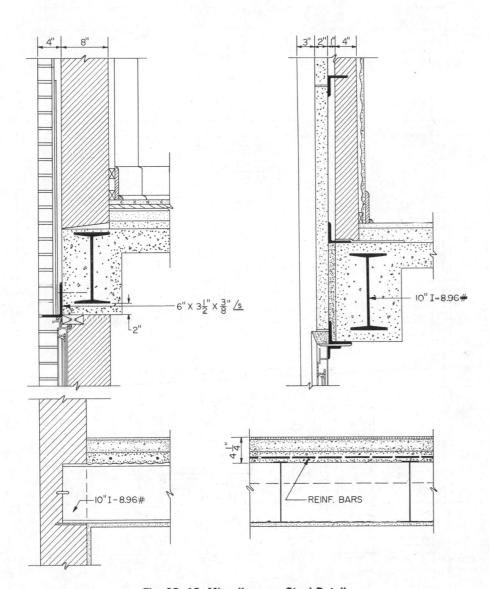

Fig. 23–18. Miscellaneous Steel Details.

23.16 Chimneys and Heating Ducts. Included in this category of details are chimneys, stacks for ventilation, fireplaces, and outdoor ovens and grills. The detailing of these features will be subject to the various building code requirements, and typical details can be found in handbooks. Figure 23–22 shows a detail of a small fireplace. A chimney with three flues, one serving for the fireplace, the second for the central heating plant, and the third as a gas range vent, can be seen in Figs. 23–19 to 23–21.

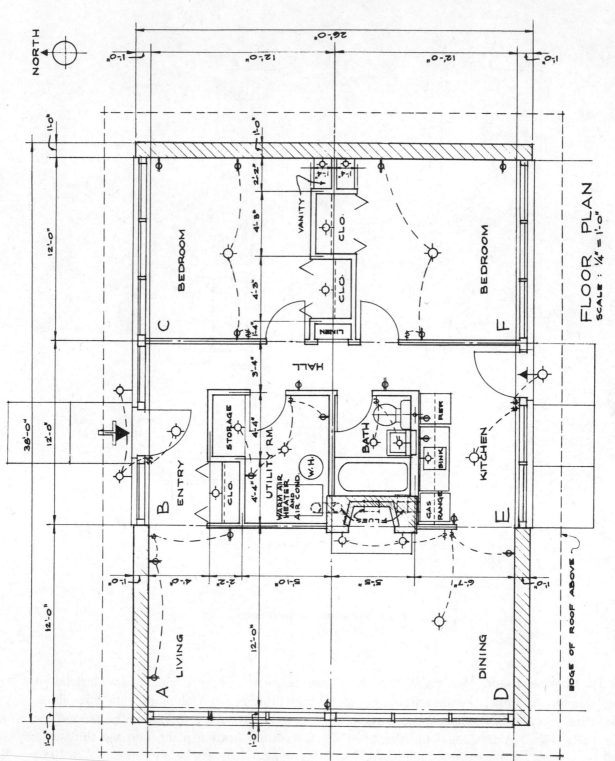

Fig. 23-19. Plan.

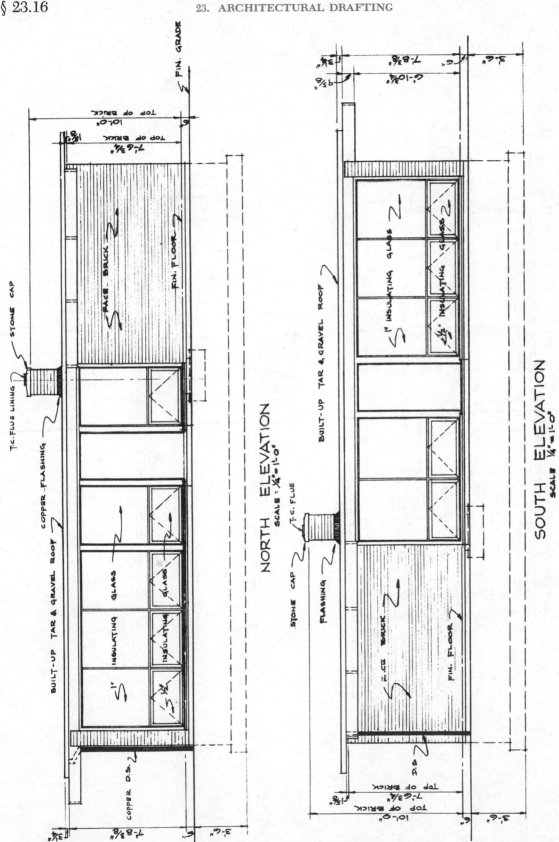

NORTH ELEVATION
SCALE: ¼"=1'-0"

SOUTH ELEVATION
SCALE ¼"=1'-0"

Fig. 23-20. Elevations.

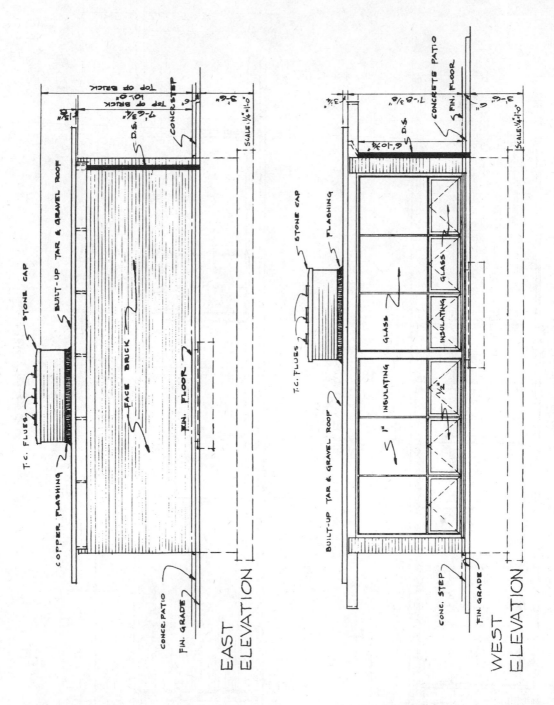

Fig. 23-21. Elevations.

23.17 Built-in Installations. Almost every residential structure will contain some built-in installations, such as kitchen cabinets, library shelves, access panels to mechanical equipment, etc. These must be carefully detailed and coordinated with the dimensions and design shown on the plan and elevation drawings.

23.18 Working Drawings. Working drawings represent the final stage of the architect's work. They include the plans, elevations, sections, and details. Before these drawings are carried to completion, many basic decisions will have been made both by the architect and by the owner. The preliminary sketches or drawings thus have served as the basis for discussions about the size of the building and the location of partitions, stairways, exits, etc. Now the draftsman is ready to proceed on the basis of these decisions and draw the final plans. These plans must be exact and comprehensive. They must give the contractor all the necessary information for the construction of the building.

The drawings in Figs. 23–19 to 23–23 represent such a set of working drawings. This set includes the plans, elevations, and sections, plus the important details drawn at a sufficiently large scale to show the construction clearly and completely. Also, the set may contain different *schedules*,* such as a ventilation schedule, lintel schedule, and door and win-

---

* A schedule is a table or chart specifying in detail the types and sizes of a certain feature on the plans. For example, a door schedule will show in tabular form the type of door (flush panel, glass, metal, etc.) and its size and the location where it is to be used. A room finish schedule similarly will show in tabular form the finish materials for every room such as the tile floor, acoustical plaster ceiling, plaster walls, etc.

dow schedule. All this information is required for construction purposes, or by city or state building codes.

The working drawings are started by drawing and completing the ground floor plan, usually at the scale of $\frac{1}{4}'' = 1'-0''$. From this plan the other floor plans, such as the basement floor plan, second floor plan, etc., are traced in outline and modified with all necessary features for the respective floor. Thus, a draftsman is able to coordinate the basement columns and girders with the load of the first floor partitions, etc.

The elevations are usually drawn after the locations of all openings on the plan have been fixed and dimensioned. Ordinarily they are drawn at the scale of $\frac{1}{4}'' = 1'-0''$, and should indicate materials that are to be used on the outside of the building. These indications are sometimes symbols, sometimes abbreviations. Examples can be found in Figs. 23–20 and 23–21. The elevations almost always include the outline of the foundations in hidden lines, so that the total height of the structure and the relationship of the floor level to the grade level can be determined from the elevations.

Working drawings in a sense represent an elaborate assembly of many parts and many materials. They should be studied from this point of view. They indicate how the different parts of the structure form the whole, how different materials join together, and how the different phases of construction should be coordinated. The student should note particularly the system by which the details, shown in Fig. 23–22, are drawn and related to the plans and elevations.

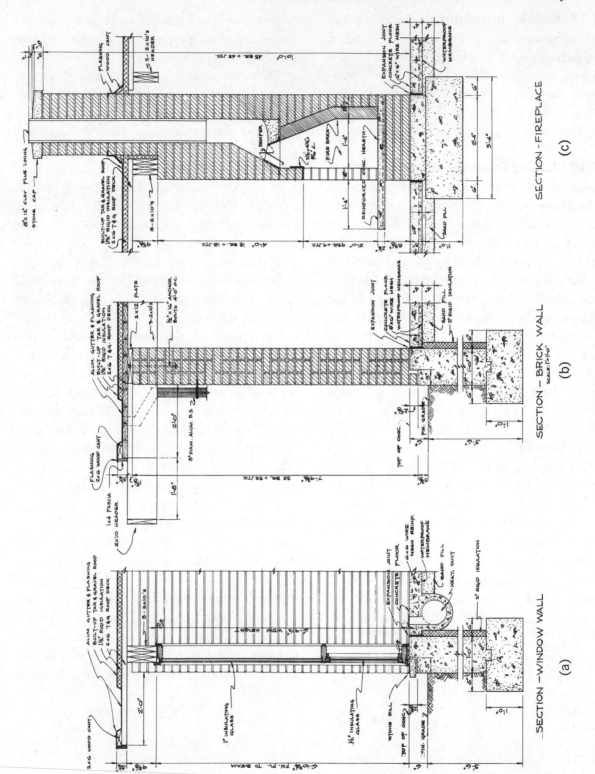

Fig. 23-22. Details.

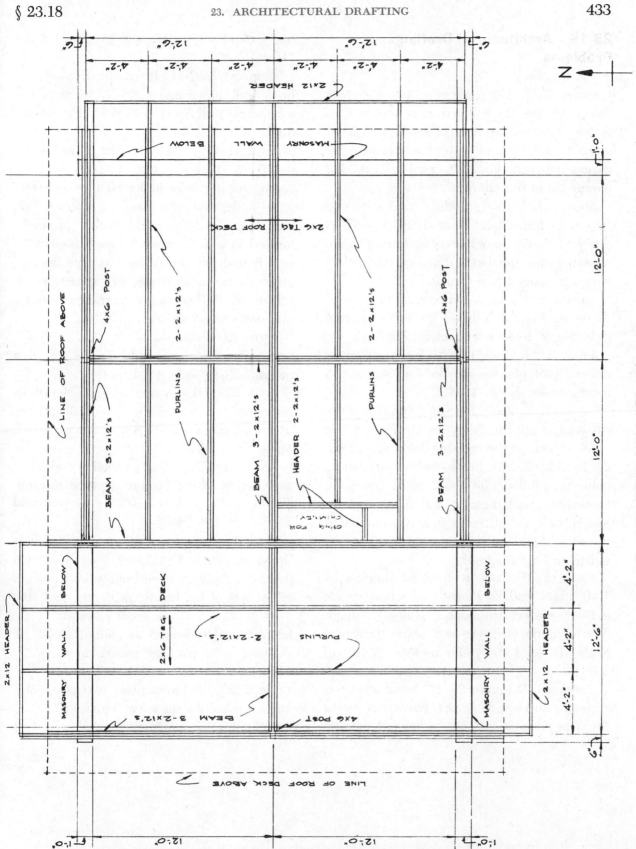

ROOF FRAMING PLAN
SCALE : ½" = 1'-0"

Fig. 23-23. Framing Plan.

23.19 Architectural Drafting Problems

PROB. 23/1. Using architectural letters as shown in Fig. 23–3 (vertical or inclined, as assigned), letter five lines of lettering $\frac{1}{8}''$ high, making lines 6″ long with approximately $\frac{1}{16}''$ spaces between. For text, use the beginning paragraph of this chapter.

PROB. 23/2. Using architectural letters as shown in Fig. 23–3 (vertical or inclined, as assigned), letter three lines of lettering $\frac{1}{4}''$ high, repeating the alphabet and making the lines 6″ long and spaced approximately $\frac{1}{8}''$ apart.

PROB. 23/3. Using architectural title letters shown in Fig. 23–5, letter the following title in letters $\frac{1}{2}''$ high: Architectural Drafting.

PROB. 23/4. Using Fig. 23–8 as an example, draw a plot plan of the building where you reside. Scale: $\frac{1}{32}'' = 1'-0''$.

PROB. 23/5. Study Fig. 23–7 and then identify and locate the following elements of the house shown on the working drawings in Figs. 23–19 to 23–23: flue lining, flashing, insulation, columns, subfloor, fireplace, joists, gravel fill, foundation wall, hearth, and heating plant. Sketch each detail freehand approximately to the scale shown on the working drawing. (Do not trace from the book.)

PROB. 23/6. Draw a detail of foundations 1′–0″ thick and 2′–4″ wide on which stands a 1′–4″ concrete foundation wall 4′–6″ high. The top of the wall is to be 8″ above the grade. Scale: $1'' = 1'-0''$ (refer to Figs. 23–9 and 23–10).

PROB. 23/7. Figure 23–12 shows a section through a brick and concrete block wall. Using the data given in Fig. 23–11, convert the sec-tion into a brick veneer wall 10″ thick. Scale: $1\frac{1}{2}'' = 1'-0''$.

PROB. 23/8. Convert section through the brick wall shown in Fig. 23–22 by including a casement-sash detail as shown in Fig. 23–16. Scale: $1\frac{1}{2}'' = 1'-0''$.

PROB. 23/9. The plan of the house shown in Fig. 23–19 consists of six areas between column center lines indicated by the letters A to F. Redesign the plan in a manner such that the "Kitchen" and "Bath," presently located in area E, are relocated properly in area B; and the "Entry" and "Utility Room," presently located in area B, are relocated properly in area E. If necessary, rearrange windows and doors to suit the new plan.

PROB. 23/10. Add a 10′–0″ × 20′–0″ carport adjoining areas C and F on the new plan of Prob. 23/9. Scale: $\frac{1}{4}'' = 1'-0''$.

PROB. 23/11. Draw front and rear elevations for completed Prob. 23/9, using wall sections of completed Prob. 23/8 in areas A and D. Scale: $\frac{1}{4}'' = 1'-0''$.

PROB. 23/12. Draw cross-section and framing details of the carport required in Prob. 23/10. Scale: $\frac{1}{4}'' = 1'-0''$ for section and $3'' = 1'-0''$ for details.

PROB. 23/13. Draw elevation of the fireplace shown on plan in Fig. 23–19. Using your own design, include an open built-in bookcase on either side of the hearth. Scale: $1'' = 1'-0''$.

PROB. 23/14. Draw detail elevation of the kitchen wall indicated in plan, Fig. 23–19, showing all appliances and cabinets. Scale: $1'' = 1'-0''$.

PROB. 23/15. Draw plan, two elevations, and a section of a single-car garage (12′–0″ × 20′–0″) of frame construction.

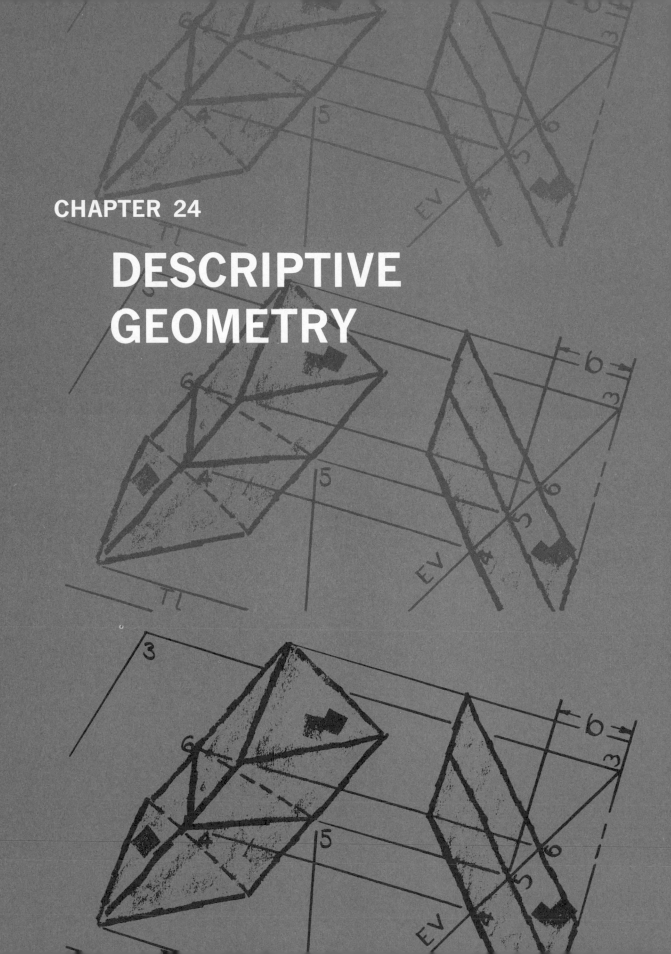

CHAPTER 24

DESCRIPTIVE
GEOMETRY

24.1 Definition. A consolidation of several dictionary definitions is: *"Descriptive Geometry** is that branch of geometry which provides a graphical solution of a three-dimensional problem by means of projections upon mutually perpendicular planes."* In Chapter 6 (and succeeding chapters) we followed that definition, perhaps without realizing it. The "three-dimensional problem" we solved was that of representing an object, such as the house of Fig. 6–2, which was solved by *views* ("projections") on "mutually perpendicular planes": the front, top, and right-side planes. Actually, people in the technical drawing field have come to mean problem solving rather than representation when they use the term "descriptive geometry." You have seen in Chap-

* For a complete discussion of this subject, see Paré, Loving, and Hill, *Descriptive Geometry*, 3rd ed. (Macmillan, 1965), or any standard descriptive geometry text.

ter 12 how auxiliary views may be used to find certain angles, Sec. 12.9, or the true length of an oblique line, Sec. 12.11. These are descriptive geometry solutions.

24.2 Angle Between Line and Plane. At Fig. 24–1 (a) we see a repetition of Fig. 12–14. By assuming a line of sight perpendicular to any view (in this case, the front view), we obtain an auxiliary view showing the true length of hip rafter 1–2.

At (b) a portion of the front wall has been added in the auxiliary view. Since this *frontal plane* appears as a line ("EV," or edge view), the true angle of 35° between line 1–2 and the plane appears and may be dimensioned as shown.

Of course, the angle between a hip rafter and a wall of a house may not be very useful information for a carpenter, but the general

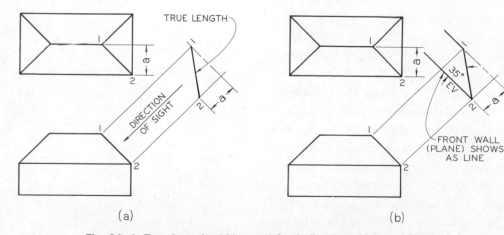

(a)   (b)

Fig. 24–1. True Length of Line and Angle Between Line and Plane.

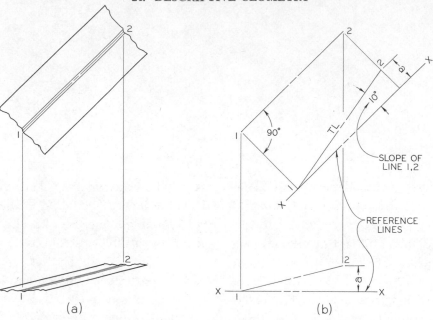

(a)   (b)

Fig. 24–2. Slope of a Line.

problem of finding the angle between a line and a plane is important in other fields. For example, in Fig. 24–2 (a) are shown the top (map) and front elevation views of a portion of a roadway. For the purpose at hand, we need be concerned only with the center line 1–2 of the roadway, as at (b).

If we introduce a horizontal reference plane (horizontal in space) through point 1, it appears as a horizontal line X–X in the front view. Now, the true angle between a line and horizontal is sometimes called the *slope* of the line. To find the slope of line 1–2 in Fig. 24–2 (b) we must construct an auxiliary view showing 1–2 in true length (TL) and the horizontal plane in edge view (EV). To do this we simply use our horizontal plane as our reference plane X–X and project at right angles to the top view, so that X–X appears perpendicular to the projection lines (and parallel to the top view). As in Fig. 24–1 (b) we now have a line true length (1–2) and a plane in edge view (X–X). The angle of 10° between them is thus true size and is the slope of line 1–2.

The civil engineer usually expresses map directions as compass bearings, Sec. 21.1, and inclination with horizontal as *percent grade* rather than slope. Thus, the previous center line 1–2 would be described, Fig. 24–3, as

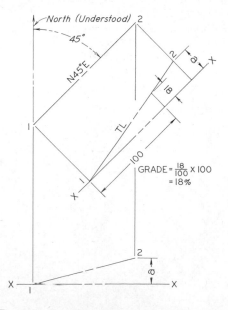

Fig. 24–3. Bearing and Percent Grade of Line.

having a bearing of N 45° E and an upgrade from point 1 of 18 percent. Note the way the grade is measured. Any convenient length (100 units here) is set off along the horizontal plane from point 1, and the corresponding distance to line 1–2 at right angles to the horizontal plane, 18 units, is measured. Since the 18 units is measured perpendicular to the horizontal plane, it is vertical *in space* (although not vertical on the paper). Thus the definition of grade is:

$$\text{Percent grade} = \frac{\text{Vertical distance}}{\text{Horizontal distance}} \times 100,$$

the two distances being measured between any convenient two points on the line.*

24.3 Edge and True-Size Views of Oblique Plane.
In Sec. 7.13 we noted how a plane surface appears when viewed from various

* In trigonometry, this proportion would be the tangent of the slope angle. Thus tan 10° = .18. (From a table of natural tangents, tan 10° = .1763.)

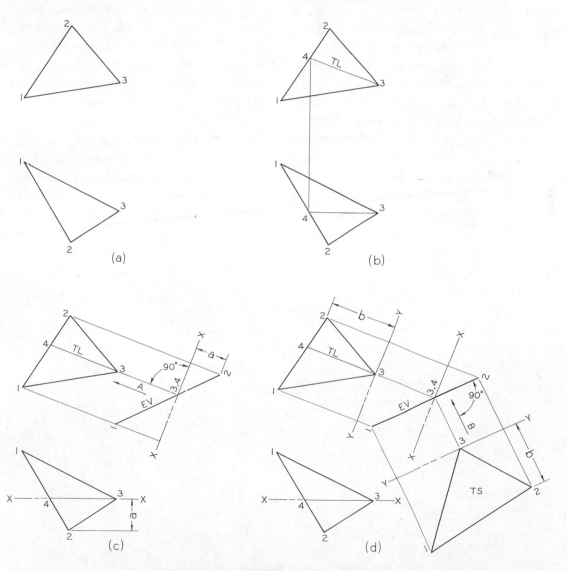

Fig. 24–4. Edge and True-Size Views of Oblique Plane.

directions. In Fig. 7–16 (a) and (b) we observed that under some circumstances a plane appears as a line—in edge view (EV). In Fig. 12–12 we observed further that when two intersecting surfaces appear as lines we can measure the true angle between them, and that this situation results when we look parallel to their line of intersection, obtaining a "point view" of this line. And, finally, in Fig. 12–15 we discovered that we could obtain an edge view of a plane surface by looking parallel to a true-length line in the surface.

Now, how can we apply what we have learned to obtain an edge view of plane 1–2–3 of Fig. 24–4 (a)? First, we observe that none of the three given edges is true length. We could construct an auxiliary view showing any one of these lines in true length, but it is simpler to add a line, as at (b), that will be true length in one of the given views. In this case we have chosen to add a horizontal line through point 3 in the front view. This line intersects line 1–2 at point 4. Point 4 is pro-

jected to the top view of 1–2, and line 3–4 is true length in the top view as indicated.

As shown at (c), we now select a line of sight A parallel to 3–4, and construct the corresponding auxiliary view. The result is a point view of line 3–4 and, accordingly, the desired edge view of plane 1–2–3.

In Sec. 12.12 we obtained a true-size (TS) view of a plane surface by looking perpendicular to its edge view and constructing the corresponding secondary auxiliary view. In our present case we can do the same thing, Fig. 24–4 (d). True-size views are, of course, very useful because all lines and angles of the true-size surface may be measured. In fact, any needed plane-geometry constructions may be performed in such a view.

24.4 Piercing Point. We may use the foregoing edge-view construction in another way: to find the point of intersection (*piercing point*) of a line and an oblique plane. Suppose, in Fig. 24–5 (a), we are given the front and top views

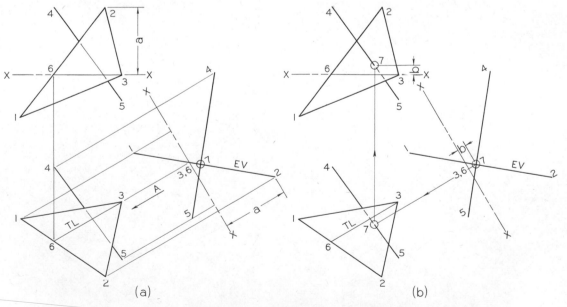

Fig. 24–5. Piercing Point of Line and Plane.

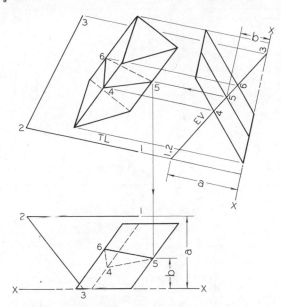

Fig. 24–6. Piercing Points—Intersection of Plane and Prism.

point 7 back to the given front and top views as at (b). Note the use of measurement b to check the accuracy of location of point 7 in the top view.

An application of the piercing-point construction is shown in Fig. 24–6. When the edge view of plane 1–2–3 is obtained in the auxiliary view, the piercing points 4, 5, and 6 of the edges of the prism with plane 1–2–3 are evident. Triangle 4–5–6 is the intersection of plane 1–2–3 and the prism.

24.5 Descriptive Geometry Problems. Problem layouts are given in Figs. 24–7 to 24–9, each designed to occupy one-fourth or one-half of Layout B (Appendix, page 474). Thus, combinations of two, three, or four problems may be assigned for one sheet. For practice, cross-section paper may be used, and rough solutions sketched freehand. But for accurate answers, of course, the problems must be laid out and solved carefully with instruments. Spacing or layout dimensions should be omitted, but dimensions for given or required measured quantities should be shown. The symbols TL, EV, and TS should also be used where they will help to clarify the method of solution.

of plane 1–2–3 and line 4–5. Note that again we do not have a true-length line. This time we have chosen to add line 3–6 in the top view, which, when projected to the front view, appears true length as indicated. This gives us the direction of line of sight A, and we obtain the edge view of plane 1–2–3. Line 4–5 intersects this edge view at point 7, the desired piercing point. If necessary, we can project

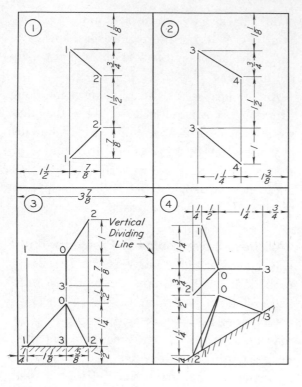

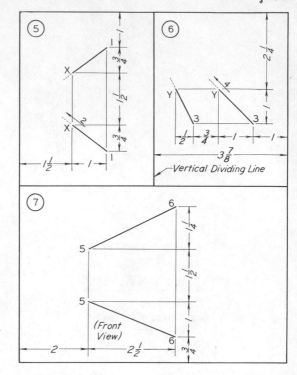

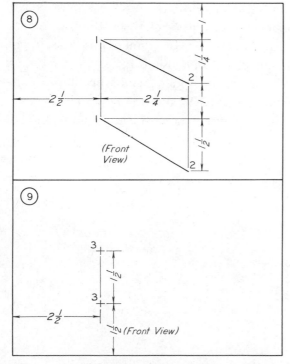

Problem Statements

1. Find and measure the true length of line 1–2.

2. Find and measure the true length of line 3–4.

3. Find and dimension the true lengths of tripod legs 0–1, 0–2, and 0–3. Scale: $1'' = 1'-0$.

4. Find and dimension the true lengths of tripod legs 0–1, 0–2, and 0–3. Scale: $\frac{3}{4}'' = 1'-0$.

5. Line 1–2 has a true length of $2''$. Complete the given views.

6. Line 3–4 has a true length of $9'-6$. Complete the given views. Scale: $\frac{1}{4}'' = 1'-0$.

7. Find and measure the true length and slope of line 5–6. Scale: $\frac{1}{8}'' = 1'-0$.

8. Find and measure the angle between line 1–2 and a frontal plane. What are the bearing and percent grade of line 1–2?

9. If a line 3–4 has a bearing of N 60° E, a downgrade from point 3 of 30%, and a true length of $3''$, complete the front and top views of 3–4.

Fig. 24–7. Descriptive Geometry Problems. Use Layout B (Appendix, page 474). Omit spacing dimensions, but show dimensions for required numerical answers. Also show symbols TL, EV, and TS where appropriate for clarifying your solutions.

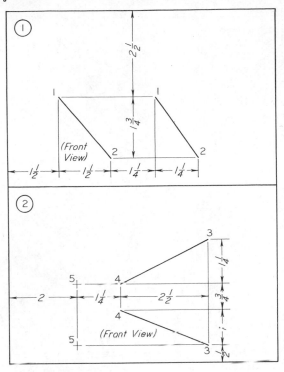

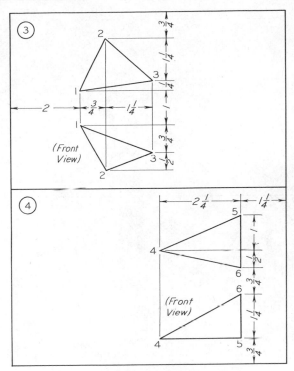

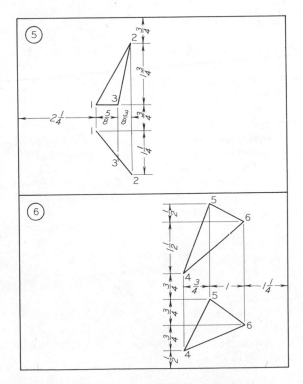

Problem Statements

1. Find the angle between line 1–2 and a profile plane. What is the bearing of line 1–2?

2. Line 3–4 represents a tunnel. Another tunnel is to be constructed starting at point 5 and meeting tunnel 3–4 at a point, 6, which is to be 600′ along 3–4 from point 3. Find the bearing, length, and grade of tunnel 5–6. Scale: $\frac{1}{4}'' = 100'$ (or $1'' = 400'$).

3. Add a horizontal line to plane 1–2–3 through point 3. Find a point view of this horizontal line and the accompanying edge view of plane 1–2–3. This edge view makes it possible to measure the true angle between plane 1–2–3 and a horizontal plane. If assigned, dimension this angle.

4. Find the true angle between plane 4–5–6 and a horizontal plane. (See the statement for problem 3.)

5. Construct a true-size view of plane 1–2–3. If assigned, calculate the area of triangle 1–2–3. Scale: $\frac{1}{4}'' = 10'$ ($1'' = 40'$).

6. Construct a true-size view of triangle 4–5–6. Measure and dimension the true angle at 4.

Fig. 24–8. Descriptive Geometry Problems. Use Layout B (Appendix, page 474). Omit spacing dimensions, but show dimensions for required numerical answers. Also show symbols TL, EV, and TS where appropriate for clarifying your solutions.

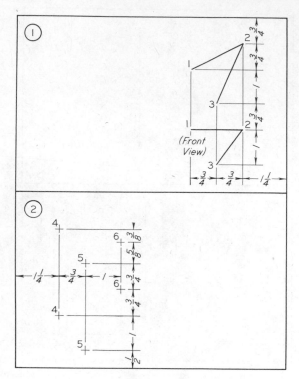

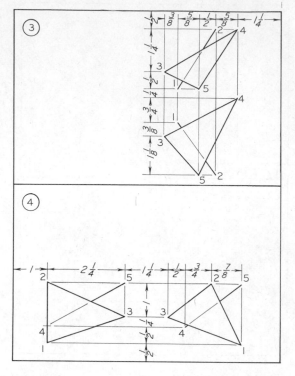

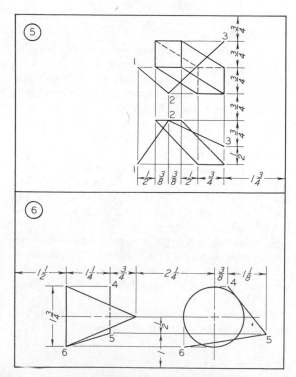

Problem Statements

1. Through the use of a true-size view of plane 1–2–3, find the bisector of angle 1–2–3 and project it back to the given front and top views. If assigned, find and measure the bearing and grade of the bisector.

2. Points 4, 5, and 6 determine a plane. Also three points determine a circle (see Fig. 5–14). Construct a true-size view of plane 4–5–6. Then find the center of the circle through the three points and project this center back to the given views.

3. Find the piercing point of line 1–2 in plane 3–4–5. If assigned, regard triangle 3–4–5 as opaque and show line 1–2 in proper visibility.

4. Find the piercing point of line 4–5 in plane 1–2–3. If assigned, regard triangle 1–2–3 as opaque, and show line 4–5 in proper visibility.

5. Obtain an edge view of plane 1–2–3 and determine in the given views the line of intersection of plane 1–2–3 and the prism.

6. Obtain the line of intersection of plane 4–5–6 and the right-circular cone. If assigned, develop a pattern of the conical surface, including the line of intersection. (See Sec. 17.17.)

Fig. 24–9. Descriptive Geometry Problems. Use Layout B (Appendix, page 474). Omit spacing dimensions, but show dimensions for required numerical answers. Also show symbols TL, EV, and TS where appropriate for clarifying your solutions.

APPENDIX

TECHNICAL TERMS

"The beginning of wisdom is to call things by their right names."

—CHINESE PROVERB

n means a *noun;*
v means a *verb*.

Acme (*n*)—Screw thread form, Secs. 14.4, 14.11.

Addendum (*n*)—Radial distance from pitch circle to top of gear tooth, Sec. 19.5.

Allen Screw (*n*)—Special set screw or cap screw with hexagon socket in head, Secs. 14.24, 14.26.

Allowance (*n*)—Minimum clearance between mating parts, Secs. 9.26, 9.27.

Alloy (*n*)—Two or more metals in combination, usually a fine metal with a baser metal.

Aluminum (*n*)—A lightweight but relatively strong metal. Often alloyed with copper to increase hardness and strength.

Angle Iron (*n*)—A structural shape whose section is a right angle, Sec. 20.3.

Anneal (*v*)—To heat and cool gradually, to reduce brittleness and increase ductility, Sec. 10.17.

Arc-weld (*v*)—To weld by electric arc. The work is usually the positive terminal.

Babbitt (*n*)—A soft alloy for bearings, mostly of tin with small amounts of copper and antimony.

Bearing (*n*)—A supporting member for a rotating shaft.

445

Bevel (*n*)—An inclined edge, not at right angle to joining surface.

Bolt Circle (*n*)—A circular center line on a drawing, containing the centers of holes about a common center, Sec. 9.17.

Bore (*v*)—To enlarge a hole with a boring bar or tool in a lathe, drill press, or boring mill, Secs. 10.11, 10.12.

Boss (*n*)—A cylindrical projection on a casting or a forging.

BOSS

Brass (*n*)—An alloy of copper and zinc.

Braze (*v*)—To join with hard solder of brass or zinc.

Brinell (*n*)—A method of testing hardness of metal.

Broach (*n*)—A long cutting tool with a series of teeth that gradually increase in size which is forced through a hole or over a surface to produce a desired shape, Sec. 10.16.

Bronze (*n*)—An alloy of eight or nine parts of copper and one part of tin.

Buff (*v*)—To finish or polish on a buffing wheel composed of fabric with abrasive powders.

Burnish (*v*)—To finish or polish by pressure upon a smooth rolling or sliding tool.

Burr (*n*)—A jagged edge on metal resulting from punching or cutting.

Bushing (*n*)—A replaceable lining or sleeve for a bearing.

Calipers (*n*)—Instrument (of several types) for measuring diameters, Sec. 10.10.

Cam (*n*)—A rotating member for changing circular motion to reciprocating motion, Sec. 19.1.

Carburize (*v*)—To heat a low-carbon steel to approximately 2000°F. in contact with material which adds carbon to the surface of the steel, and to cool slowly in preparation for heat treatment, Sec. 10.17.

Caseharden (*v*)—To harden the outer surface of a carburized steel by heating and then quenching, Sec. 10.17.

Castellate (*v*)—To form like a castle, as a castellated shaft or nut.

Casting (*n*)—A metal object produced by pouring molten metal into a mold, Sec. 10.2.

Cast Iron (*n*)—Iron melted and poured into molds, Sec. 10.2.

Center Drill (*n*)—A special drill to produce bearing holes in the ends of a workpiece to be mounted between centers. Also called a "combined drill and countersink," Table 5 (Appendix, page 457).

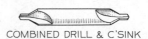

COMBINED DRILL & C'SINK

Chamfer (*n*)—A narrow inclined surface along the intersection of two surfaces.

CHAMFER

Chase (*v*)—To cut threads with an external cutting tool, Secs. 10.11, 14.3.

Cheek (*n*)—The middle portion of a three-piece flask used in molding, Sec. 10.2.

Chill (*v*)—To harden the outer surface of cast iron by quick cooling, as in a metal mold.

Chip (*v*)—To cut away metal with a cold chisel.

Chuck (*n*)—A mechanism for holding a rotating tool or workpiece.

Coin (*v*)—To form a part in one stamping operation.

Cold Rolled Steel (CRS) (*n*)—Open hearth or Bessemer steel, containing 0.12% to 0.20% carbon, which has been rolled while cold to produce a smooth, quite accurate stock.

Collar (*n*)—A round flange or ring fitted on a shaft to prevent sliding.

COLLAR

Colorharden (*v*)—Same as *caseharden*, except that it is done to a shallower depth, usually for appearance only.

Cope (*n*)—The upper portion of a flask used in molding, Sec. 10.2.

Core (*v*)—To form a hollow portion in a casting by using a dry-sand core or a green-sand core in a mold, Sec. 10.2.

Coreprint (*n*)—A projection on a pattern which forms an opening in the sand to hold the end of a core, Sec. 10.2.

Cotter Pin (*n*)—A split pin used as a fastener, usually to prevent a nut from unscrewing.

Counterbore (*v*)—To enlarge an end of a hole cylindrically with a *counterbore*, Sec. 10.12.

COUNTERBORE

Countersink (*v*)—To enlarge an end of a hole conically, usually with a *countersink*, Sec. 10.12.

COUNTERSINK

Crown (*n*)—A raised contour, as on the surface of a pulley.

Cyanide (*v*)—To surface-harden steel by heating in contact with a cyanide salt, followed by quenching.

Dedendum (*n*)—Distance from pitch circle to bottom of tooth space, Sec. 19.5.

Development (*n*)—Drawing of the surface of an object unfolded or rolled out on a plane, Sec. 17.1.

Diametral Pitch (*n*)—Number of gear teeth per inch of pitch diameter, Sec. 19.5.

Die (*n*)—(1) Hardened metal piece shaped to cut or form a required shape in a sheet of metal by pressing it against a mating die. (2) Also used for cutting small threads. In a sense is opposite to a tap, Sec. 14.12.

Die Casting (*n*)—Process of forcing molten metal under pressure into metal dies or molds, producing a very accurate and smooth casting, Sec. 10.7.

Die Stamping (*n*)—Process of cutting or forming a piece of sheet metal with a die.

Dog (*n*)—A small auxiliary clamp used to prevent work from rotating in relation to the face plate of a lathe, Sec. 10.11.

Dowel (*n*)—A cylindrical pin, commonly used between two contacting flat surfaces to prevent sliding.

DOWEL

Draft (*n*)—The tapered shape of the parts of a pattern to permit it to be easily withdrawn from the sand or, on a forging, to permit it to be easily withdrawn from the dies, Secs. 10.2, 10.6.

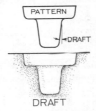

PATTERN

DRAFT

DRAFT

Drag (*n*)—Lower portion of a flask used in molding, Sec. 10.2.

Draw (*v*)—To stretch or otherwise to deform metal. Also to temper steel.

Drill (*v*)—To cut a cylindrical hole with a drill. A *blind hole* does not go through the piece, Sec. 10.12.

Drill Press (*n*)—A machine for drilling and other hole-forming operations, Sec. 10.12.

Drop Forge (*v*)—To form a piece while hot between dies in a drop hammer or with great pressure, Sec. 10.6.

Face (*v*)—To finish a surface at right angles, or nearly so, to the center line of rotation on a lathe, Sec. 10.11.

FAO—Finish all over, Sec. 9.7.

Feather Key (*n*)—A flat key which is sunk partly in a shaft and partly in a hub, permitting the hub to slide lengthwise of the shaft, Sec. 14.28.

File (*v*)—To finish or smooth with a file.

Fillet (*n*)—An interior rounded intersection between two surfaces, Sec. 10.3.

Fin (*n*)—A thin extrusion of metal at the intersection of dies or sand molds, Secs. 10.2, 10.6.

Fit (*n*)—Degree of tightness or looseness between two mating parts, as a *loose fit*, a *snug fit*, or a *tight fit*, Secs. 9.26, 9.28.

Fixture (*n*)—A special device for holding the work in a machine tool, *but not for guiding the cutting tool.*

Flange (*n*)—A relatively thin rim around a piece.

FLANGE

Flash (*n*)—Same as *fin*.

Flask (*n*)—A box made of two or more parts for holding the sand in sand molding, Sec. 10.2.

Flute (*n*)—Groove, as on twist drills, reamers, and taps.

Forge (*v*)—To force metal while it is hot to take on a desired shape by hammering or pressing.

Galvanize (*v*)—To cover a surface with a thin layer of molten alloy, composed mainly of zinc, to prevent rusting.

Gasket (*n*)—A thin piece of rubber, metal, or some other material, placed between surfaces to make a tight joint.

Gate (*n*)—The opening in a sand mold at the bottom of the *sprue* through which the molten metal passes to enter the cavity or mold, Sec. 10.2.

Graduate (*v*)—To set off accurate divisions on a scale or dial.

Grind (*v*)—To remove metal by means of an abrasive wheel, often made of carborundum. Used chiefly where accuracy is required, Sec. 10.15.

Harden (*v*)—To heat steel above a critical temperature and then quench in water or oil, Sec. 10.17.

Heat-treat (*v*)—To change the properties of metals by heating and then cooling, Sec. 10.17.

Interchangeable (*adj.*)—Refers to a part made to limit dimensions so that it will fit any mating part similarly manufactured, Sec. 9.26.

Jig (*n*)—A device *for guiding a tool* in cutting a piece. Usually it holds the work in position.

Journal (*n*)—Portion of a rotating shaft supported by a bearing.

Kerf (*n*)—Groove or cut made by a saw.

KERF

Key (*n*)—A small piece of metal sunk partly into both shaft and hub to prevent rotation, Sec. 14.28.

Keyseat (*n*)—A slot or recess in a shaft to hold a key, Sec. 14.28.

KEYSEAT

Keyway (*n*)—A slot in a hub or portion surrounding a shaft to receive a key, Sec. 14.28.

KEYWAY

Knurl (*v*)—To impress a pattern of dents in a turned surface with a knurling tool to produce a better hand grip, Sec. 10.11.

Lap (*v*)—To produce a very accurate finish by sliding contact with a *lap*, or piece of wood, leather, or soft metal impregnated with abrasive powder.

Lathe (*n*)—A machine used to shape metal or other materials by rotating against a tool, Sec. 10.11.

Lug (*n*)—An irregular projection of metal, but not round as in the case of a *boss*, usually with a hole in it for a bolt or screw.

Malleable Casting (*n*)—A casting which has been made less brittle and tougher by annealing.

Mill (*v*)—To remove material by means of a rotating cutter on a milling machine, Sec. 10.14.

Mold (*n*)—The mass of sand or other material which forms the cavity into which molten metal is poured, Sec. 10.2.

MS (*n*)—Machinery steel, sometimes called *mild steel*, with a small percentage of carbon. Cannot be hardened.

Neck (*v*)—To cut a groove called a *neck* around a cylindrical piece.

NECK

Normalize (*v*)—To heat steel above its critical temperature and then to cool it in air, Sec. 10.17.

Pack-harden (*v*)—To *carburize*, then to *case-harden*, Sec. 10.17.

Pad (*n*)—A slight projection, usually to provide a bearing surface around one or more holes.

PAD

Pattern (*n*)—A model, usually of wood, used in forming a mold for a casting. In sheet metal work a pattern is called a *development*, Sec. 10.2.

Peen (*v*)—To hammer into shape with a ball-peen hammer.

Pickle (*v*)—To clean forgings or castings in dilute sulfuric acid.

Pinion (*n*)—The smaller of two mating gears, Sec. 19.5.

Pitch Circle (*n*)—An imaginary circle corresponding to the circumference of the friction gear from which the spur gear was derived, Sec. 19.6.

Plane (*v*)—To remove material by means of the *planer*.

Planish (*v*)—To impart a planished surface to sheet metal by hammering with a smooth-surfaced hammer.

Plate (*v*)—To coat a metal piece with another metal, such as chrome or nickel, by electrochemical methods.

Polish (*v*)—To produce a highly finished or polished surface by friction, using a very fine abrasive.

Profile (*v*)—To cut any desired outline by moving a small rotating cutter, usually with a master template as a guide.

Punch (*v*)—To cut an opening of a desired shape with a rigid tool having the same shape, by pressing the tool through the work.

Quench (*v*)—To immerse a heated piece of metal in water or oil in order to harden it.

Rack (*n*)—A flat bar with gear teeth in a straight line to engage with teeth in a gear.

Ream (*v*)—To enlarge a finished hole slightly to give it greater accuracy, with a *reamer*.

Relief (*n*)—An offset of surfaces to provide clearance for machining.

RELIEF

Rib (*n*)—A relatively thin flat member acting as a brace or support.

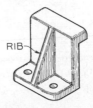
RIB

Rivet (*v*)—To connect with rivets, or to clench over the end of a pin by hammering.

Round (*n*)—An exterior rounded intersection of two surfaces, Sec. 10.3.

SAE—Society of Automotive Engineers.

Sandblast (*v*)—To blow sand at high velocity with compressed air against castings or forgings to clean them.

Scleroscope (*n*)—An instrument for measuring hardness of metals.

Scrape (*v*)—To remove metal by scraping with a hand scraper, usually to fit a bearing.

Shape (*v*)—To remove metal from a piece with a *shaper*, Sec. 10.13.

Shear (*v*)—To cut metal by means of shearing with two blades in sliding contact.

Sherardize (*v*)—To galvanize a piece with a coating of zinc by heating it in a drum with zinc powder, to a temperature of 575° to 850° F.

Shim (*n*)—A thin piece of metal or other material used as a spacer in adjusting two parts.

Solder (*v*)—To join with solder, usually composed of lead and tin.

Spin (*v*)—To form a rotating piece of sheet metal into a desired shape by pressing it with a smooth tool against a rotating form.

Spline (*n*)—A keyway, usually one of a series cut around a shaft or hole.

SPLINED HOLE

Spotface (*v*)—To produce a round *spot* or bearing surface around a hole, usually with a *spotfacer*. The spotface may be on top of a boss or it may be sunk into the surface, Sec. 10.12.

SPOTFACE

Sprue (*n*)—A hole in the sand leading to the *gate* which leads to the mold, through which the metal enters, Sec. 10.2.

Steel Casting (*n*)—Like cast-iron casting except that in the furnace scrap steel has been added to the casting.

Swage (*v*)—To hammer metal into shape while it is held over a *swage*, or die, which fits in a hole in the *swage block*, or anvil.

Sweat (*v*)—To fasten metal together by the use of solder between the pieces and by the application of heat and pressure.

Tap (*v*)—To cut relatively small internal threads with a *tap*, Sec. 14.12.

Taper (*n*)—Conical form given to a shaft or a hole. Also refers to the slope of a plane surface, Sec. 9.16.

Taper pin (*n*)—A small tapered pin for fastening, usually to prevent a collar or hub from rotating on a shaft.

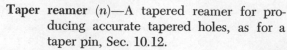

TAPER PIN

Taper reamer (*n*)—A tapered reamer for producing accurate tapered holes, as for a taper pin, Sec. 10.12.

Temper (*v*)—To reheat hardened steel to bring it to a desired degree of hardness, Sec. 10.17.

Template or **Templet** (*n*)—A guide or pattern used to mark out the work, guide the tool in cutting it, or check the finished product.

Tin (*n*)—A silvery metal used in alloys and for coating other metals, such as tin plate.

Tolerance (*n*)—Total amount of variation permitted in limit dimensions of a part, Secs. 9.26, 9.27.

Trepan (*v*)—To cut a circular groove in the flat surface at one end of a hole.

Tumble (*v*)—To clean rough castings or forgings in a revolving drum filled with scrap metal.

Turn (*v*)—To produce, on a lathe, a surface parallel to the center line, Sec. 10.11.

Twist Drill (*n*)—A drill for use in a drill press, Sec. 10.12.

Undercut (*n*)—A recessed cut or a cut with inwardly sloping sides.

UNDERCUT

Upset (*v*)—To form a head or enlarged end on a bar or rod by pressure or by hammering between dies.

Web (*n*)—A thin flat part joining larger parts. Also known as a *rib*.

Weld (*v*)—To unite metal pieces by pressure or fusion welding processes, Sec. 10.9.

Woodruff Key (*n*)—A semicircular flat key, Sec. 14.28.

WOODRUFF KEYS

Wrought Iron (*n*)—Iron of low carbon content useful because of its toughness, ductility, and malleability.

VISUAL AIDS FOR TECHNICAL DRAWING

The following visual aids are suggested. However, the instructor should preview each visual aid before using it in class, in order to determine its suitability for the subject area and for the level of students. All motion pictures listed are on 16 mm film. The sources for the aids listed below are as follows:

DA   Du Art Film Laboratories, U.S. Government Film Services, 245 W. 55th St., New York, N.Y. 10019

JH   Jam Handy Organization, 2821 E. Grand Blvd., Detroit, Mich. 48211

McG   McGraw-Hill Book Co., Text Film Division, 330 W. 42nd St., New York, N.Y. 10036

P   Purdue University, Audio Visual Center, West Lafayette, Ind. 47907

PSU   Pennsylvania State University, Audio Visual Aids Library, University Park, Pa. 16802

UC   University of California, Extension Media Center, 2223 Fulton St., Berkeley, Calif. 94720

Chapter 1   The Graphic Language

The Draftsman—10 min. sound movie—P.
The Language of Drawing—10 min. sound movie —McG.
According to Plan—9 min. sound movie—McG.

Chapter 2   Freehand Sketching

Freehand Drafting—12 min. silent movie—P.
Pictorial Sketching—11 min. sound movie—McG.

Chapter 3   Mechanical Drawing

Use of T-Squares and Triangles—20 min. silent movie—P.

Chapter 4   Lettering

Capital Letters—21 min. sound movie—P.
Lettering Instructional Materials—22 min. sound movie—PSU.
Lower-Case Letters—17 min. sound movie—P.
Technical Lettering—filmstrips (a series of five) —JH.

Chapter 5   Geometry of Technical Drawing

Applied Geometry—16 min. silent movie—P.

Chapter 6   Views of Objects

Behind the Shop Drawing—18 min. sound movie—P.

Multiview Drawing—24 min. silent movie—P.

Orthographic Projection—18 min. sound movie—McG.

Reading a Three-View Drawing—9 min. sound movie—DA.

Shape Description—31 min. sound movie—P.

Shape Description, Parts I and II—11 and 8 min., respectively, sound movies—McG.

Visualizing an Object—10 min. sound movie—DA.

Chapter 7   Techniques and Applications

Basic Reproduction Processes in the Graphic Arts—25 min. sound movie—PSU.

Drafting Tips—28 min. sound movie—PSU.

Chapter 8   Inking

Tracing with Ink—32 min. silent movie—P.

Chapter 9   Dimensioning

Principal Dimensions, References, Surfaces and Tolerances—12 min. sound movie—DA.

Selection of Dimensions—18 min. sound movie—McG.

Size Description—13 min. sound movie—McG.

Chapter 10   Shop Processes*

Cutting a Keyway on the End of a Finished Shaft—13 min. sound movie—PSU.

Drawings and the Shop—15 min. sound movie—McG.

The Drill Press—10 min. sound movie—DA.

Fixed Gages—16 min. sound movie—P.

The Lathe—15 min. sound movie—DA.

* For thread cutting, see list for Chapter 14.

Laying Out Small Castings—16 min. sound movie—PSU.

Machining a Cast Iron Rectangular Block—25 min. sound movie—PSU.

Machining a Tool Steel V-Block—21 min. sound movie—PSU.

The Milling Machine—15 min. sound movie—DA.

Precision Layout and Measuring—11 min. sound movie—PSU.

The Shaper—15 min. sound movie—DA.

Shop Drawing—22 min. sound movie—PSU.

Shop Procedures—17 min. sound movie—McG.

Shop Work—27 min. silent movie—P.

Chapter 11   Sectional Views

Sections and Conventions—15 min. sound movie—McG.

Sectional Views—15 min. silent movie—P.

Sectional Views and Projections, Finish Marks—15 min. sound movie—DA.

Chapter 12   Auxiliary Views

Auxiliary Views—17 min. silent movie—P.

Auxiliary Views, Parts I and II—11 and 10 min., respectively, sound movies—McG.

Auxiliary Views: Single Auxiliaries—23 min. sound movie—McG.

Auxiliary Views: Double Auxiliaries—13 min. sound movie—McG.

Chapter 14   Threads and Fasteners

Cutting an Internal Acme Thread—22 min. sound movie—PSU.

Cutting an External Acme Thread—12 min. sound movie—PSU.

Cutting an External National Fine Thread—11 min. sound movie—P.

Cutting Threads with Taps and Dies—20 min. sound movie—P.

Screw Threads—22 min. silent movie—P.

Chapter 15   Working Drawings

Concepts and Principles of Functional Drafting—20 min. sound movie—PSU.

Design for Manufacture—29 min. sound movie —PSU.

Chapter 16   Pictorial Drawings

Discovering Perspective—14 min. sound movie —UC.

Perspective Drawing—10 min. sound movie— UC.

Pictorial Drawing (Isometric)—21 min. silent movie—P.

Pictorial Sketching—11 min. sound movie—McG.

Chapter 17   Developments and Intersections

Development of Surfaces—22 min. silent movie —P.

Finding the Line of Intersection of Two Solids— 22 min. sound movie—PSU.

Intersection of Surfaces—9 min. silent movie—P.

Oblique Cones and Transition Developments— 11 min. sound movie—McG.

Simple Developments—11 min. sound movie— McG.

Chapter 19   Cams and Gears

Cutting a Short Rack—18 min. sound movie— DA.

Cutting Teeth on a Worm Gear—18 min. sound movie—DA.

Principles of Gearing: An Introduction—19 min. sound movie—DA.

Chapter 20   Structural Drawing

Structural Drawing—20 min. silent movie—P.

Reinforced Concrete Construction—30 min. sound movie—PSU.

Chapter 21   Map Drafting

Introduction to Map Projection—20 min. sound movie—PSU.

Chapter 22   Electrical Drafting

Elements of Electrical Circuits—11 min. sound movie—PSU.

Ohm's Law—19 min. sound movie—PSU.

Series and Parallel Circuits—11 min. sound movie —PSU.

Chapter 23   Architectural Drafting

Introduction to Architecture—9 min. sound movie—PSU.

TABLES

1. USA STANDARD UNIFIED AND AMERICAN THREAD SERIES[1]
(See pages 250, 251)

Nominal Diameter	Coarse[2] NC UNC		Fine[2] NF UNF		Extra Fine[3] NEF UNEF	
	Threads per In.	Tap Drill[4]	Threads per In.	Tap Drill[4]	Threads per In.	Tap Drill[4]
0 (.060)			80	$\frac{3}{64}$		
1 (.073)	64	No. 53	72	No. 53		
2 (.086)	56	No. 50	64	No. 50		
3 (.099)	48	No. 47	56	No. 45		
4 (.112)	40	No. 43	48	No. 42		
5 (.125)	40	No. 38	44	No. 37		
6 (.138)	32	No. 36	40	No. 33		
8 (.164)	32	No. 29	36	No. 29		
10 (.190)	24	No. 25	32	No. 21		
12 (.216)	24	No. 16	28	No. 14	32	No. 13
$\frac{1}{4}$	20	No. 7	28	No. 3	32	$\frac{7}{32}$
$\frac{5}{16}$	18	F	24	I	32	$\frac{9}{32}$
$\frac{3}{8}$	16	$\frac{5}{16}$	24	Q	32	$\frac{11}{32}$
$\frac{7}{16}$	14	U	20	$\frac{25}{64}$	28	$\frac{13}{32}$
$\frac{1}{2}$	13	$\frac{27}{64}$	20	$\frac{29}{64}$	28	$\frac{15}{32}$
$\frac{9}{16}$	12	$\frac{31}{64}$	18	$\frac{33}{64}$	24	$\frac{33}{64}$
$\frac{5}{8}$	11	$\frac{17}{32}$	18	$\frac{37}{64}$	24	$\frac{37}{64}$
$\frac{11}{16}$					24	$\frac{41}{64}$
$\frac{3}{4}$	10	$\frac{21}{32}$	16	$\frac{11}{16}$	20	$\frac{45}{64}$
$\frac{13}{16}$					20	$\frac{49}{64}$
$\frac{7}{8}$	9	$\frac{49}{64}$	14	$\frac{13}{16}$	20	$\frac{53}{64}$
$\frac{15}{16}$					20	$\frac{57}{64}$

NOTE: USA Standards referred to in the Appendix may be obtained from the United States of America Standards Institute, 10 East 40th Street, New York, N.Y. 10016.

[1] USAS Bl.1—1960. For 8-, 12-, and 16-pitch thread series, see this standard.

1. USA STANDARD UNIFIED AND AMERICAN THREAD SERIES (Continued)
(See pages 250, 251)

Nominal Diameter	Coarse[2] NC UNC		Fine[2] NF UNF		Extra Fine[3] NEF UNEF	
	Threads per In.	Tap Drill[4]	Threads per In.	Tap Drill[4]	Threads per In.	Tap Drill[4]
1	8	7/8	12	59/64	20	61/64
1 1/16					18	1
1 1/8	7	63/64	12	1 3/64	18	1 5/64
1 3/16					18	1 9/64
1 1/4	7	1 7/64	12	1 11/64	18	1 3/16
1 5/16					18	1 17/64
1 3/8	6	1 7/32	12	1 19/64	18	1 5/16
1 7/16					18	1 3/8
1 1/2	6	1 11/32	12	1 27/64	18	1 7/16
1 9/16					18	1 1/2
1 5/8					18	1 9/16
1 11/16					18	1 5/8
1 3/4	5	1 9/16			16	1 11/16
2	4 1/2	1 25/32			16	1 15/16
2 1/4	4 1/2	2 1/32				
2 1/2	4	2 1/4				
2 3/4	4	2 1/2				
3	4	2 3/4				
3 1/4	4	3				
3 1/2	4	3 1/4				
3 3/4	4	3 1/2				
4	4	3 3/4				

[2] Classes 1A, 2A, 3A, 1B, 2B, 3B, 2, and 3.
[3] Classes 2A, 2B, 2, and 3.
[4] For approximately 75% full depth of thread. For decimal sizes of numbered and lettered drills, see page 456.

2. DECIMAL EQUIVALENTS OF TWIST DRILLS[1]

(See page 179)

All dimensions are given in inches. See decimal equivalents table on page 473.

Size	Drill Diameter	Size	Drill Diameter	Size	Drill Diameter	Size	Drill Diameter	Size	Drill Diameter
1	.2280	17	.1730	33	.1130	49	.0730	65	.0350
2	.2210	18	.1695	34	.1110	50	.0700	66	.0330
3	.2130	19	.1660	35	.1100	51	.0670	67	.0320
4	.2090	20	.1610	36	.1065	52	.0635	68	.0310
5	.2055	21	.1590	37	.1040	53	.0595	69	.0292
6	.2040	22	.1570	38	.1015	54	.0550	70	.0280
7	.2010	23	.1540	39	.0995	55	.0520	71	.0260
8	.1990	24	.1520	40	.0980	56	.0465	72	.0250
9	.1960	25	.1495	41	.0960	57	.0430	73	.0240
10	.1935	26	.1470	42	.0935	58	.0420	74	.0225
11	.1910	27	.1440	43	.0890	59	.0410	75	.0210
12	.1890	28	.1405	44	.0860	60	.0400	76	.0200
13	.1850	29	.1360	45	.0820	61	.0390	77	.0180
14	.1820	30	.1285	46	.0810	62	.0380	78	.0160
15	.1800	31	.1200	47	.0785	63	.0370	79	.0145
16	.1770	32	.1160	48	.0760	64	.0360	80	.0135

LETTER SIZES

A	.234	G	.261	L	.290	Q	.332	V	.377
B	.238	H	.266	M	.295	R	.339	W	.386
C	.242	I	.272	N	.302	S	.348	X	.397
D	.246	J	.277	O	.316	T	.358	Y	.404
E	.250	K	.281	P	.323	U	.368	Z	.413
F	.257								

[1] Drills designated in fractions are also frequently used, and are available in diameters $1/64''$ to $1\frac{3}{4}''$ in $1/64''$ increments, $1\frac{3}{4}''$ to $2\frac{1}{4}''$ in $1/32''$ increments, and $2\frac{1}{4}''$ to $3\frac{1}{2}''$ in $1/16''$ increments. Drills larger than $3\frac{1}{2}''$ are seldom used, and are regarded as special drills.

3. SQUARE AND ACME THREADS[1]

(See pages 241, 245, 247)

Size	Threads per Inch	Size	Threads per Inch	Size	Threads per Inch	Size	Threads per Inch
$3/8$	12	$7/8$	5	2	$2\frac{1}{2}$	$3\frac{1}{2}$	$1\frac{1}{3}$
$7/16$	10	1	5	$2\frac{1}{4}$	2	$3\frac{3}{4}$	$1\frac{1}{3}$
$1/2$	10	$1\frac{1}{8}$	4	$2\frac{1}{2}$	2	4	$1\frac{1}{3}$
$9/16$	8	$1\frac{1}{4}$	4	$2\frac{3}{4}$	2	$4\frac{1}{4}$	$1\frac{1}{3}$
$5/8$	8	$1\frac{1}{2}$	3	3	$1\frac{1}{2}$	$4\frac{1}{2}$	1
$3/4$	6	$1\frac{3}{4}$	$2\frac{1}{2}$	$3\frac{1}{4}$	$1\frac{1}{2}$	over $4\frac{1}{2}$	1

[1]See Table 4 (page 457) for USA Standard General Purpose Acme Threads.

4. USA STANDARD GENERAL PURPOSE ACME THREADS[1]
(See pages 241, 247)

Size	Threads per Inch	Size	Threads per Inch	Size	Threads per Inch	Size	Threads per Inch
¼	16	¾	6	1½	4	3	2
5/16	14	⅞	6	1¾	4	3½	2
⅜	12	1	5	2	4	4	2
7/16	12	1⅛	5	2¼	3	4½	2
½	10	1¼	5	2½	3	5	2
⅝	8	1⅜	4	2¾	3	. . .	. .

[1] USAS B1.5—1952.

5. SHAFT CENTER SIZES
(See Fig. 14-4)

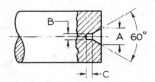

Shaft Diameter				Shaft Diameter			
D	A	B	C	D	A	B	C
3/16 to 7/32	5/64	3/64	1/16	1⅛ to 1 15/32	5/16	5/32	5/32
¼ to 11/32	3/32	3/64	1/16	1½ to 1 31/32	⅜	5/32	5/32
⅜ to 17/32	⅛	1/16	5/64	2 to 2 31/32	7/16	7/32	3/16
9/16 to 25/32	3/16	5/64	3/32	3 to 3 31/32	½	7/32	7/32
13/16 to 1 3/32	¼	3/32	3/32	4 and over	9/16	7/32	7/32

6. USA STANDARD TAPER PIPE THREADS[1]
(See page 252)

Nominal Pipe Size[2]	Outside Diameter	Threads per Inch	Tap Drill[3]	Nominal Pipe Size[2]	Outside Diameter	Threads per Inch	Tap Drill[3]
⅛	0.405	27	11/32	2	2.375	11½	2 7/32
¼	0.540	18	7/16	2½	2.875	8	2 ⅝
⅜	0.675	18	37/64	3	3.500	8	3 ¼
½	0.840	14	23/32	3½	4.000	8	3 ¾
¾	1.050	14	59/64	4	4.500	8	
1	1.315	11½	1 5/32	5	5.563	8	
1¼	1.660	11½	1 ½	6	6.625	8	
1½	1.900	11½	1 47/64	8	8.625	8	

[1] From USAS B2.1—1960.   [2] Sizes 1/16 and 10 to 24 are omitted.   [3] Suggested sizes; not standard.

7. USA STANDARD SQUARE AND HEXAGON BOLTS[1] AND NUTS[2] AND HEXAGON-HEAD CAP SCREWS[3]

Boldface type indicates product features unified dimensionally with British and Canadian standards.
All dimensions are in inches.
For thread series, minimum thread lengths, and bolt lengths, see Sec. 14.20.

Nominal Size D Body Dia. of Bolt		REGULAR BOLTS					HEAVY BOLTS		
		Width Across Flats W		Height H			Width Across Flats W	Height H	
		Sq.	Hex.	Sq. (Unfin.)	Hex. (Unfin.)	Hex. Cap Scr.[3] (Fin.)		Hex. (Unfin.)	Hex. Screw (Fin.)
¼	0.2500	⅜	7/16	11/64	11/64	5/32	....	....	....
5/16	0.3125	½	½	13/64	7/32	13/64	....	....	....
⅜	0.3750	9/16	9/16	¼	¼	15/64	....	....	....
7/16	0.4375	⅝	⅝	19/64	19/64	9/32	....	....	....
½	0.5000	¾	¾	21/64	11/32	5/16	⅞	11/32	5/16
9/16	0.5625	....	13/16	....	....	23/64	....	....	....
⅝	0.6250	15/16	15/16	27/64	27/64	25/64	1 1/16	27/64	25/64
¾	0.7500	1⅛	1⅛	½	½	15/32	1¼	½	15/32
⅞	0.8750	1 5/16	1 5/16	19/32	37/64	35/64	1 7/16	37/64	35/64
1	1.0000	1½	1½	21/32	43/64	39/64	1⅝	43/64	39/64
1⅛	1.1250	1 11/16	1 11/16	¾	¾	11/16	1 13/16	¾	11/16
1¼	1.2500	1⅞	1⅞	27/32	27/32	25/32	2	27/32	25/32
1⅜	1.3750	2 1/16	2 1/16	29/32	29/32	27/32	2 3/16	29/32	27/32
1½	1.5000	2¼	2¼	1	1	15/16	2⅜	1	15/16
1¾	1.7500	....	2⅝	....	1 5/32	1 3/32	2¾	1 5/32	1 3/32
2	2.0000	....	3	....	1 11/32	1 7/32	3⅛	1 11/32	1 7/32
2¼	2.2500	....	3⅜	....	1½	1⅜	3½	1½	1⅜
2½	2.5000	....	3¾	....	1 21/32	1 17/32	3⅞	1 21/32	1 17/32
2¾	2.7500	....	4⅛	....	1 13/16	1 11/16	4¼	1 13/16	1 11/16
3	3.0000	....	4½	....	2	1⅞	4⅝	2	1⅞
3¼	3.2500	....	4⅞	....	2 3/16	....	....	....	....
3½	3.5000	....	5¼	....	2 5/16	....	....	....	....
3¾	3.7500	....	5⅝	....	2½	....	....	....	....
4	4.0000	....	6	....	2 11/16	....	....	....	....

[1] USAS B18.2.1—1965.
[2] USAS B18.2.2—1965.
[3] Hexagon cap screws and finished hexagon bolts are combined as a single product.

7. USA STANDARD SQUARE AND HEXAGON HEAD BOLTS AND NUTS AND HEXAGON–HEAD CAP SCREWS (continued)

See USAS B18.2.2—1965 for jam nuts, slotted nuts, thick nuts, thick slotted nuts, and castle nuts. For methods of drawing bolts and nuts and hexagon-head cap screws, see Figs. 14–26 to 14–29 and 14-31.

NOMINAL SIZE D BODY DIA. OF BOLT		REGULAR NUTS					HEAVY NUTS			
		WIDTH ACROSS FLATS W		THICKNESS T			WIDTH ACROSS FLATS W.	THICKNESS T		
		SQ.	HEX.	SQ. (UNFIN.)	HEX. FLAT (UNFIN.)	HEX. (FIN.)		SQ. (UNFIN.)	HEX. FLAT (UNFIN.)	HEX. (FIN.)
1/4	0.2500	7/16	7/16	7/32	7/32	7/32	1/2	1/4	15/64	15/64
5/16	0.3125	9/16	1/2	17/64	17/64	17/64	9/16	5/16	19/64	19/64
3/8	0.3750	5/8	9/16	21/64	21/64	21/64	11/16	3/8	23/64	23/64
7/16	0.4375	3/4	11/16	3/8	3/8	3/8	3/4	7/16	27/64	27/64
1/2	0.5000	13/16	3/4	7/16	7/16	7/16	7/8*	1/2	31/64	31/64
9/16	0.5625	....	7/8	....	31/64	31/64	15/16	....	35/64	35/64
5/8	0.6250	1	15/16	35/64	35/64	35/64	1 1/16*	5/8	39/64	39/64
3/4	0.7500	1 1/8	1 1/8	21/32	41/64	41/64	1 1/4*	3/4	47/64	47/64
7/8	0.8750	1 5/16	1 5/16	49/64	3/4	3/4	1 7/16*	7/8	55/64	55/64
1	1.0000	1 1/2	1 1/2	7/8	55/64	55/64	1 5/8*	1	63/64	63/64
1 1/8	1.1250	1 11/16	1 11/16	1	1	31/32	1 13/16*	1 1/8	1 1/8	1 7/64
1 1/4	1.2500	1 7/8	1 7/8	1 3/32	1 3/32	1 1/16	2*	1 1/4	1 1/4	1 7/32
1 3/8	1.3750	2 1/16	2 1/16	1 13/64	1 13/64	1 11/64	2 3/16*	1 3/8	1 3/8	1 11/32
1 1/2	1.5000	2 1/4	2 1/4	1 5/16	1 5/16	1 9/32	2 3/8*	1 1/2	1 1/2	1 15/32
1 5/8	1.6250	....	....	....	....	....	2 9/16	....	....	1 19/32
1 3/4	1.7500	....	....	....	....	....	2 3/4	....	1 3/4	1 23/32
1 7/8	1.8750	....	....	....	....	....	2 15/16	....	....	1 27/32
2	2.0000	....	....	....	....	....	3 1/8	....	2	1 31/32
2 1/4	2.2500	....	....	....	....	....	3 1/2	....	2 1/4	2 13/64
2 1/2	2.5000	....	....	....	....	....	3 7/8	....	2 1/2	2 29/64
2 3/4	2.7500	....	....	....	....	....	4 1/4	....	2 3/4	2 45/64
3	3.0000	....	....	....	....	....	4 5/8	....	3	2 61/64
3 1/4	3.2500	....	....	....	....	....	5	....	3 1/4	3 3/16
3 1/2	3.5000	....	....	....	....	....	5 5/8	....	3 1/2	3 7/16
3 3/4	3.7500	....	....	....	....	....	5 3/4	....	3 3/4	3 11/16
4	4.0000	....	....	....	....	....	6 1/8	....	4	3 15/16

* Product feature not unified for heavy square nut.

8. USA STANDARD SLOTTED[1] AND SOCKET HEAD[2] CAP SCREWS[3]
(See page 257)

FLAT HEAD ROUND HEAD FILLISTER HEAD SOCKET HEAD

Lengths of screws, L: Consecutive lengths vary $\frac{1}{8}''$ for lengths $\frac{1}{4}''$ to $1''$, $\frac{1}{4}''$ for lengths $1''$ to $4''$, and $\frac{1}{2}''$ for lengths $4''$ to $6''$.

Threads: Coarse or Fine Thread Series. Class 3A fit.

NOMINAL SIZE D	FLAT HEAD[1]	ROUND HEAD[1]		FILLISTER HEAD[1]		SOCKET HEAD[2]		
	A	B	C	E	F	G	J	S
0 (.060)	....	....	....	....	....	.096	.05	.054
1 (.073)	....	....	....	....	....	.118	$\frac{1}{16}$	.066
2 (.086)	....	....	....	....	....	.140	$\frac{5}{64}$	.077
3 (.099)	....	....	....	....	....	.161	$\frac{5}{64}$	.089
4 (.112)	....	....	....	....	....	.183	$\frac{3}{32}$	.101
5 (.125)	....	....	....	....	....	.205	$\frac{3}{32}$	.112
6 (.138)	....	....	....	....	....	.226	$\frac{7}{64}$	.124
8 (.164)	....	....	....	....	....	.270	$\frac{9}{64}$	.148
10 (.190)	....	....	....	....	....	$\frac{5}{16}$	$\frac{5}{32}$	.171
$\frac{1}{4}$	$\frac{1}{2}$	$\frac{7}{16}$	.191	$\frac{3}{8}$	$\frac{11}{64}$	$\frac{3}{8}$	$\frac{3}{16}$	.225
$\frac{5}{16}$	$\frac{5}{8}$	$\frac{9}{16}$	.245	$\frac{7}{16}$	$\frac{13}{64}$	$\frac{15}{32}$	$\frac{1}{4}$	.281
$\frac{3}{8}$	$\frac{3}{4}$	$\frac{5}{8}$	.273	$\frac{9}{16}$	$\frac{1}{4}$	$\frac{9}{16}$	$\frac{5}{16}$	.337
$\frac{7}{16}$	$\frac{13}{16}$	$\frac{3}{4}$	$\frac{21}{64}$	$\frac{5}{8}$	$\frac{19}{64}$	$\frac{21}{32}$	$\frac{3}{8}$	.394
$\frac{1}{2}$	$\frac{7}{8}$	$\frac{13}{16}$	.355	$\frac{3}{4}$	$\frac{21}{64}$	$\frac{3}{4}$	$\frac{3}{8}$	.450
$\frac{9}{16}$	1	$\frac{15}{16}$	.409	$\frac{13}{16}$	$\frac{3}{8}$	....	....	....
$\frac{5}{8}$	$1\frac{1}{8}$	1	$\frac{7}{16}$	$\frac{7}{8}$	$\frac{27}{64}$	$\frac{15}{16}$	$\frac{1}{2}$	.562
$\frac{3}{4}$	$1\frac{3}{8}$	$1\frac{1}{4}$	$\frac{35}{64}$	1	$\frac{1}{2}$	$1\frac{1}{8}$	$\frac{5}{8}$	.675
$\frac{7}{8}$	$1\frac{5}{8}$	....	....	$1\frac{1}{8}$	$\frac{19}{32}$	$1\frac{5}{16}$	$\frac{3}{4}$	.787
1	$1\frac{7}{8}$	....	....	$1\frac{5}{16}$	$\frac{21}{32}$	$1\frac{1}{2}$	$\frac{3}{4}$	.900
$1\frac{1}{8}$	$2\frac{1}{16}$	....	....	....	....	$1\frac{11}{16}$	$\frac{7}{8}$	1.012
$1\frac{1}{4}$	$2\frac{5}{16}$	....	....	....	....	$1\frac{7}{8}$	$\frac{7}{8}$	1.125
$1\frac{3}{8}$	$2\frac{9}{16}$	....	....	....	....	$2\frac{1}{16}$	1	1.237
$1\frac{1}{2}$	$2\frac{13}{16}$	....	....	....	....	$2\frac{1}{4}$	1	1.350

[1] USAS B18.6.2—1956.
[2] USAS B18.3—1961.
[3] For hexagon-head screws, see Sec. 14.24 and Table 7 (page 458).

9. USA STANDARD MACHINE SCREWS[1]
(See pages 258, 259)

Threads: Coarse or Fine Thread Series, Class 2 fit. *Thread length:* On screws 2″ long and shorter, the threads extend to within two threads of the head; on longer screws the thread length is $1\frac{3}{4}''$.

Heads may be recessed instead of slotted.

For slot proportions, see Fig. 14-33. Points have plain sheared ends and are not chamfered.

All dimensions are in inches. See decimal equivalents table on page 473.

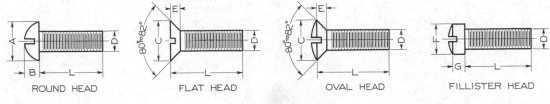

ROUND HEAD   FLAT HEAD   OVAL HEAD   FILLISTER HEAD

DIAMETER	ROUND HEAD		FLAT AND OVAL HEADS		FILLISTER HEAD		DIAMETER	ROUND HEAD		FLAT AND OVAL HEADS		FILLISTER HEAD	
D	A	B	C	E	F	G	D	A	B	C	E	F	G
0 (.060)	.113	.053	.119	.035	.096	.045	12 (.216)	.408	.153	.438	.132	.357	.148
1 (.073)	.138	.061	.146	.043	.118	.053	$\frac{1}{4}$	.472	.175	.507	.153	.414	.170
2 (.086)	.162	.069	.172	.051	.140	.062	$\frac{5}{16}$	.590	.216	.635	.191	.518	.211
3 (.099)	.187	.078	.199	.059	.161	.070	$\frac{3}{8}$	.708	.256	.762	.230	.622	.253
4 (.112)	.211	.086	.225	.067	.183	.079	$\frac{7}{16}$	.750	.328	.812	.223	.625	.265
5 (.125)	.236	.095	.252	.075	.205	.088	$\frac{1}{2}$	.813	.355	.875	.223	.750	.297
6 (.138)	.260	.103	.279	.083	.226	.096	$\frac{9}{16}$	.938	.410	1.000	.260	.812	.336
8 (.164)	.309	.120	.332	.100	.270	.113	$\frac{5}{8}$	1.000	.438	1.125	.298	.875	.375
10 (.190)	.359	.137	.385	.116	.313	.130	$\frac{3}{4}$	1.250	.547	1.375	.372	1.000	.441

[1] USAS B18.6.3—1962, which see for other style heads—Truss, Binding, Pan, Hexagon, and 100° Flat Heads.

10. SQUARE AND FLAT KEYS, PLAIN TAPER KEYS, AND GIB HEAD KEYS
(See pages 260, 261)

Plain taper square and flat keys have the same dimensions as the plain parallel stock keys, with the addition of the taper on top. Gib head taper square and flat keys have the same dimensions as the plain taper keys, with the addition of the gib head.

Stock lengths for plain taper and gib head taper keys: The minimum stock length equals 4W, and the maximum equals 16W. The increments of increase of length equal 2W.

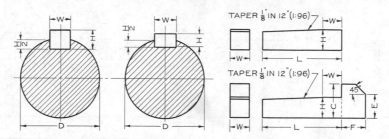

Shaft Diameters	Square Stock Key	Flat Stock Key	Gib Head Taper Stock Key					
			Square			Flat		
			Height	Length	Height to Chamfer	Height	Length	Height to Chamfer
D	W = H	W × H	C	F	E	C	F	E
$\frac{1}{2}$ to $\frac{9}{16}$	$\frac{1}{8}$	$\frac{1}{8} \times \frac{3}{32}$	$\frac{1}{4}$	$\frac{7}{32}$	$\frac{5}{32}$	$\frac{3}{16}$	$\frac{1}{8}$	$\frac{1}{8}$
$\frac{5}{8}$ to $\frac{7}{8}$	$\frac{3}{16}$	$\frac{3}{16} \times \frac{1}{8}$	$\frac{5}{16}$	$\frac{9}{32}$	$\frac{7}{32}$	$\frac{1}{4}$	$\frac{3}{16}$	$\frac{5}{32}$
$\frac{15}{16}$ to $1\frac{1}{4}$	$\frac{1}{4}$	$\frac{1}{4} \times \frac{3}{16}$	$\frac{7}{16}$	$\frac{11}{32}$	$\frac{11}{32}$	$\frac{5}{16}$	$\frac{1}{4}$	$\frac{3}{16}$
$1\frac{5}{16}$ to $1\frac{3}{8}$	$\frac{5}{16}$	$\frac{5}{16} \times \frac{1}{4}$	$\frac{9}{16}$	$\frac{13}{32}$	$\frac{13}{32}$	$\frac{3}{8}$	$\frac{5}{16}$	$\frac{1}{4}$
$1\frac{7}{16}$ to $1\frac{3}{4}$	$\frac{3}{8}$	$\frac{3}{8} \times \frac{1}{4}$	$\frac{11}{16}$	$\frac{15}{32}$	$\frac{15}{32}$	$\frac{7}{16}$	$\frac{3}{8}$	$\frac{5}{16}$
$1\frac{13}{16}$ to $2\frac{1}{4}$	$\frac{1}{2}$	$\frac{1}{2} \times \frac{3}{8}$	$\frac{7}{8}$	$\frac{19}{32}$	$\frac{5}{8}$	$\frac{5}{8}$	$\frac{1}{2}$	$\frac{7}{16}$
$2\frac{5}{16}$ to $2\frac{3}{4}$	$\frac{5}{8}$	$\frac{5}{8} \times \frac{7}{16}$	$1\frac{1}{16}$	$\frac{23}{32}$	$\frac{3}{4}$	$\frac{3}{4}$	$\frac{5}{8}$	$\frac{1}{2}$
$2\frac{7}{8}$ to $3\frac{1}{4}$	$\frac{3}{4}$	$\frac{3}{4} \times \frac{1}{2}$	$1\frac{1}{4}$	$\frac{7}{8}$	$\frac{7}{8}$	$\frac{7}{8}$	$\frac{3}{4}$	$\frac{5}{8}$
$3\frac{3}{8}$ to $3\frac{3}{4}$	$\frac{7}{8}$	$\frac{7}{8} \times \frac{5}{8}$	$1\frac{1}{2}$	1	1	$1\frac{1}{16}$	$\frac{7}{8}$	$\frac{3}{4}$
$3\frac{7}{8}$ to $4\frac{1}{2}$	1	$1 \times \frac{3}{4}$	$1\frac{3}{4}$	$1\frac{3}{16}$	$1\frac{3}{16}$	$1\frac{1}{4}$	1	$1\frac{3}{16}$
$4\frac{3}{4}$ to $5\frac{1}{2}$	$1\frac{1}{4}$	$1\frac{1}{4} \times \frac{7}{8}$	2	$1\frac{7}{16}$	$1\frac{7}{16}$	$1\frac{1}{2}$	$1\frac{1}{4}$	1
$5\frac{3}{4}$ to 6	$1\frac{1}{2}$	$1\frac{1}{2} \times 1$	$2\frac{1}{2}$	$1\frac{3}{4}$	$1\frac{3}{4}$	$1\frac{3}{4}$	$1\frac{1}{2}$	$1\frac{1}{4}$

11. USA STANDARD WOODRUFF KEYS[1]

(See page 260)

All dimensions are in inches. See decimal equivalents table on page 473.

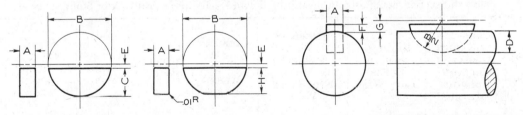

Key No.[2]	Nominal Sizes				Maximum Sizes			Key No.[2]	Nominal Sizes				Maximum Sizes		
	A × B	E	F	G	H	D	C		A × B	E	F	G	H	D	C
204	1/16 × 1/2	3/64	1/32	5/64	.194	.1718	.203	808	1/4 × 1	1/16	1/8	3/16	.428	.3130	.438
304	3/32 × 1/2	3/64	3/64	3/32	.194	.1561	.203	809	1/4 × 1 1/8	5/64	1/8	13/64	.475	.3590	.484
305	3/32 × 5/8	1/16	3/64	7/64	.240	.2031	.250	810	1/4 × 1 1/4	5/64	1/8	13/64	.537	.4220	.547
404	1/8 × 1/2	3/64	1/16	7/64	.194	.1405	.203	811	1/4 × 1 3/8	3/32	1/8	7/32	.584	.4690	.594
405	1/8 × 5/8	1/16	1/16	1/8	.240	.1875	.250	812	1/4 × 1 1/2	7/64	1/8	15/64	.631	.5160	.641
406	1/8 × 3/4	1/16	1/16	1/8	.303	.2505	.313	1008	5/16 × 1	1/16	5/32	7/32	.428	.2818	.438
505	5/32 × 5/8	1/16	5/64	9/64	.240	.1719	.250	1009	5/16 × 1 1/8	5/64	5/32	15/64	.475	.3278	.484
506	5/32 × 3/4	1/16	5/64	9/64	.303	.2349	.313	1010	5/16 × 1 1/4	5/64	5/32	15/64	.537	.3908	.547
507	5/32 × 7/8	1/16	5/64	9/64	.365	.2969	.375	1011	5/16 × 1 3/8	3/32	5/32	8/32	.584	.4378	.594
606	3/16 × 3/4	1/16	3/32	5/32	.303	.2193	.313	1012	5/16 × 1 1/2	7/64	5/32	17/64	.631	.4848	.641
607	3/16 × 7/8	1/16	3/32	5/32	.365	.2813	.375	1210	3/8 × 1 1/4	5/64	3/16	17/64	.537	.3595	.547
608	3/16 × 1	1/16	3/32	5/32	.428	.3443	.438	1211	3/8 × 1 3/8	3/32	3/16	9/32	.584	.4065	.594
609	3/16 × 1 1/8	5/64	3/32	11/64	.475	.3903	.484	1212	3/8 × 1 1/2	7/64	3/16	19/64	.631	.4535	.641
807	1/4 × 7/8	1/16	1/8	3/16	.365	.2500	.375	....	....	...	...	...	....	....	....

[1] USAS B17f—1930 (Reaffirmed 1955).

[2] Key numbers indicate nominal key dimensions. The last two digits give the nominal diameter B in eighths of an inch, and the digits before the last two give the nominal width A in thirty-seconds of an inch.

12. WOODRUFF KEY SIZES FOR DIFFERENT SHAFT DIAMETERS [1]

Shaft Diameter In.	Key Numbers	Shaft Diameter In.	Key Numbers
5/16 to 3/8	204	1 to 1 3/16	606, 607, 608, 609
7/16 to 1/2	304, 305	1 1/4 to 1 7/16	807, 808, 809
9/16 to 3/4	404, 405, 406	1 1/2 to 1 3/4	810, 811, 812
13/16 to 15/16	505, 506, 507	1 13/16 to 2 1/8	1011, 1012
		2 3/16 to 2 1/2	1211, 1212

[1] Suggested sizes; not standard.

13. STANDARDS FOR WIRE GAGES[1]

Dimensions of sizes in decimal parts of an inch. See decimal equivalents table on page 473.
The difference between the Stubs' Iron Wire Gage and the Stubs' Steel Wire Gage should be noted, the first being commonly known as the English Standard Wire, or Birmingham Gage, which designates the Stubs' soft wire sizes, and the second being used in measuring drawn steel wire or drill rods of Stubs' make.

No. of Wire	American or Brown & Sharpe for Non-Ferrous Metals	Birmingham, or Stubs' Iron Wire	American S. & W. Co.'s (Washburn & Moen) Std. Steel Wire	American S. & W. Co.'s Music Wire	Imperial Wire	Stubs' Steel Wire	Steel Manu-facturers' Sheet Gage	No. of Wire
7–0's	.651354	....	.4900	....	.500	....	.....	7–0's
6–0's	.580049	....	.4615	.004	.464	....	.....	6–0's
5–0's	.516549	.500	.4305	.005	.432	....	.....	5–0's
4–0's	.460	.454	.3938	.006	.400	....	.....	4–0's
000	.40964	.425	.3625	.007	.372	....	.....	000
00	.3648	.380	.3310	.008	.348	....	.....	00
0	.32486	.340	.3065	.009	.324	....	.....	0
1	.2893	.300	.2830	.010	.300	.227	.....	1
2	.25763	.284	.2625	.011	.276	.219	.....	2
3	.22942	.259	.2437	.012	.252	.212	.2391	3
4	.20431	.238	.2253	.013	.232	.207	.2242	4
5	.18194	.220	.2070	.014	.212	.204	.2092	5
6	.16202	.203	.1920	.016	.192	.201	.1943	6
7	.14428	.180	.1770	.018	.176	.199	.1793	7
8	.12849	.165	.1620	.020	.160	.197	.1644	8
9	.11443	.148	.1483	.022	.144	.194	.1495	9
10	.10189	.134	.1350	.024	.128	.191	.1345	10
11	.090742	.120	.1205	.026	.116	.188	.1196	11
12	.080808	.109	.1055	.029	.104	.185	.1046	12
13	.071961	.095	.0915	.031	.092	.182	.0897	13
14	.064084	.083	.0800	.033	.080	.180	.0747	14
15	.057068	.072	.0720	.035	.072	.178	.0673	15
16	.05082	.065	.0625	.037	.064	.175	.0598	16
17	.045257	.058	.0540	.039	.056	.172	.0538	17
18	.040303	.049	.0475	.041	.048	.168	.0478	18
19	.03589	.042	.0410	.043	.040	.164	.0418	19
20	.031961	.035	.0348	.045	.036	.161	.0359	20
21	.028462	.032	.0317	.047	.032	.157	.0329	21
22	.025347	.028	.0286	.049	.028	.155	.0299	22
23	.022571	.025	.0258	.051	.024	.153	.0269	23
24	.0201	.022	.0230	.055	.022	.151	.0239	24
25	.0179	.020	.0204	.059	.020	.148	.0209	25
26	.01594	.018	.0181	.063	.018	.146	.0179	26
27	.014195	.016	.0173	.067	.0164	.143	.0164	27
28	.012641	.014	.0162	.071	.0149	.139	.0149	28
29	.011257	.013	.0150	.075	.0136	.134	.0135	29
30	.010025	.012	.0140	.080	.0124	.127	.0120	30
31	.008928	.010	.0132	.085	.0116	.120	.0105	31
32	.00795	.009	.0128	.090	.0108	.115	.0097	32
33	.00708	.008	.0118	.095	.0100	.112	.0090	33
34	.006304	.007	.0104	....	.0092	.110	.0082	34
35	.005614	.005	.0095	....	.0084	.108	.0075	35
36	.005	.004	.0090	....	.0076	.106	.0067	36
37	.004453	....	.0085	....	.0068	.103	.0064	37
38	.003965	....	.0080	....	.0060	.101	.0060	38
39	.003531	....	.0075	....	.0052	.099	.....	39
40	.003144	....	.0070	....	.0048	.097	.....	40

[1] Courtesy Brown & Sharpe Mfg. Co.

14. USA STANDARD PLAIN WASHERS[1]

For parts lists, etc., give inside diameter, outside
diameter, and the thickness; for example,
0.344 × 0.688 × 0.065 TYPE A PLAIN WASHER.

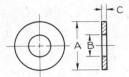

PREFERRED SIZES OF TYPE A PLAIN WASHERS[2]

NOMINAL WASHER SIZE[3]			INSIDE DIA.	OUTSIDE DIA	NOMINAL THICKNESS
			A	B	C
....	....		0.078	0.188	0.020
....	....		0.094	0.250	0.020
....	....		0.125	0.312	0.032
No. 6	0.138		0.156	0.375	0.049
No. 8	0.164		0.188	0.438	0.049
No. 10	0.190		0.219	0.500	0.049
3/16	0.188		0.250	0.562	0.049
No. 12	0.216		0.250	0.562	0.065
1/4	0.250	N	0.281	0.625	0.065
1/4	0.250	W	0.312	0.734	0.065
5/16	0.312	N	0.344	0.688	0.065
5/16	0.312	W	0.375	0.875	0.083
3/8	0.375	N	0.406	0.812	0.065
3/8	0.375	W	0.438	1.000	0.083
7/16	0.438	N	0.469	0.922	0.065
7/16	0.438	W	0.500	1.250	0.083
1/2	0.500	N	0.531	1.062	0.095
1/2	0.500	W	0.562	1.375	0.109
9/16	0.562	N	0.594	1.156	0.095
9/16	0.562	W	0.625	1.469	0.109
5/8	0.625	N	0.656	1.312	0.095
5/8	0.625	W	0.688	1.750	0.134
3/4	0.750	N	0.812	1.469	0.134
3/4	0.750	W	0.812	2.000	0.148
7/8	0.875	N	0.938	1.750	0.134
7/8	0.875	W	0.938	2.250	0.165
1	1.000	N	1.062	2.000	0.134
1	1.000	W	1.062	2.500	0.165
1 1/8	1.125	N	1.250	2.250	0.134
1 1/8	1.125	W	1.250	2.750	0.165
1 1/4	1.250	N	1.375	2.500	0.165
1 1/4	1.250	W	1.375	3.000	0.165
1 3/8	1.375	N	1.500	2.750	0.165
1 3/8	1.375	W	1.500	3.250	0.180
1 1/2	1.500	N	1.625	3.000	0.165
1 1/2	1.500	W	1.625	3.500	0.180
1 5/8	1.625		1.750	3.750	0.180
1 3/4	1.750		1.875	4.000	0.180
1 7/8	1.875		2.000	4.250	0.180
2	2.000		2.125	4.500	0.180
2 1/4	2.250		2.375	4.750	0.220
2 1/2	2.500		2.625	5.000	0.238
2 3/4	2.750		2.875	5.250	0.259
3	3.000		3.125	5.500	0.284

[1] From USAS B27.2—1965. For complete listings, see the Standard.

[2] Preferred sizes are for the most part from series previously designated "Standard Plate" and "SAE." Where common sizes existed in the two series, the SAE size is designated "N" (narrow) and the Standard Plate "W" (wide).

[3] Nominal washer sizes are intended for use with comparable nominal screw or bolt sizes.

15. USA STANDARD LOCK WASHERS[1]

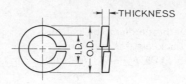

For parts lists, etc., give nominal size and series; for example, ¼″ REGULAR LOCK WASHER

PREFERRED SERIES

Nominal Washer Size[2]		Inside Dia. Min.	Regular		Extra Duty		Hi-Collar	
			Outside Dia. Max.	Thickness Min.	Outside Dia. Max.	Thickness Min.	Outside Dia. Max.	Thickness Min.
No. 2	0.086	0.088	0.172	0.020	0.208	0.027	. . . .	. . . .
No. 3	0.099	0.101	0.195	0.025	0.239	0.034	. . . .	. . . .
No. 4	0.112	0.115	0.209	0.025	0.253	0.034	0.173	0.022
No. 5	0.125	0.128	0.236	0.031	0.300	0.045	0.202	0.030
No. 6	0.138	0.141	0.250	0.031	0.314	0.045	0.216	0.030
No. 8	0.164	0.168	0.293	0.040	0.375	0.057	0.267	0.047
No. 10	0.190	0.194	0.334	0.047	0.434	0.068	0.294	0.047
No. 12	0.216	0.221	0.377	0.056	0.497	0.080	. . . .	. . . .
¼	0.250	0.255	0.489	0.062	0.535	0.084	0.365	0.078
⁵⁄₁₆	0.312	0.318	0.586	0.078	0.622	0.108	0.460	0.093
⅜	0.375	0.382	0.683	0.094	0.741	0.123	0.553	0.125
⁷⁄₁₆	0.438	0.446	0.779	0.109	0.839	0.143	0.647	0.140
½	0.500	0.509	0.873	0.125	0.939	0.162	0.737	0.172
⁹⁄₁₆	0.562	0.572	0.971	0.141	1.041	0.182	. . . .	. . . .
⅝	0.625	0.636	1.079	0.156	1.157	0.202	0.923	0.203
¹¹⁄₁₆	0.688	0.700	1.176	0.172	1.258	0.221	. . . .	. . . .
¾	0.750	0.763	1.271	0.188	1.361	0.241	1.111	0.218
¹³⁄₁₆	0.812	0.826	1.367	0.203	1.463	0.261	. . . .	. . . .
⅞	0.875	0.890	1.464	0.219	1.576	0.285	1.296	0.234
¹⁵⁄₁₆	0.938	0.954	1.560	0.234	1.688	0.308	. . . .	. . . .
1	1.000	1.017	1.661	0.250	1.799	0.330	1.483	0.250
1¹⁄₁₆	1.062	1.080	1.756	0.266	1.910	0.352	. . . .	. . . .
1⅛	1.125	1.144	1.853	0.281	2.019	0.375	1.669	0.313
1³⁄₁₆	1.188	1.208	1.950	0.297	2.124	0.396	. . . .	. . . .
1¼	1.250	1.271	2.045	0.312	2.231	0.417	1.799	0.313
1⁵⁄₁₆	1.312	1.334	2.141	0.328	2.335	0.438	. . . .	. . . .
1⅜	1.375	1.398	2.239	0.344	2.439	0.458	2.041	0.375
1⁷⁄₁₆	1.438	1.462	2.334	0.359	2.540	0.478	. . . .	. . . .
1½	1.500	1.525	2.430	0.375	2.638	0.496	2.170	0.375

[1] From USAS B27.1—1965. For complete listing, see the Standard.
[2] Nominal washer sizes are intended for use with comparable nominal screw or bolt sizes.

16. GRAPHICAL SYMBOLS FOR ELECTRICAL DIAGRAMS[1]

SWITCHES

DISCONNECT	CIRCUIT INTERRUPTER	CIRCUIT BREAKER	LIMIT			
			SPRING RETURN		NEUTRAL POSITION	MAINTAINED POSITION
			NORMALLY OPEN	NORMALLY CLOSED		
					N P	
			HELD CLOSED	HELD OPEN		

LIQUID LEVEL		VACUUM & PRESSURE		TEMPERATURE ACTUATED		FLOW (AIR, WATER, ETC)	
NORMALLY OPEN	NORMALLY CLOSED	NORMALLY OPEN	NORMALLY CLOSED	NORMALLY OPEN	NORMALLY CLOSED	NORMALLY OPEN	NORMALLY CLOSED

SPEED (PLUGGING)	ANTI-PLUG	SELECTOR		FOOT	
		PREFERRED PUSH BUTTON TYPE	ALTERNATE DRUM TYPE	NORMALLY CLOSED	NORMALLY OPEN
F	F	1 2 3	1 2 3		
	F R				

PUSH BUTTONS

SINGLE CIRCUIT		DOUBLE CIRCUIT	MUSHROOM HEAD	MAINTAINED CONTACT
NORMALLY OPEN	NORMALLY CLOSED			

TIMER CONTACTS. CONTACT ACTION RETARDED WHEN COIL IS:				GENERAL CONTACTS. STARTERS, RELAYS, ETC		
ENERGIZED		DE-ENERGIZED		OVERLOAD THERMAL	NORMALLY OPEN	NORMALLY CLOSED
NORMALLY OPEN	NORMALLY CLOSED	NORMALLY OPEN	NORMALLY CLOSED			

CONDUCTORS		COILS			
NOT CONNECTED	CONNECTED	RELAYS TIMERS, ETC.	OVERLOAD THERMAL	SOLENOID	CONTROL TRANSFORMER
					H1 H3 H2 H4
					X1 X2

(CONTINUED)

[1] Reproduced by permission from JIC Electrical Standards for Industrial Equipment, 1957, Joint Industrial Conference, National Machine Tool Builders Association, 2139 Wisconsin Avenue, Washington, D.C. 20007.

16. GRAPHICAL SYMBOLS FOR ELECTRICAL DIAGRAMS (continued)

COILS (CONTINUED)			
AUTO TRANSFORMER	REACTORS		ADJUSTABLE

	IRON CORE	AIR CORE	

(SHOWN WITH IRON CORE)

RECTIFIERS		MOTORS		LOCATION OF RELAY CONTACTS
HALF WAVE	FULL WAVE	THREE PHASE	D. C. TYPES	
			FIELDS	ARMATURE

ICR (2-3-4)

NUMBERS IN PARENTHESIS DESIGNATE THE LOCATION OF RELAY CONTACTS. A LINE UNDERNEATH A LOCATION NUMBER SIGNIFIES A NORMALLY CLOSED CONTACT

RESISTORS

FIXED	TAPPED	POTENTIOMETER OR RHEOSTAT
RES		
H		
HEATING ELEMENT		

DENOTE PURPOSE

ELECTRONIC TUBES

COLD CATHODE	DIODE	TRIODE	TETRODE	PENTODE	IGNITRON	PHOTO-CELL
VOLTAGE REG.					DOT IN ANY TUBE DENOTES GAS	

MISCELLANEOUS

FUSE (POWER OR CONTROL CIRCUIT)	HORN, SIREN, ETC.	BELL OR BUZZER	PLUG AND RECEPTACLE	METER SHUNT	METER
					V M
					AM

THERMOCOUPLES	LAMPS	BATTERY	GROUND	CAPACITOR	
	PUSH TO TEST			FIXED	ADJUSTABLE
	R DENOTE COLOR BY LETTER	+			

17. GRAPHICAL ELECTRICAL WIRING SYMBOLS FOR ARCHITECTURAL AND ELECTRICAL LAYOUT DRAWINGS[1]

LIGHTING OUTLETS

Ceiling	Wall	
		Surface or Pendant Incandescent Mercury Vapor or Similar Lamp Fixture
		Recessed Incandescent Mercury Vapor or Similar Lamp Fixture
		Surface or Pendant Individual Fluorescent Fixture
		Recessed Individual Fluorescent Fixture
		Surface or Pendant Continuous-Row Fluorescent Fixture
		* Recessed Continuous-Row Fluorescent Fixture
		** Bare-Lamp Fluorescent Strip
		Surface or Pendant Exit Light
		Recessed Exit Light
		Blanked Outlet
		Junction Box
		Outlet Controlled by Low-Voltage Switching When Relay is Installed in Outlet Box

* In the case of combination continuous-row fluorescent and incandescent spotlights, use combinations of the above standard symbols.

** In the case of continuous-row bare-lamp fluorescent strip above an area-wide diffusing means, show each fixture run; using the standard symbol; indicate area of diffusing means and type of light shading and/or drawing notation.

RECEPTACLE OUTLETS

Ungrounded	Grounding	
		Single Receptacle Outlet
		Duplex Receptacle Outlet
		Triplex Receptacle Outlet
		Quadruplex Receptacle Outlet
		Duplex Receptacle Outlet—Split Wired
		Triplex Receptacle Outlet—Split Wired
		* Single Special-Purpose Receptacle Outlet
		* Duplex Special-Purpose Receptacle Outlet
		Range Outlet
		Special-Purpose Connection or Provision For Connection. Use subscript Letters To Indicate Function (DW—Dishwasher; CD—Clothes Dryer, etc.)
		Multi-Outlet Assembly. (Extend arrows to limit of installation. Use appropriate symbol to indicate type of outlet. Also indicate spacing of outlets as x inches.)
		Clock Hanger Receptacle
		Fan Hanger Receptacle
		Floor Single Receptacle Outlet
		Floor Duplex Receptacle Outlet
		* Floor Special-Purpose Outlet
		Floor Telephone Outlet—Public
		Floor Telephone Outlet—Private

* Use numeral or letter either within the symbol or as a subscript alongside the symbol keyed to explanation in the drawing list of symbols to indicate type of receptacle or usage.

SWITCH OUTLETS

S	Single-Pole Switch
S_2	Double-Pole Switch
S_3	Three-Way Switch
S_4	Four-Way Switch
S_K	Key-Operated Switch
S_P	Switch and Pilot Lamp
S_L	Switch for Low-Voltage Switching System
S_{LM}	Master Switch for Low-Voltage Switching System
S	Switch and Single Receptacle
S	Switch and Double Receptacle
S_D	Door Switch
S_T	Time Switch
S_{CB}	Circuit Breaker Switch
S_{MC}	Momentary Contact Switch or Pushbutton For Other Than Signalling System
S	Ceiling Pull Switch

SIGNALLING SYSTEM OUTLETS RESIDENTIAL OCCUPANCIES

	Pushbutton
	Buzzer
	Bell
	Combination Bell-Buzzer
CH	Chime
	Annunicator
D	Electric Door Opener
M	Maid's Signal Plug
	Interconnection Box
BT	Bell-Ringing Transformer
	Outside Telephone
	Interconnecting Telephone
R	Radio Outlet
TV	Television Outlet

CIRCUITING

Wiring Method Identification By Notation On Drawing Or In Specifications

—————— Wiring Concealed in Ceiling or Wall

— — — Wiring Concealed in Floor

- - - - - - Wiring Exposed

Note: Use heavy-weight line to identify service and feeders. Indicate empty conduit by notation CO (Conduit only)

Branch Circuit Home Run to Panel Board. Number of arrows indicates number of circuits. (A numeral at each arrow may be used to identify circuit number.) Note: Any circuit without further identification indicates two-wire circuit. For a greater number of wires, indicate with cross lines, e.g.:

—/// 3 wires; —//// 4 wires, etc.

Unless indicated otherwise, the wire size of the circuit is the minimum size required by the specification.
Identify different functions of wiring system, e.g., signalling system by notation or other means.

——————o Wiring Turned Up

——————● Wiring Turned Down

[1] USAS Y32.9—1962. © The Institute of Electrical and Electronics Engineers, United Engineering Center, 345 East 47th Street, New York, N.Y. 10017. Reproduced by permission.

18. STANDARD WELDING SYMBOLS[1]

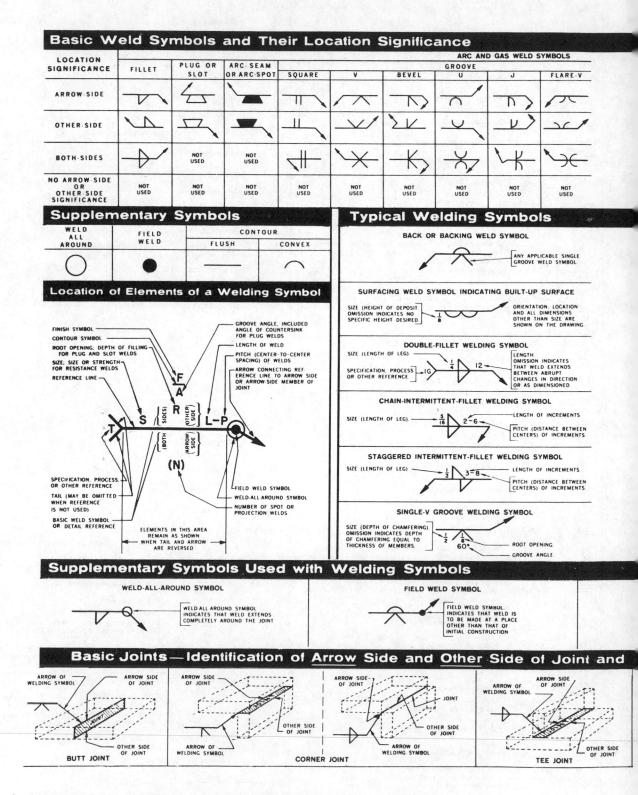

Basic Weld Symbols and Their Location Significance

LOCATION SIGNIFICANCE	FILLET	PLUG OR SLOT	ARC-SEAM OR ARC-SPOT	ARC AND GAS WELD SYMBOLS					
				GROOVE					
				SQUARE	V	BEVEL	U	J	FLARE-V
ARROW-SIDE									
OTHER-SIDE									
BOTH-SIDES		NOT USED	NOT USED						
NO ARROW-SIDE OR OTHER-SIDE SIGNIFICANCE	NOT USED	NOT USED	NOT USED	NOT USED	NOT USED	NOT USED	NOT USED	NOT USED	NOT USED

Supplementary Symbols

WELD ALL AROUND	FIELD WELD	CONTOUR	
		FLUSH	CONVEX

Location of Elements of a Welding Symbol

FINISH SYMBOL
CONTOUR SYMBOL
ROOT OPENING, DEPTH OF FILLING FOR PLUG AND SLOT WELDS
SIZE; SIZE OR STRENGTH FOR RESISTANCE WELDS
REFERENCE LINE

GROOVE ANGLE, INCLUDED ANGLE OF COUNTERSINK FOR PLUG WELDS
LENGTH OF WELD
PITCH (CENTER-TO-CENTER SPACING) OF WELDS
ARROW CONNECTING REFERENCE LINE TO ARROW SIDE OR ARROW SIDE MEMBER OF JOINT

F
A
R
S (SIDES) (OTHER SIDE) L-P
(BOTH) (ARROW SIDE)
(N)

SPECIFICATION, PROCESS, OR OTHER REFERENCE
TAIL (MAY BE OMITTED WHEN REFERENCE IS NOT USED)
BASIC WELD SYMBOL OR DETAIL REFERENCE

FIELD WELD SYMBOL
WELD-ALL-AROUND SYMBOL
NUMBER OF SPOT OR PROJECTION WELDS

ELEMENTS IN THIS AREA REMAIN AS SHOWN WHEN TAIL AND ARROW ARE REVERSED

Typical Welding Symbols

BACK OR BACKING WELD SYMBOL
ANY APPLICABLE SINGLE GROOVE WELD SYMBOL.

SURFACING WELD SYMBOL INDICATING BUILT-UP SURFACE
SIZE (HEIGHT OF DEPOSIT OMISSION INDICATES NO SPECIFIC HEIGHT DESIRED.) $\frac{1}{8}$
ORIENTATION, LOCATION AND ALL DIMENSIONS OTHER THAN SIZE ARE SHOWN ON THE DRAWING.

DOUBLE-FILLET WELDING SYMBOL
SIZE (LENGTH OF LEG). $\frac{1}{4}$ 12
SPECIFICATION, PROCESS OR OTHER REFERENCE. IG
LENGTH OMISSION INDICATES THAT WELD EXTENDS BETWEEN ABRUPT CHANGES IN DIRECTION OR AS DIMENSIONED.

CHAIN-INTERMITTENT-FILLET WELDING SYMBOL
SIZE (LENGTH OF LEG) $\frac{5}{16}$ 2-6
LENGTH OF INCREMENTS.
PITCH (DISTANCE BETWEEN CENTERS) OF INCREMENTS.

STAGGERED INTERMITTENT-FILLET WELDING SYMBOL
SIZE (LENGTH OF LEG) $\frac{1}{2}$ 3-8
LENGTH OF INCREMENTS.
PITCH (DISTANCE BETWEEN CENTERS) OF INCREMENTS.

SINGLE-V GROOVE WELDING SYMBOL
SIZE (DEPTH OF CHAMFERING) OMISSION INDICATES DEPTH OF CHAMFERING EQUAL TO THICKNESS OF MEMBERS. $\frac{1}{2}$ $\frac{1}{8}$ 60°
ROOT OPENING.
GROOVE ANGLE.

Supplementary Symbols Used with Welding Symbols

WELD-ALL-AROUND SYMBOL
WELD ALL AROUND SYMBOL INDICATES THAT WELD EXTENDS COMPLETELY AROUND THE JOINT.

FIELD WELD SYMBOL
FIELD WELD SYMBOL INDICATES THAT WELD IS TO BE MADE AT A PLACE OTHER THAN THAT OF INITIAL CONSTRUCTION.

Basic Joints—Identification of Arrow Side and Other Side of Joint and

ARROW OF WELDING SYMBOL
ARROW SIDE OF JOINT
OTHER SIDE OF JOINT
BUTT JOINT

ARROW SIDE OF JOINT
OTHER SIDE OF JOINT
ARROW OF WELDING SYMBOL
CORNER JOINT

ARROW SIDE OF JOINT
JOINT
OTHER SIDE OF JOINT
ARROW OF WELDING SYMBOL
CORNER JOINT

ARROW OF WELDING SYMBOL
ARROW SIDE OF JOINT
OTHER SIDE OF JOINT
TEE JOINT

[1] AWS A2.0-58, American Welding Society, 345 East 47th Street, New York, N.Y. 10017.

18. STANDARD WELDING SYMBOLS (continued)

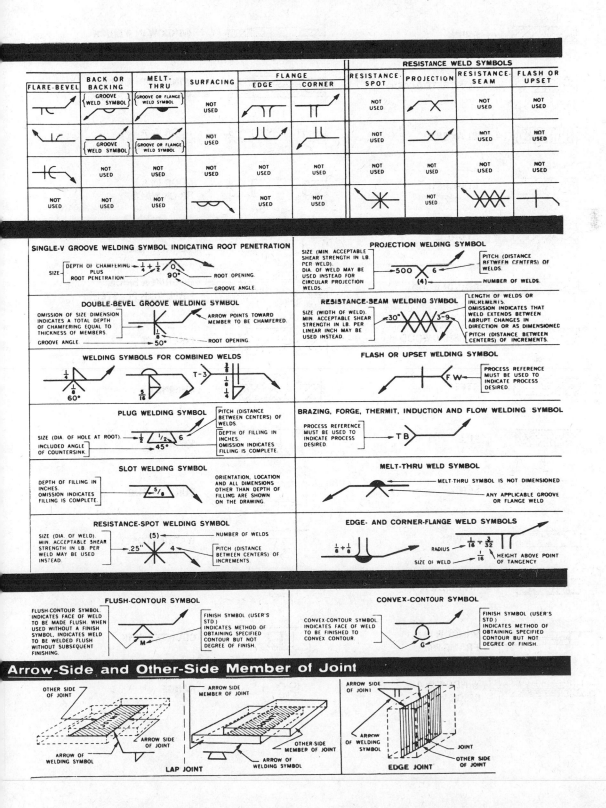

RESISTANCE WELD SYMBOLS

FLARE-BEVEL	BACK OR BACKING	MELT-THRU	SURFACING	FLANGE EDGE	FLANGE CORNER	RESISTANCE-SPOT	PROJECTION	RESISTANCE-SEAM	FLASH OR UPSET
	GROOVE WELD SYMBOL	GROOVE OR FLANGE WELD SYMBOL	NOT USED			NOT USED		NOT USED	NOT USED
	GROOVE WELD SYMBOL	GROOVE OR FLANGE WELD SYMBOL	NOT USED			NOT USED		NOT USED	NOT USED
	NOT USED	NOT USED	NOT USED	NOT USED	NOT USED	NOT USED	NOT USED	NOT USED	NOT USED
NOT USED	NOT USED	NOT USED		NOT USED	NOT USED		NOT USED		

SINGLE-V GROOVE WELDING SYMBOL INDICATING ROOT PENETRATION

SIZE — DEPTH OF CHAMFERING $\frac{1}{4} + \frac{1}{2}$ — ROOT PENETRATION — PLUS — 90° — GROOVE ANGLE — ROOT OPENING

DOUBLE-BEVEL GROOVE WELDING SYMBOL

OMISSION OF SIZE DIMENSION INDICATES A TOTAL DEPTH OF CHAMFERING EQUAL TO THICKNESS OF MEMBERS.
ARROW POINTS TOWARD MEMBER TO BE CHAMFERED.
GROOVE ANGLE — 50° — $\frac{1}{8}$ — ROOT OPENING

WELDING SYMBOLS FOR COMBINED WELDS

$\frac{1}{4}$ — $\frac{1}{8}$ — 60° — $\frac{5}{16}$ — T-3 — $\frac{1}{8}$

PLUG WELDING SYMBOL

SIZE (DIA. OF HOLE AT ROOT) — $\frac{1}{2}$ — $\frac{1}{2}$" — 6 — 45°
INCLUDED ANGLE OF COUNTERSINK.
PITCH (DISTANCE BETWEEN CENTERS) OF WELDS.
DEPTH OF FILLING IN INCHES. OMISSION INDICATES FILLING IS COMPLETE.

SLOT WELDING SYMBOL

DEPTH OF FILLING IN INCHES. OMISSION INDICATES FILLING IS COMPLETE.
$\frac{5}{8}$
ORIENTATION, LOCATION AND ALL DIMENSIONS OTHER THAN DEPTH OF FILLING ARE SHOWN ON THE DRAWING.

RESISTANCE-SPOT WELDING SYMBOL

SIZE (DIA. OF WELD) MIN. ACCEPTABLE SHEAR STRENGTH IN LB. PER WELD MAY BE USED INSTEAD.
(5) — NUMBER OF WELDS
.25" — 4 — PITCH (DISTANCE BETWEEN CENTERS) OF INCREMENTS.

PROJECTION WELDING SYMBOL

SIZE (MIN. ACCEPTABLE SHEAR STRENGTH IN LB. PER WELD). DIA. OF WELD MAY BE USED INSTEAD FOR CIRCULAR PROJECTION WELDS.
500 — 6 — PITCH (DISTANCE BETWEEN CENTERS) OF WELDS.
(4) — NUMBER OF WELDS.

RESISTANCE-SEAM WELDING SYMBOL

SIZE (WIDTH OF WELD). MIN. ACCEPTABLE SHEAR STRENGTH IN LB. PER LINEAR INCH MAY BE USED INSTEAD.
30° — 3-9
LENGTH OF WELDS OR INCREMENTS. OMISSION INDICATES THAT WELD EXTENDS BETWEEN ABRUPT CHANGES IN DIRECTION OR AS DIMENSIONED.
PITCH (DISTANCE BETWEEN CENTERS) OF INCREMENTS.

FLASH OR UPSET WELDING SYMBOL

FW — PROCESS REFERENCE MUST BE USED TO INDICATE PROCESS DESIRED.

BRAZING, FORGE, THERMIT, INDUCTION AND FLOW WELDING SYMBOL

PROCESS REFERENCE MUST BE USED TO INDICATE PROCESS DESIRED.
TB

MELT-THRU WELD SYMBOL

MELT-THRU SYMBOL IS NOT DIMENSIONED
ANY APPLICABLE GROOVE OR FLANGE WELD

EDGE- AND CORNER-FLANGE WELD SYMBOLS

$\frac{1}{8} + \frac{1}{8}$
RADIUS — $\frac{1}{16}$ T $\frac{3}{32}$
SIZE OF WELD — $\frac{1}{16}$ — HEIGHT ABOVE POINT OF TANGENCY

FLUSH-CONTOUR SYMBOL

FLUSH-CONTOUR SYMBOL INDICATES FACE OF WELD TO BE MADE FLUSH. WHEN USED WITHOUT A FINISH SYMBOL, INDICATES WELD TO BE WELDED FLUSH WITHOUT SUBSEQUENT FINISHING.
M
FINISH SYMBOL (USER'S STD.) INDICATES METHOD OF OBTAINING SPECIFIED CONTOUR BUT NOT DEGREE OF FINISH.

CONVEX-CONTOUR SYMBOL

CONVEX-CONTOUR SYMBOL INDICATES FACE OF WELD TO BE FINISHED TO CONVEX CONTOUR.
G
FINISH SYMBOL (USER'S STD.) INDICATES METHOD OF OBTAINING SPECIFIED CONTOUR BUT NOT DEGREE OF FINISH.

Arrow-Side and Other-Side Member of Joint

OTHER SIDE OF JOINT — ARROW SIDE OF JOINT — ARROW OF WELDING SYMBOL
LAP JOINT

ARROW SIDE MEMBER OF JOINT — OTHER-SIDE MEMBER OF JOINT — ARROW OF WELDING SYMBOL

ARROW SIDE OF JOINT — ARROW OF WELDING SYMBOL — JOINT — OTHER SIDE OF JOINT
EDGE JOINT

19. ARCHITECTURAL SYMBOLS

	BRICK			WINDOW IN A BRICK VENEER WALL
	CONCRETE BLOCK			WINDOW IN A BRICK WALL WITH PLASTER
	CLAY TILE			DOOR IN A FRAME WALL
	CONCRETE			DOOR IN A BRICK WALL
	STONE			DOOR IN A BRICK VENEER WALL
	ROUGH WOOD			DOOR IN A BRICK WALL WITH PLASTER
	FINISHED WOOD			FIREPLACE
	STEEL			CHIMNEY
	FILL			STAIRS
	SAND			BUILT-IN-TUB
	GRAVEL			WATER CLOSET
	INSULATION			LAVATORY
	WINDOW IN A FRAME WALL			COUNTER SINK
	WINDOW IN A BRICK WALL			RANGE

20. DECIMAL EQUIVALENTS

Decimal Measurements may be set off directly on drawings with the aid of an engineers scale, Sec. 3.22.

		$\frac{1}{64}$	.015625			$\frac{33}{64}$	.515625
	$\frac{1}{32}$		.03125		$\frac{17}{32}$		.53125
		$\frac{3}{64}$	.046875			$\frac{35}{64}$	.546875
$\frac{1}{16}$			.0625	$\frac{9}{16}$			.5625
		$\frac{5}{64}$	.078125			$\frac{37}{64}$	.578125
	$\frac{3}{32}$		.09375		$\frac{19}{32}$		.59375
		$\frac{7}{64}$	.109375			$\frac{39}{64}$	.609375
$\frac{1}{8}$			.125	$\frac{5}{8}$			.625
		$\frac{9}{64}$	.140625			$\frac{41}{64}$	.640625
	$\frac{5}{32}$		.15625		$\frac{21}{32}$		.65625
		$\frac{11}{64}$	.171875			$\frac{43}{64}$	.671875
$\frac{3}{16}$			.1875	$\frac{11}{16}$			.6875
		$\frac{13}{64}$	.203125			$\frac{45}{64}$	.703125
	$\frac{7}{32}$		.21875		$\frac{23}{32}$		.71875
		$\frac{15}{64}$	.234375			$\frac{47}{64}$	.734375
$\frac{1}{4}$			.25	$\frac{3}{4}$			.75
		$\frac{17}{64}$	.265625			$\frac{49}{64}$	.765625
	$\frac{9}{32}$		.28125		$\frac{25}{32}$		.78125
		$\frac{19}{64}$	.296875			$\frac{51}{64}$	.796875
$\frac{5}{16}$			.3125	$\frac{13}{16}$			.8125
		$\frac{21}{64}$	.328125			$\frac{53}{64}$	.828125
	$\frac{11}{32}$		.34375		$\frac{27}{32}$		.84375
		$\frac{23}{64}$	.359375			$\frac{55}{64}$	.859375
$\frac{3}{8}$			.375	$\frac{7}{8}$			.875
		$\frac{25}{64}$	.390625			$\frac{57}{64}$	.890625
	$\frac{13}{32}$		.40625		$\frac{29}{32}$		.90625
		$\frac{27}{64}$	.421875			$\frac{59}{64}$	.921875
$\frac{7}{16}$			.4375	$\frac{15}{16}$			.9375
		$\frac{29}{64}$	.453125			$\frac{61}{64}$	.953125
	$\frac{15}{32}$		.46875		$\frac{31}{32}$		.96875
		$\frac{31}{64}$	.484375			$\frac{63}{64}$	.984375
$\frac{1}{2}$			.5	1			1.

SHEET LAYOUTS

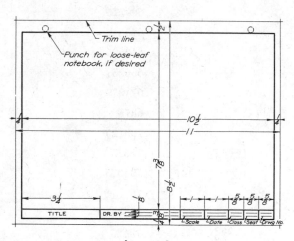

Layout A

See Fig. 3–43 showing steps in drawing this layout.

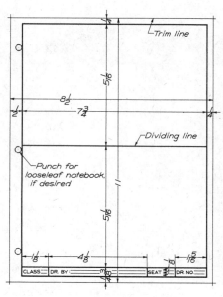

Layout B

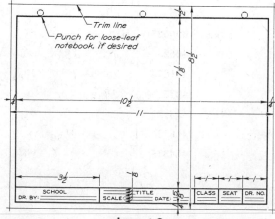

Layout C

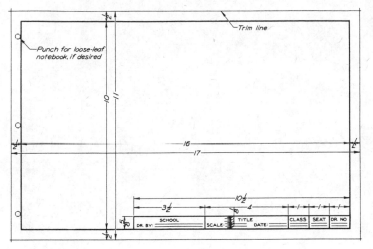

Layout D

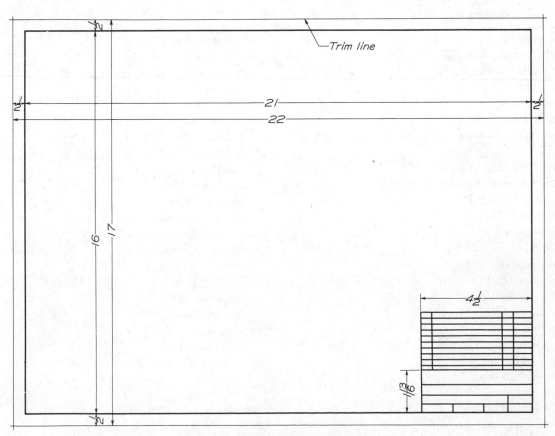

NO.	PART NAME	REQ'D	MATERIAL
4	VISE SCREW	1	C.R.S.
3	SLIDING JAW	1	C.I.
2	SECONDARY BASE	1	C.I.
1	VISE BASE	1	C.I.

SCHOOL OR COMPANY
CITY, STATE

DRAWING TITLE

DR. BY:

**Block Title and
Parts List for Layout E**

Layout E
See Block Title and Parts List Above.

INDEX

Numbers listed are page numbers

476

SHEET LAYOUTS

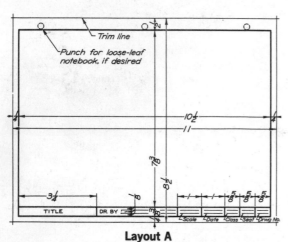

Layout A

See Fig. 3–43 showing steps in drawing this layout.

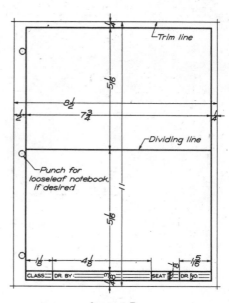

Layout B

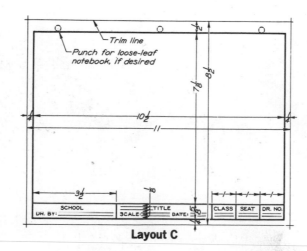

Layout C